Introduction

This book is designed for Year 9 students and is part of a series covering Key Stage 3 Mathematics. Each textbook in the series is accompanied by an extensive Teacher's Resource including additional material. This allows the series to be used with the full ability range of students.

The series builds on the National Numeracy Strategy in primary schools and its extension into Key Stage 3. It is designed to support the style of teaching and the lesson framework to which students will be accustomed.

This book is presented as a series of double-page spreads, each of which is designed to be a teaching unit. The left-hand page covers the material to be taught and the right-hand page provides examples for the students to work through. Each chapter ends with a review exercise covering all its content. Further worksheets, tests and ICT materials are provided in the Teacher's Resource.

An important feature of the left-hand pages is the Tasks, which are printed in boxes. These are intended to be carried out by the student in mid-lesson. Their aim is twofold: in the first place they give the students practice on what they have just been taught, allowing them to consolidate their understanding. However, the tasks then extend the ideas and raise questions, setting the agenda for the later part of the lesson. Further guidance on the Tasks is available in the Teacher's Resource.

Another key feature of the left-hand pages is the Discussion Points. These are designed to help teachers engage their students in whole class discussion. Teachers should see the icon as an opportunity and an invitation.

Several other symbols and instructions are used in this book. These are explained on the 'How to use this book' page for students opposite. The symbol indicates to the teacher that there is additional ICT material directly linked to that unit of work. This is referenced in the teaching notes for that unit in the Teacher's Resource.

The order of the chapters in this book ensures that the subject is developed logically, at each stage building on previous knowledge. The Teacher's Resource includes a Scheme of Work based on this order. However, teachers are of course free to vary the order to meet their own circumstances and needs.

The first 22 chapters complete the new work needed for the Key Stage 3 National Test. Chapter 23 consists of questions on earlier work for students preparing for this National Test. The remaining two chapters are designed for use after the students have taken the National Test. Chapter 24 introduces them to some of the financial calculations they will meet in everyday life. Chapter 25 consists of Investigations in preparation for GCSE coursework.

The authors would like to thank all those who helped in preparing this book, particularly those involved with the writing of materials for the accompanying Teacher's Resource.

Roger Porkess 2002
Series Editor

Catherine Berry • Margaret Bland

Anthony Eccles • Dave Faulkner

Sophie Goldie • Tim Holmes

Simon Jowett • Habib Rashid

Leonie Turner • Brandon Wilshaw

SERIES EDITOR: Roger Porkess

Hodder & Stoughton

A MEMBER OF THE HODDER HEADLINE GROUP

Acknowledgements

Every effort has been made to trace and acknowledge ownership of copyright. The publishers will be glad to make suitable arrangements with any copyright holder whom it has not been possible to contact.

Illustrations were drawn by Maggie Brand, Tom Cross, Jeff Edwards and Joe McEwan.

Photos supplied by Hodder Picture Library (page 8); Richard Cummins/CORBIS (page 22); Action Plus Sports Images (pages 24, 103); Dr A. J. Finch (page 47); Simon Jowett (page 77); Photostore (page 103); Bob Krist/CORBIS (page 146); Ruth Nossek (pages 156, 157); Gianni Dagli Orti/CORBIS (page 197)

Cover design and page design by Julie Martin.

Orders: please contact Bookpoint Ltd, 130 Milton Park, Abingdon, Oxon OX14 4SB. Telephone: (44) 01235 827720, Fax (44) 01235 400454. Lines are open from 9.00–6.00, Monday to Saturday, with a 24 hour message answering service. You can also order through our website at www.hodderheadline.co.uk

British Library Cataloguing in Publication Data
A catalogue record for this title is available from The British Library

ISBN 0 340 779772

First published 2002
Impression number 10 9 8 7 6 5 4 3
Year 2008 2007 2006 2005 2004 2003

Cover photo from Jacey, Debut Art

Typeset by Tech-Set Ltd, Gateshead, Tyne & Wear.
Printed in Italy for Hodder & Stoughton Educational, a division of Hodder Headline, 338 Euston Road, London NW1 3BH by Printer Trento.

How to use this book

This symbol means that you will need to think carefully about a point. Your teacher may ask you to join in a discussion about it.

This symbol next to a question means that you are allowed (and indeed expected) to use your calculator for this question.

This symbol means exactly the opposite – you are not allowed to use your calculator for this question.

This is a warning sign. It is used where a common mistake, or misunderstanding, is being described. It is also used to identify questions which are slightly more difficult or which require a little more thought. It should be read as 'caution'.

This is the ICT symbol. It should alert your teacher to the fact that there is some additional material in the accompanying Teacher's Resource using ICT for this unit of work.

Each chapter of work in this book is divided into a series of double-page spreads – or units of work. The left-hand page is the teaching page, and the right-hand page involves an exercise and sometimes additional activities or investigations to do with that topic.

You will also come across the following features in the units of work:

The tasks give you the opportunity to work alone, in pairs or in small groups on an activity in the lesson. It gives you the chance to practise what you have just been taught, and to discuss ideas and raise questions about the topic.

Do the right thing!

These boxes give you a set of step-by-step instructions on how to carry out a particular technique in maths, usually to do with shape work.

Do you remember?

These boxes give you the chance to review work that you have covered in the previous year.

This downline at the edge of the page indicates that this is a review (or revision) of material which you have already met.

Contents

1 Units

Why is being able to measure accurately important?
Think of five examples. In each case say why accuracy is important.

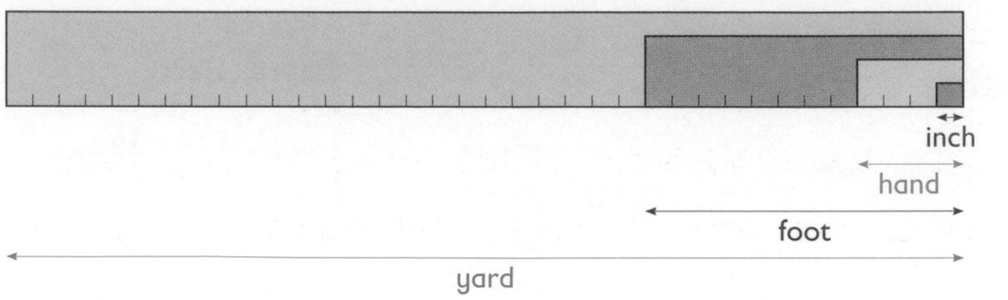

In 1196 standardised units of measurement were first recorded. During the reign of Edward I (1272–1307), an inch was defined as 'three grains of barley, dry and round'.

1 inch

Task

Work in pairs for this Task.

1. Both measure the width of your desk in hands. Compare your results.

   Does your answer agree with your friend's?

   What are the disadvantages of measuring in this way?

Length	Estimate	
Your own height		hands
Length of your desk		inches
Width of exercise book		inches
Length of your arm		feet
Length of pen		inches

2. Copy and complete the table. *Estimate* the measurements required.
3. Take one hand to be 4 inches.
   Convert **(a)** your height **(b)** the length of your desk to inches.

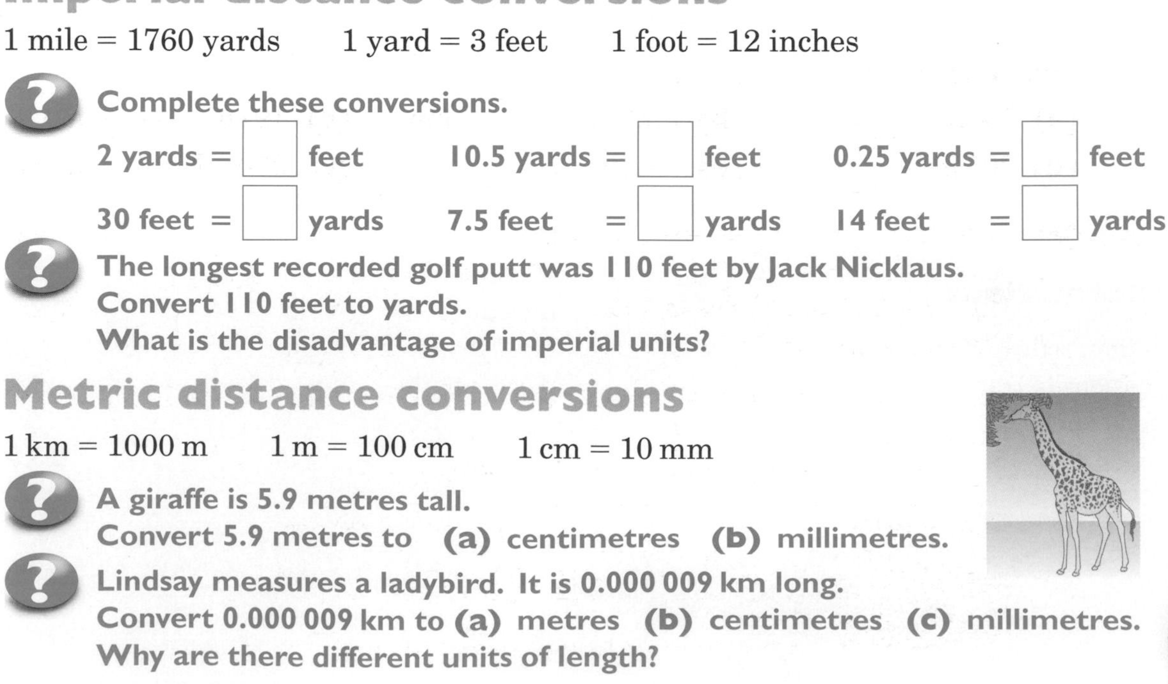

Imperial distance conversions

1 mile = 1760 yards   1 yard = 3 feet   1 foot = 12 inches

Complete these conversions.

2 yards = ☐ feet   10.5 yards = ☐ feet   0.25 yards = ☐ feet

30 feet = ☐ yards   7.5 feet = ☐ yards   14 feet = ☐ yards

The longest recorded golf putt was 110 feet by Jack Nicklaus.
Convert 110 feet to yards.
What is the disadvantage of imperial units?

Metric distance conversions

1 km = 1000 m   1 m = 100 cm   1 cm = 10 mm

A giraffe is 5.9 metres tall.
Convert 5.9 metres to (a) centimetres (b) millimetres.

Lindsay measures a ladybird. It is 0.000 009 km long.
Convert 0.000 009 km to (a) metres (b) centimetres (c) millimetres.
Why are there different units of length?

Exercise

1 Convert these measurements from feet to inches.

(a) 6 feet **(b)** 4 feet **(c)** $\frac{1}{2}$ foot **(d)** $5\frac{1}{2}$ feet

1 foot = 12 inches.

2 Convert these measurements from inches to feet.

(a) 24 inches **(b)** 36 inches **(c)** 60 inches **(d)** 18 inches

3 Convert these measurements from miles to yards.

(a) 2 miles **(b)** 3 miles **(c)** 10 miles **(d)** $\frac{1}{2}$ mile

1 mile = 1760 yards.

4 Furlongs are used to measure distances in horseracing.

1 mile = 8 furlongs.

A racecourse is 3 miles long.
There are 4 furlongs to go in the race.

How far have the horses run?
Give your answer in

(a) furlongs
(b) yards.

5 The heights of horses are measured in hands.
The winner of the race is St. Paddy. He is 19 hands tall.

(a) How many inches is 19 hands?
(b) How many feet is 19 hands?

1 hand = 4 inches
12 inches = 1 foot.

6 Stones and pounds (lb) are measures of weight.
Convert these measurements to stones.

(a) 28 lb **(b)** 140 lb **(c)** 210 lb **(d)** 217 lb

1 stone = 14 lb.

7 A smaller measure of weight is the ounce (oz).
Convert these measurements from pounds to ounces.

(a) 2 pounds **(b)** 5 pounds **(c)** 6.5 pounds **(d)** $\frac{1}{4}$ pound

1 pound (lb) = 16 ounces (oz).

8 Oliver has ten glasses.
Each glass holds 250 ml of liquid.
How many litres of cola does
Oliver need to fill all ten glasses?

1000 ml = 1 l.

9 Copy and complete these conversions.

(a) 5 km = ☐ m **(b)** 3500 m = ☐ km **(c)** 19 cm = ☐ mm

(d) 305 g = ☐ kg **(e)** 3.6 litres = ☐ ml **(f)** 250 ml = ☐ litres

10 **(a)** The sides of a square are 1 m long.
This is the same as 100 cm or 1000 mm.
The area of the square is 1 m^2.
Write this in **(i)** cm^2 **(ii)** mm^2.

1 m
100 cm
1000 mm

(b) The edges of a cube are all 1 m so the volume of the cube is 1 m^3.
Write this in **(i)** cm^3 **(ii)** mm^3.

(c) Copy and complete this conversion table.

Length	1 m	100 cm	1000 mm
Area	1 m^2	$100^2 = 10\,000$ cm^2	mm^2
Volume	1 m^3	cm^3	mm^3

Imperial–metric conversions

Alan and Jim are marking out a football pitch.

How many metres does Alan mark for the posts?

Task

The goalkeeper is not happy …
The posts have to be 24 feet apart.

1 How many inches are there in 24 feet?

2 Approximately how many centimetres is this?

3 Change your answer to part **2** into metres.

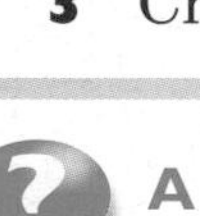

A more accurate conversion is to use 1 inch = 2.54 cm.
Calculate a more accurate distance to mark the goal posts.

Avonford Town fans travel far and wide to see their team play.

Convert these distances to miles.
(a) 16 km **(b)** 80 km **(c)** 20 km **(d)** 10 km **(e)** 25 km
When are the conversions easy? When are they more difficult?

A conversion graph can help with more difficult conversions.

Task

Drawing a graph to convert miles to kilometres.

1 Draw x and y axes. Put miles on the x axis.
Use a scale of one square for 5 miles, one square for 5 km.

2 Convert 50 miles to km. Draw the graph.

3 Use your graph to convert these distances to miles.
(a) 25 km **(b)** 55 km **(c)** 70 km **(d)** 76 km

4 Use your graph to convert these distances to kilometres.
(a) 8 miles **(b)** 26 miles **(c)** 37 miles **(d)** 46 miles

A French car is driving on the M6. The speed limit is 70 m.p.h.
What is this in km h^{-1}?

Exercise

Approximate conversions	Weight		Volume		Distance or length			
Metric	1 kg	28 g	1 litre	4.5 litres	8 km	1 m	30 cm	2.5 cm
Imperial	2.2 pounds	1 ounce	1.75 pints	1 gallon	5 miles	39 inches	1 foot	1 inch

1 Sylvia drives across France. She has enough petrol for 18 miles.
(a) The next petrol station is 28 km away. Convert 28 km to miles.
(b) A more accurate conversion is that 1 mile = 1.609 km.
Use this conversion to calculate the distance to the next petrol station.
(c) Sylvia gets to the petrol station. She fills her petrol tank.
The tank takes 9 gallons. How many litres is 9 gallons?

2 Georgina is 5 feet 4 inches tall.
Convert 5 feet 4 inches to centimetres.
Does the coat fit Georgina?

3 Look at the recipe for Caramel Pear.
Convert the recipe to metric units.

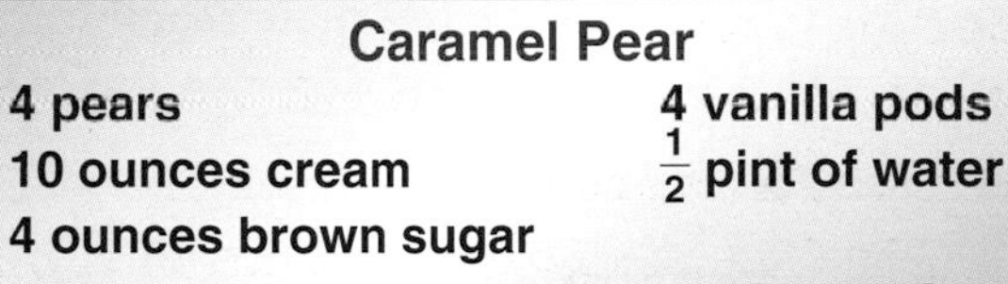

4 Joanne makes a wooden car in DT.
She uses an old plan marked in imperial units.
(a) How long is the car in centimetres?
Joanne has two choices of wood:
60 mm thick or 80 mm thick.
(b) Which is closer to $\frac{1}{4}$ inch?
(c) Convert 6.6 pounds to kilograms.

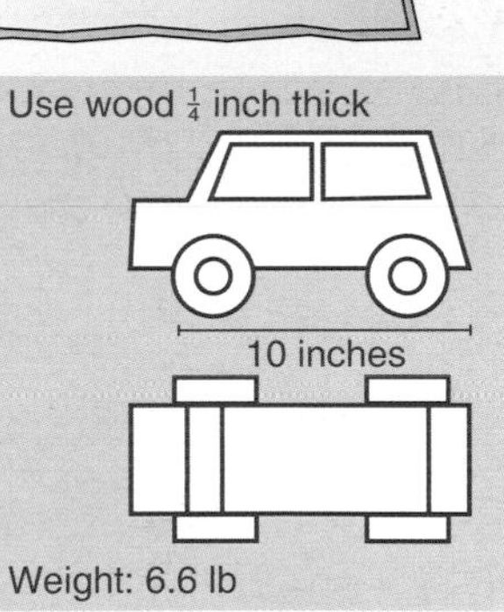

5 **(a)** Convert 10 kg to pounds.
(b) Draw a graph to convert kilograms to pounds.
Put kilograms on the x axis. Use a scale of one square to 1 kg, one square to 1 pound. Go up to 10 kg.
(c) Use your graph to convert these weights to pounds.
(i) 3 kg **(ii)** 7 kg **(iii)** 8 kg **(iv)** 9.5 kg
(d) Use your graph to convert these weights to kg.
(i) 5 pounds **(ii)** 8 pounds **(iii)** 2.6 pounds **(iv)** 15.5 pounds

Activity

Joanne gets some crisp boxes.
Each box is a cube of side 50 cm.
Work out the volume of a box in cm^3 and in m^3.

Joanne makes scenery for a school play.
She uses crisp boxes to make a crate.
The crate has to be a cube. The length of *each* side is 200 cm.
How many crisp boxes does she need?
What is the volume of the crate in cm^3 and in m^3.

Time

Katie has 20 minutes before she goes out. Will she have time to listen to all of her CD?

Sophie catches a train from Norwich to Peterborough.
She draws a time line for the journey.

Norwich 11 45
Peterborough 13 25

1145 1200 1300 1325

How can the time line be used to work out the journey time?
How else could Sophie work out the journey time?

Task

Sophie visits a friend in Newcastle. She goes from Norwich to Peterborough first. Then she changes trains and goes from Peterborough to Newcastle.

Station			
Norwich	1112	1145	1249
Peterborough	1235	1325	1414

Station			
Peterborough	1246	1315	1419
Newcastle	1457	1540	1634

1 Look at the timetables.
Work out three possible journeys for Sophie. How long does each journey take?

2 What is the disadvantage of catching the 1145 from Norwich?
Why might Sophie choose to catch this train anyway?

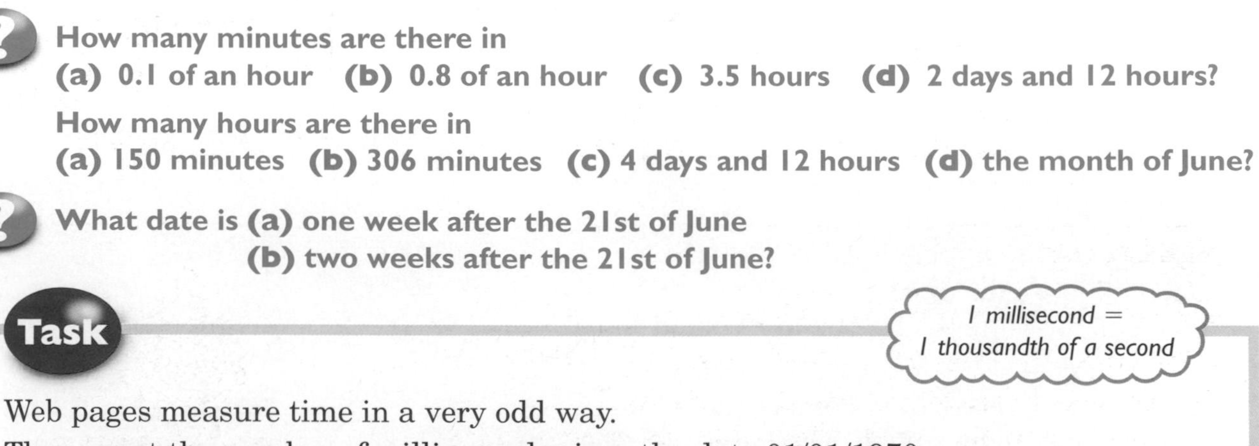

How many minutes are there in
(a) 0.1 of an hour (b) 0.8 of an hour (c) 3.5 hours (d) 2 days and 12 hours?

How many hours are there in
(a) 150 minutes (b) 306 minutes (c) 4 days and 12 hours (d) the month of June?

What date is (a) one week after the 21st of June
(b) two weeks after the 21st of June?

Task

1 millisecond = 1 thousandth of a second

Web pages measure time in a very odd way.
They count the number of milliseconds since the date 01/01/1970.
What date and time is 942 929 101 089 milliseconds after 01/01/1970?

Exercise

1 **(a)** How long does it take to play this CD single?
(b) Michael has 15 minutes before he goes out.
How long will he have left after listening to this CD?

THE FEVER CD single	
DONNY	
The Fever	3.36
Radio Edit	4.15
Dance Mix	6.48

2 Sophie is travelling back from Newcastle to Norwich.

Station			
Newcastle	0831	0900	0938
Peterborough	1053	1112	1219

Station		
Peterborough	1146	1253
Norwich.............	1321	1429

(a) Use the timetables to work out three possible journeys for Sophie.
(b) How long does each journey take?

3 Huyen goes to the airport to meet her pen pal.
The flight is delayed by 140 minutes.
(a) Write 140 minutes in hours and minutes.
(b) The flight was due at 2045. What time is it expected now?

4 Copy and complete the table.

Flight from	Time due	Minutes late	Time now due
Barcelona	2055	100	
New York		135	2340
Sydney	2205		0035
Tokyo	2240	195	

5 How many minutes are there in
(a) 0.15 of an hour **(b)** 2.3 hours **(c)** $\frac{3}{5}$ of an hour?

6 Work out the total minutes in the following sums.
(a) 3.7 hours + 10 minutes **(b)** 1.4 hours + 50 minutes
(c) 3.3 hours + 40 minutes **(d)** 1.7 hours + 30 minutes

7 John has four weeks to rehearse with his band.
What date is it now?

Porridge Dog
LIVE! on 24th Feb

8 Add two weeks to each of these dates.
(a) 20 March **(b)** 24 September **(c)** 24 December

Activity

DJ Jive has three records to play:
American Dream (3:35) Dancer (4:45) Salsoul (3:20)
Dance records are often 'mixed' (overlapped).
DJ Jive has 10 minutes to play all three records in full.
Find two ways he can mix them.

Speed

The Eurostar train travels at a speed of 150 mph.
How far can it travel in 2.5 hours?

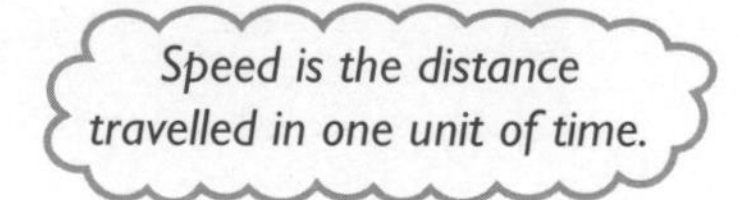

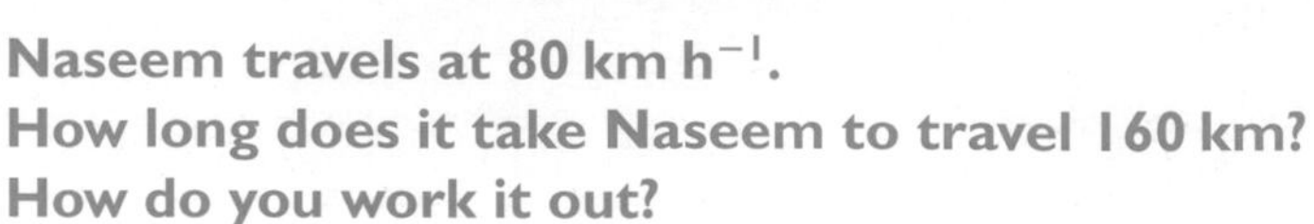

Naseem travels at 80 km h^{-1}.
How long does it take Naseem to travel 160 km?
How do you work it out?

John goes out on his bike.
He cycles 40 km in 2 hours.
Work out his speed.

80 km h^{-1} means that 80 km are travelled in one hour.

Speed, distance travelled and time taken are connected by these formulae

$$\text{Speed} = \frac{\text{Distance}}{\text{Time}} \qquad \text{Distance} = \text{Speed} \times \text{Time} \qquad \text{Time} = \frac{\text{Distance}}{\text{Speed}}$$

Task

Damon is driving at 30 km h^{-1}.

1 How many metres per hour is 30 km h^{-1}?

2 How many metres per second is 30 km h^{-1}?

3 A tree is blocking the road.
The tree is 50 m away from the car.
How long does it take to travel 50 m at a speed of 30 km h^{-1}?

Do you think Damon has time to stop before he hits the tree?

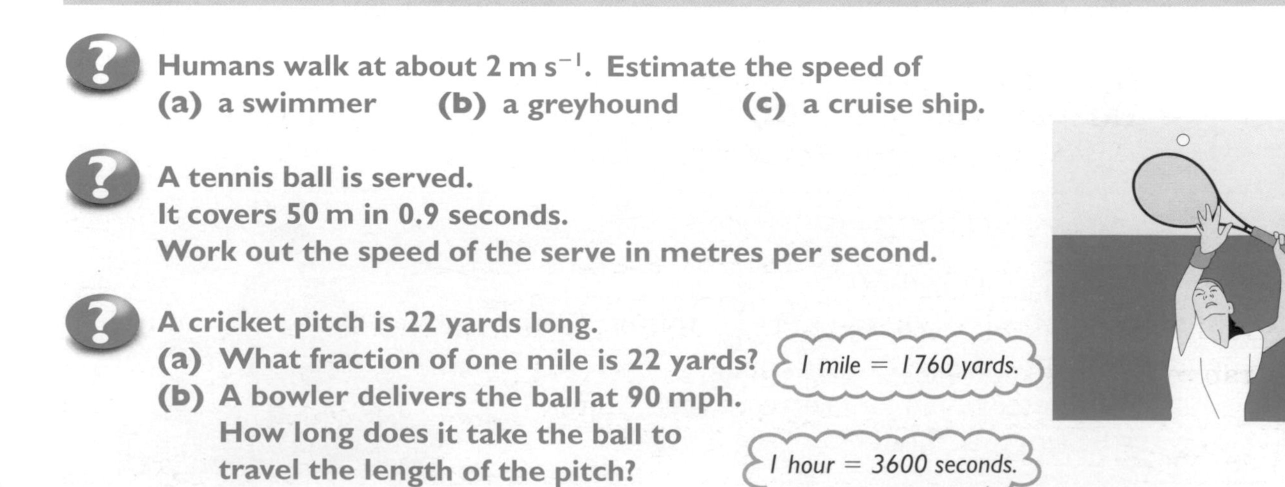

Humans walk at about 2 m s^{-1}. Estimate the speed of
(a) a swimmer (b) a greyhound (c) a cruise ship.

A tennis ball is served.
It covers 50 m in 0.9 seconds.
Work out the speed of the serve in metres per second.

A cricket pitch is 22 yards long.
(a) What fraction of one mile is 22 yards?
(b) A bowler delivers the ball at 90 mph.
How long does it take the ball to travel the length of the pitch?

1 mile = 1760 yards.

1 hour = 3600 seconds.

Exercise

1. John is driving at $90\ \mathrm{km\,h^{-1}}$. How long does it take him to drive
   (a) 90 km **(b)** 270 km **(c)** 45 km **(d)** 135 km?

2. Caroline is driving at $100\ \mathrm{km\,h^{-1}}$. How far does she travel in
   (a) 1 hour **(b)** 3 hours **(c)** 30 minutes **(d)** 1 hour 15 minutes?

3. Work out the speed of each of these trains.
   (a) An express train travelling 320 km in 2 hours
   (b) A local train travelling 40 km in 30 minutes
   (c) A freight train travelling 90 km in 1 hour 30 minutes

4. Alice goes out on her bike. She cycles at $20\ \mathrm{km\,h^{-1}}$.
   How many *minutes* does it take her to cycle
   (a) 10 km **(b)** 5 km **(c)** 4 km **(d)** 2 km?

5. Tim is driving at $60\ \mathrm{km\,h^{-1}}$. How far does he travel in
   (a) 20 minutes **(b)** 40 minutes **(c)** 45 minutes?

6. An express train travels at $180\ \mathrm{km\,h^{-1}}$.
   (a) How many kilometres does the train travel in
   (i) one minute **(ii)** one second?
   (b) How many metres does the train travel in one second?

7. Sebastian sprints 100 m in 10 seconds. Work out Sebastian's speed in
   (a) metres per second **(b)** metres per hour **(c)** kilometres per hour.

8. Alana sees a fork of lightning.
   5 seconds later she hears the thunder.
   How far away is the lightning?

   Sound travels at 331 metres per second.

9. There are yellow mileposts every mile along railway lines.
   Simon is travelling on a train.
   He times how long it takes between two mileposts.
   His train takes 36 seconds to cover one mile.
   How fast is the train travelling in mph?

Activity

The speed of music is measured in 'beats per minute' (bpm).

1. A funk record's speed is 120 bpm. It lasts for 3 minutes. How many beats does it have?
2. A rap record has 450 beats. It lasts for 5 minutes. What is its speed in bpm?
3. A drum and bass record has a speed of 170 bpm. The record has 850 beats.
   How long does it last?

Finishing off

Now that you have finished this chapter you should be able to:

- use metric and imperial units
- convert between units
- make estimates of conversions
- choose units sensibly
- solve problems involving time
- work with speed, distance and time taken.

Review exercise

1 John wants to know if a desk will fit in his bedroom.
The space he has is 4 feet long. The desk is 50 inches long.
Will it fit? (Show your working.)

2 What fraction of a mile is 440 yards?

Museum Entrance
440 yards

3 A cricket pitch is 22 yards long. How many feet is 22 yards?

4 A shot putt weighs 22 pounds. Convert 22 pounds to kilograms.

5 Copy and complete these conversions.

(a) 5 stones =	□ pounds	**(b)** 20 feet =	□ inches
(c) 48 ounces =	□ pounds	**(d)** 35 pounds =	□ stones
(e) 18 inches =	□ feet	**(f)** 5 pounds =	□ ounces
(g) 5250 g =	□ kg	**(h)** 435 g =	□ kg
(i) 0.07 kg =	□ g	**(j)** 342 m =	□ cm
(k) 0.2 m =	□ cm	**(l)** 56 cm =	□ m

6 The world's largest butterfly is the Queen Alexandra Birdwing.
It has a wing span of 28 cm. What is this in inches?

7 On a football pitch, the penalty area extends 54 feet from the goal.
Convert 54 feet to metres.

8 Joanne wants to buy some trousers.
They are 80 cm long. Joanne's size is 30 inches.
(a) Convert 30 inches to cm.
(b) Will the trousers fit Joanne?

9 Convert each of these distances to sensible units. Show your working.
(a) The distance from London to New York is 556 400 000 cm.
(b) Mount Everest is 8 848 000 mm high.
(c) The River Nile is 6 695 000 000 mm long.

10 An ostrich egg weighs 1.5 kg. What is this in pounds?

11

> **Rule 10a**
>
> The top of the basketball net must be 10 feet from the ground.

Christophe is measuring where to put a basketball net.
His tape measure is marked in metres.
How many metres is 10 feet?

12 Copy and complete this table.

Depart Exeter	1237	1336	1530
Arrive Reading	1416	1534	1716
Length of journey			

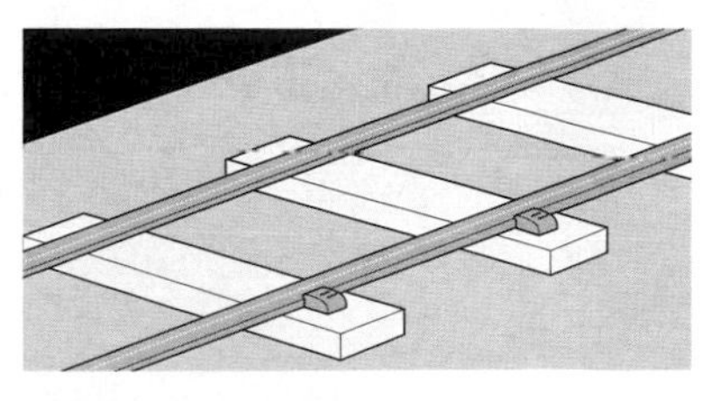

13 Convert these times to minutes.

(a) 3.5 hours **(b)** 4.1 hours **(c)** 1.2 hours **(d)** 0.05 hours

14 **(a)** Susan drives 170 miles in 4 hours 15 minutes. What is her speed?

(b) Jasmine walks 5 km in 45 minutes. What is her speed in km h^{-1}?

15 **(a)** Naomi cycles at 24 km h^{-1} for 36 minutes.
How far does she cycle (in km)?

(b) Kushal walks at 6 km h^{-1} for 20 minutes.
How far does he walk (in km)?

16 **(a)** Elizabeth walks 4 km at a speed of 6 km h^{-1}.
How long does her walk take?

(b) Jed cycles 8 km at a speed of 20 km h^{-1}.
How long does his ride take?

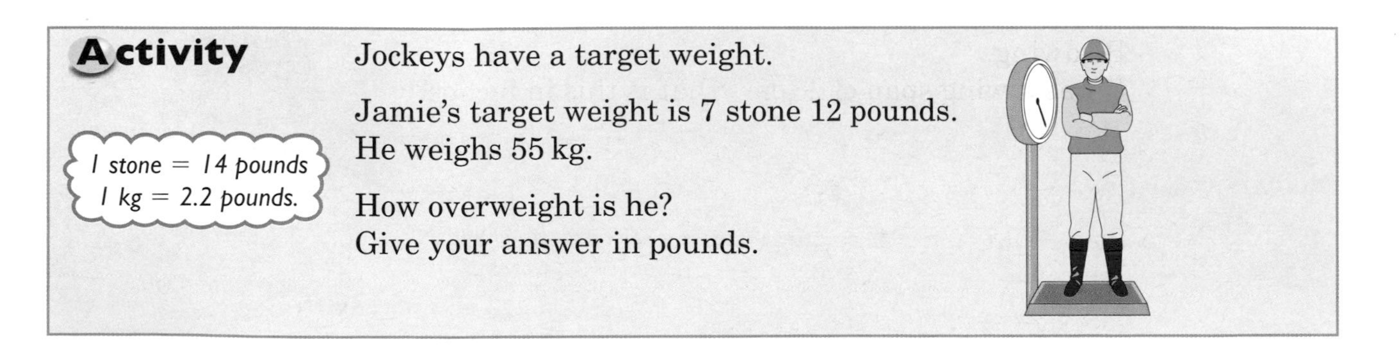

Activity

Jockeys have a target weight.

Jamie's target weight is 7 stone 12 pounds.
He weighs 55 kg.

How overweight is he?
Give your answer in pounds.

1 stone = 14 pounds
1 kg = 2.2 pounds.

Activity

Elaine's car will travel 60 miles on 1 gallon of petrol.

(a) How many *kilometres* can she travel on one gallon of petrol?

(b) How many kilometres can she travel on one *litre* of petrol?

2 Algebraic expressions

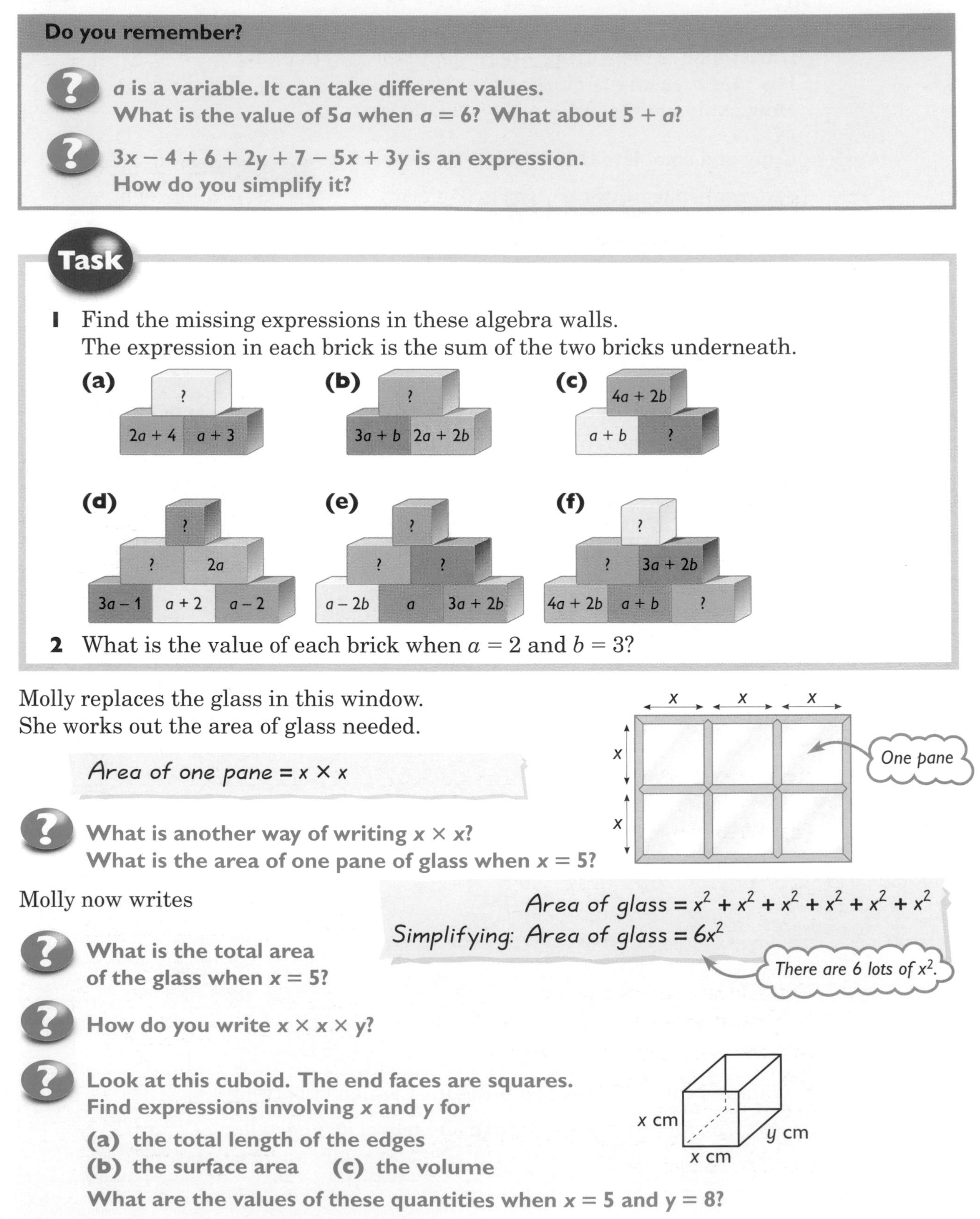

Expressions

Do you remember?

a is a variable. It can take different values.
What is the value of $5a$ when $a = 6$? What about $5 + a$?

$3x - 4 + 6 + 2y + 7 - 5x + 3y$ is an expression.
How do you simplify it?

Task

1 Find the missing expressions in these algebra walls.
The expression in each brick is the sum of the two bricks underneath.

2 What is the value of each brick when $a = 2$ and $b = 3$?

Molly replaces the glass in this window.
She works out the area of glass needed.

Area of one pane = $x \times x$

What is another way of writing $x \times x$?
What is the area of one pane of glass when $x = 5$?

Molly now writes

Area of glass = $x^2 + x^2 + x^2 + x^2 + x^2 + x^2$
Simplifying: Area of glass = $6x^2$

What is the total area of the glass when $x = 5$?

How do you write $x \times x \times y$?

Look at this cuboid. The end faces are squares.
Find expressions involving x and y for
(a) the total length of the edges
(b) the surface area (c) the volume
What are the values of these quantities when $x = 5$ and $y = 8$?

Exercise

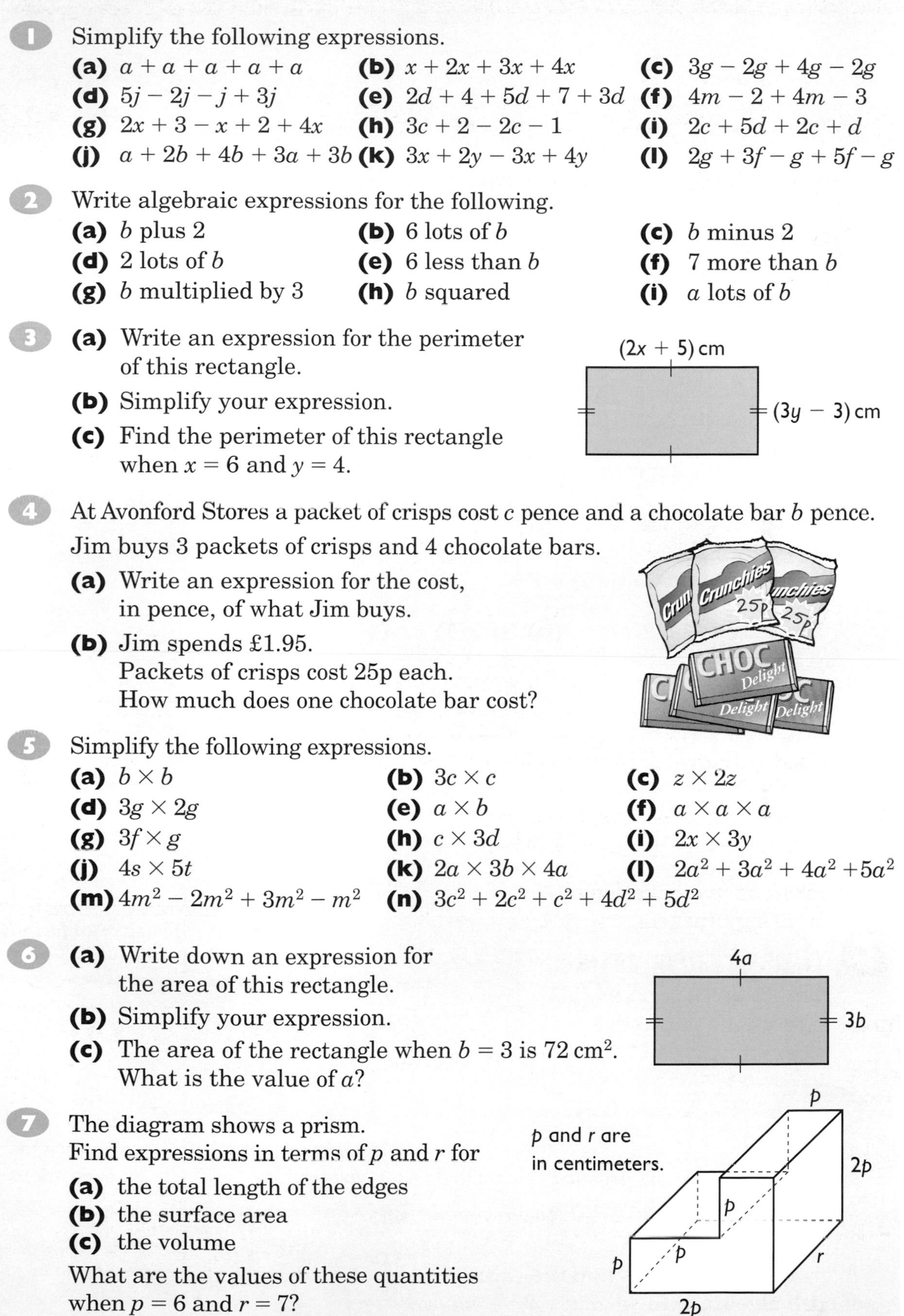

1 Simplify the following expressions.

(a) $a + a + a + a + a$ **(b)** $x + 2x + 3x + 4x$ **(c)** $3g - 2g + 4g - 2g$
(d) $5j - 2j - j + 3j$ **(e)** $2d + 4 + 5d + 7 + 3d$ **(f)** $4m - 2 + 4m - 3$
(g) $2x + 3 - x + 2 + 4x$ **(h)** $3c + 2 - 2c - 1$ **(i)** $2c + 5d + 2c + d$
(j) $a + 2b + 4b + 3a + 3b$ **(k)** $3x + 2y - 3x + 4y$ **(l)** $2g + 3f - g + 5f - g$

2 Write algebraic expressions for the following.

(a) b plus 2 **(b)** 6 lots of b **(c)** b minus 2
(d) 2 lots of b **(e)** 6 less than b **(f)** 7 more than b
(g) b multiplied by 3 **(h)** b squared **(i)** a lots of b

3 **(a)** Write an expression for the perimeter of this rectangle.
(b) Simplify your expression.
(c) Find the perimeter of this rectangle when $x = 6$ and $y = 4$.

4 At Avonford Stores a packet of crisps cost c pence and a chocolate bar b pence.
Jim buys 3 packets of crisps and 4 chocolate bars.
(a) Write an expression for the cost, in pence, of what Jim buys.
(b) Jim spends £1.95.
Packets of crisps cost 25p each.
How much does one chocolate bar cost?

5 Simplify the following expressions.

(a) $b \times b$ **(b)** $3c \times c$ **(c)** $z \times 2z$
(d) $3g \times 2g$ **(e)** $a \times b$ **(f)** $a \times a \times a$
(g) $3f \times g$ **(h)** $c \times 3d$ **(i)** $2x \times 3y$
(j) $4s \times 5t$ **(k)** $2a \times 3b \times 4a$ **(l)** $2a^2 + 3a^2 + 4a^2 + 5a^2$
(m) $4m^2 - 2m^2 + 3m^2 - m^2$ **(n)** $3c^2 + 2c^2 + c^2 + 4d^2 + 5d^2$

6 **(a)** Write down an expression for the area of this rectangle.
(b) Simplify your expression.
(c) The area of the rectangle when $b = 3$ is 72 cm^2.
What is the value of a?

7 The diagram shows a prism.
Find expressions in terms of p and r for
(a) the total length of the edges
(b) the surface area
(c) the volume
What are the values of these quantities when $p = 6$ and $r = 7$?

Brackets

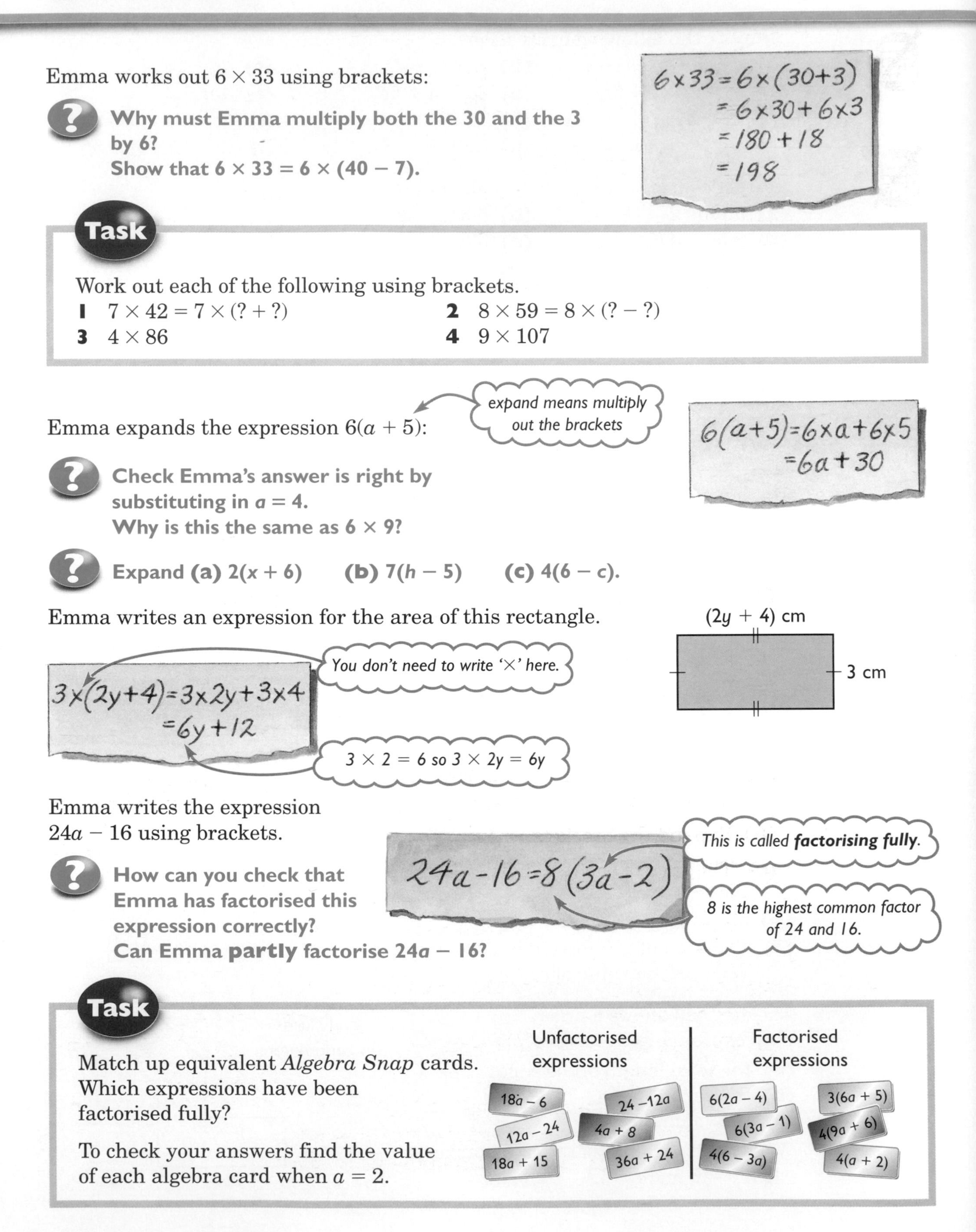

Emma works out 6×33 using brackets:

$6 \times 33 = 6 \times (30 + 3)$
$= 6 \times 30 + 6 \times 3$
$= 180 + 18$
$= 198$

Why must Emma multiply both the 30 and the 3 by 6?
Show that $6 \times 33 = 6 \times (40 - 7)$.

Task

Work out each of the following using brackets.

1 $7 \times 42 = 7 \times (? + ?)$

2 $8 \times 59 = 8 \times (? - ?)$

3 4×86

4 9×107

Emma expands the expression $6(a + 5)$:

expand means multiply out the brackets

$6(a+5) = 6 \times a + 6 \times 5$
$= 6a + 30$

Check Emma's answer is right by substituting in $a = 4$.
Why is this the same as 6×9?

Expand (a) $2(x + 6)$ (b) $7(h - 5)$ (c) $4(6 - c)$.

Emma writes an expression for the area of this rectangle.

$3 \times (2y+4) = 3 \times 2y + 3 \times 4$
$= 6y + 12$

You don't need to write '×' here.

$3 \times 2 = 6$ so $3 \times 2y = 6y$

Emma writes the expression $24a - 16$ using brackets.

$24a - 16 = 8(3a - 2)$

*This is called **factorising fully**.*

8 is the highest common factor of 24 and 16.

How can you check that Emma has factorised this expression correctly?
Can Emma partly factorise $24a - 16$?

Task

Match up equivalent *Algebra Snap* cards.
Which expressions have been factorised fully?

To check your answers find the value of each algebra card when $a = 2$.

Unfactorised expressions	Factorised expressions
$18a - 6$	$6(2a - 4)$
$24 - 12a$	$3(6a + 5)$
$12a - 24$	$6(3a - 1)$
$4a + 8$	$4(9a + 6)$
$18a + 15$	$4(6 - 3a)$
$36a + 24$	$4(a + 2)$

Exercise

1 Work out the following by expanding the brackets.

(a) $7 \times 56 = 7 \times (50 + 6)$ **(b)** $40 \times 65 = 4 \times (70 - 5)$

(c) $8 \times 34 = 8 \times (40 - 6)$ **(d)** $8 \times 99 = 8 \times (100 - 1)$

2 Use brackets to work out the following.

(a) 8×36 **(b)** 9×57 **(c)** 6×89

(d) 99×240 **(e)** 999×240 **(f)** 6002×25

3 Expand the following brackets.

(a) $3(a + 2)$ **(b)** $5(2 + b)$ **(c)** $7(c - 1)$

(d) $4(1 - d)$ **(e)** $11(5 - e)$ **(f)** $3(f - 2)$

(g) $3(2x + 4)$ **(h)** $5(2g - 3)$ **(i)** $6(2m + 3)$

4 Factorise the following expressions fully.

(a) $6y + 12$ **(b)** $2d + 2$ **(c)** $12f - 18$

(d) $24 - 18p$ **(e)** $15 + 25b$ **(f)** $48x - 30$

(g) $5x + 5y$ **(h)** $10x + 5y$ **(i)** $15x - 35y$

5 **(a)** Write an expression using brackets for the area of this rectangle.

(b) Expand the brackets.

(c) What is the area of the rectangle when $x = 6$?

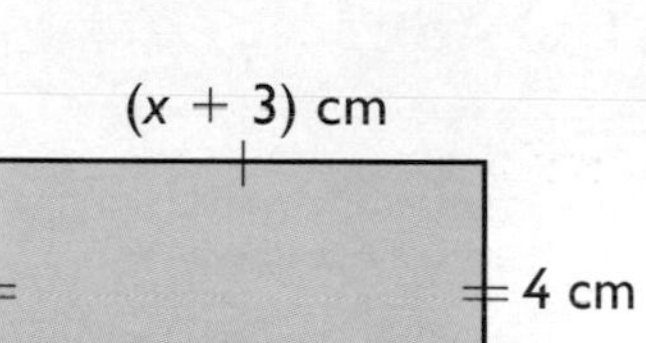

6 **(a)** Write an expression using brackets for the area of this rectangle.

(b) Expand the brackets.

(c) The area of the rectangle is $72\ \text{cm}^2$.

(i) Write an equation using your expression.

(ii) Solve your equation to find the value of p.

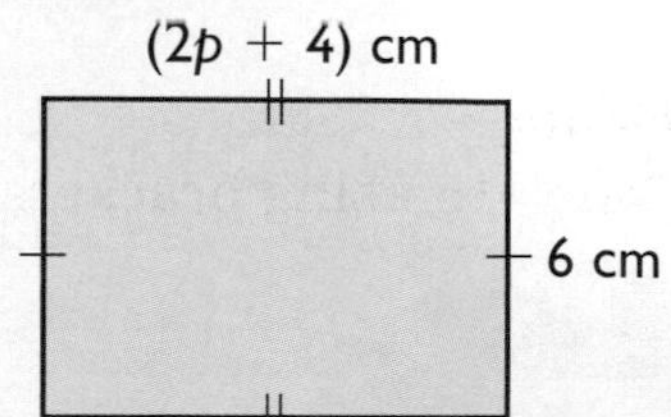

7 **(a)** Write an expression for the perimeter of this triangle.

(b) Expand the brackets and simplify your answer.

(c) The perimeter of the triangle is 39 cm.

(i) Write an equation using your expression.

(ii) Solve your equation to find the value of b.

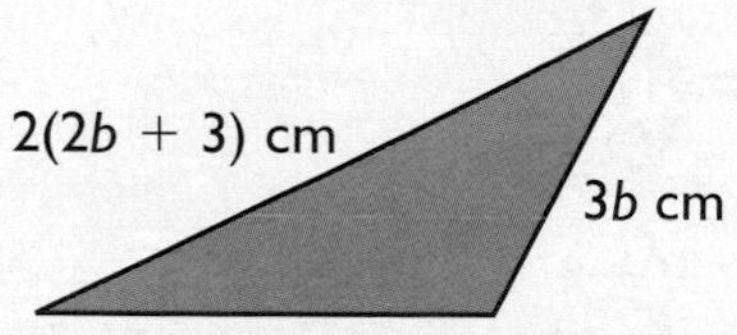

Indices

Do you remember?

$2 \times 2 \times 2 \times 2 \times 2$ can be written 2^5.

This is said '2 to the power of 5' or just '2 to the 5'

When you write a number this way it is in **index** form.

In 2^5, 2 is the **base** and 5 is the **index** (or **power**).

Task

Solve the clues to complete this cross-number.

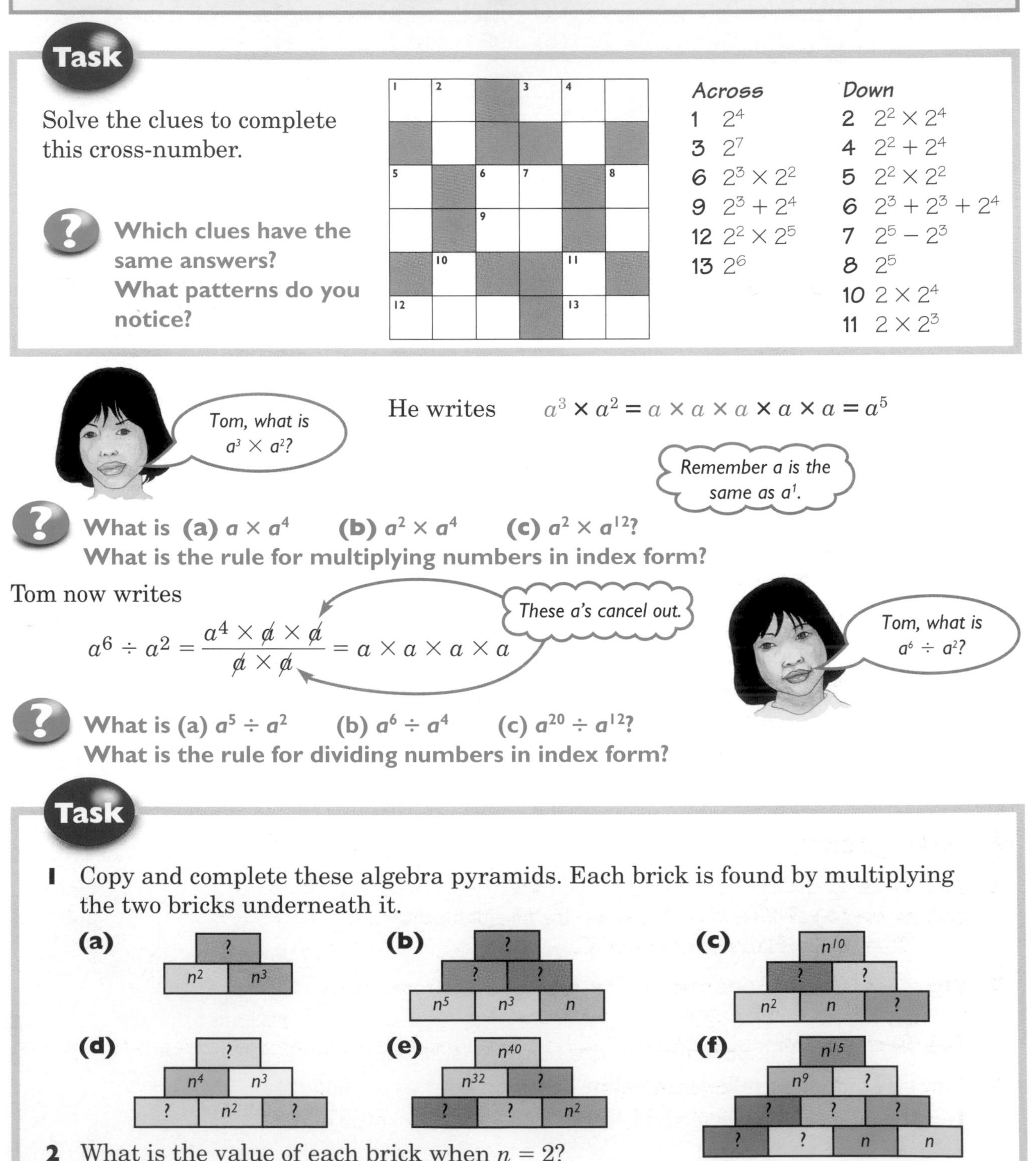

Which clues have the same answers?
What patterns do you notice?

Across	Down
1 2^4	2 $2^2 \times 2^4$
3 2^7	4 $2^2 + 2^4$
6 $2^3 \times 2^2$	5 $2^2 \times 2^2$
9 $2^3 + 2^4$	6 $2^3 + 2^3 + 2^4$
12 $2^2 \times 2^5$	7 $2^5 - 2^3$
13 2^6	8 2^5
	10 2×2^4
	11 2×2^3

Tom, what is $a^3 \times a^2$?

He writes $a^3 \times a^2 = a \times a \times a \times a \times a = a^5$

Remember a is the same as a^1.

What is (a) $a \times a^4$ (b) $a^2 \times a^4$ (c) $a^2 \times a^{12}$?
What is the rule for multiplying numbers in index form?

Tom now writes

$$a^6 \div a^2 = \frac{a^4 \times \cancel{a} \times \cancel{a}}{\cancel{a} \times \cancel{a}} = a \times a \times a \times a$$

These a's cancel out.

Tom, what is $a^6 \div a^2$?

What is (a) $a^5 \div a^2$ (b) $a^6 \div a^4$ (c) $a^{20} \div a^{12}$?
What is the rule for dividing numbers in index form?

Task

1 Copy and complete these algebra pyramids. Each brick is found by multiplying the two bricks underneath it.

(a) top: ?; bottom: n^2, n^3

(b) top: ?; middle: ?, ?; bottom: n^5, n^3, n

(c) top: n^{10}; middle: ?, ?; bottom: n^2, n, ?

(d) top: ?; middle: n^4, n^3; bottom: ?, n^2, ?

(e) top: n^{40}; middle: n^{32}, ?; bottom: ?, ?, n^2

(f) top: n^{15}; second row: n^9, ?; third row: ?, ?, ?; bottom: ?, ?, n, n

2 What is the value of each brick when $n = 2$?

Exercise

1 Write the following as single powers. The first one has been done for you.

(a) (i) $n^3 \times n^4 = n \times n \times n \times n \times n \times n \times n = n^7$

(ii) $n^2 \times n^2 = (?) \times (?) = n^?$ **(iii)** $n^2 \times n^4 = (?) \times (?) = n^?$

(iv) $n^6 \times n^8 = (?) \times (?) = n^?$ **(v)** $n^4 \times n = (?) \times (?) = n^?$

(b) What do you notice?

2 Write the following as single powers. The first one has been done for you.

(a) (i) $n^5 \div n^2 = \frac{n \times n \times n \times \not{n} \times \not{n}}{\not{n} \times \not{n}} = n^3$

(ii) $n^5 \div n = \frac{(?)}{(?)} = n^?$ **(iii)** $n^4 \div n^2 = \frac{(?)}{(?)} = n^?$

(iv) $n^8 \div n^2 = \frac{(?)}{(?)} = n^?$ **(v)** $n^4 \div n^3 = \frac{(?)}{(?)} = n^?$

(b) What do you notice

3 **(a)** Copy and complete these algebra pyramids.

Each brick is found by multiplying together the two bricks underneath it.

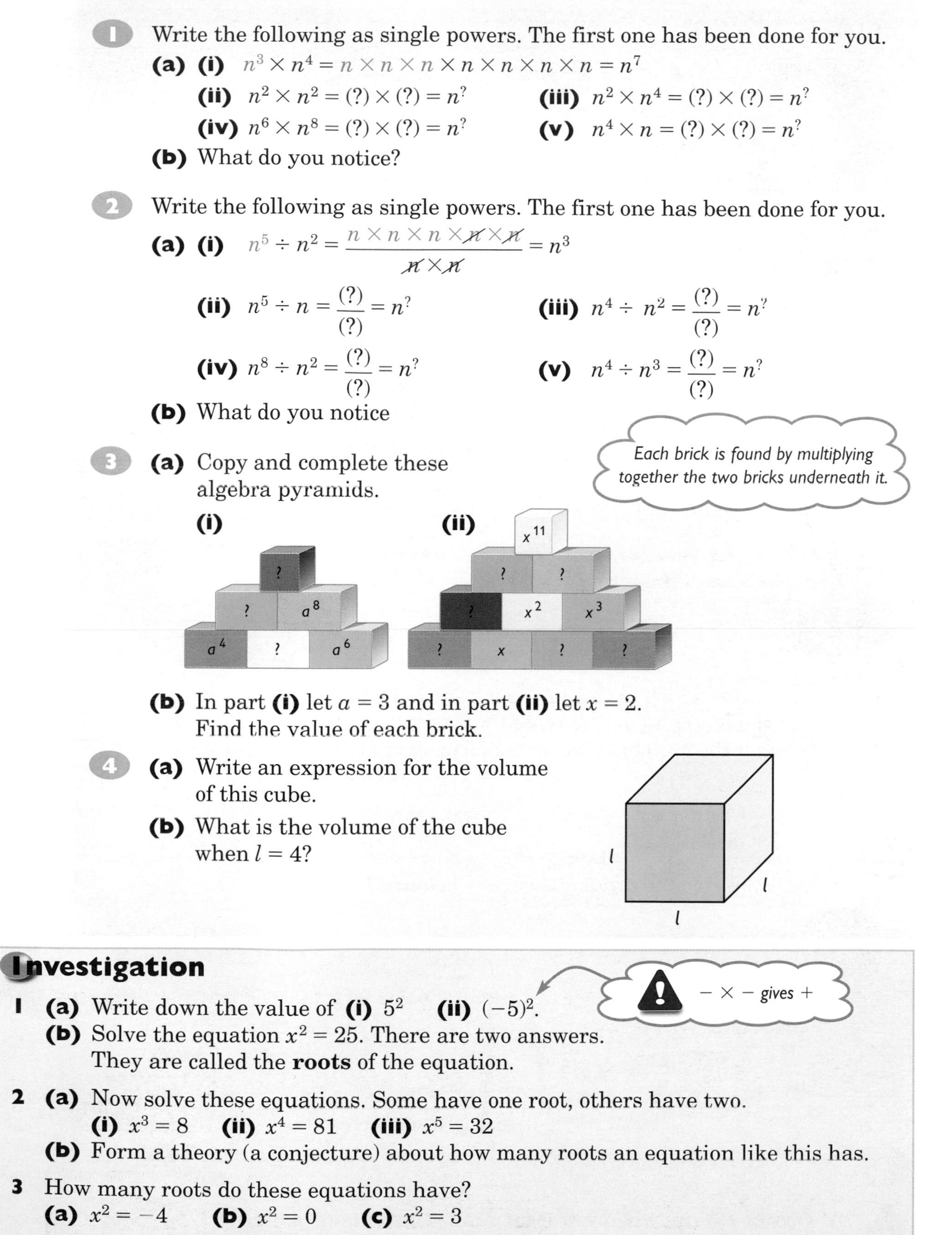

(b) In part **(i)** let $a = 3$ and in part **(ii)** let $x = 2$. Find the value of each brick.

4 **(a)** Write an expression for the volume of this cube.

(b) What is the volume of the cube when $l = 4$?

Investigation

! $- \times -$ gives +

1 **(a)** Write down the value of **(i)** 5^2 **(ii)** $(-5)^2$.

(b) Solve the equation $x^2 = 25$. There are two answers. They are called the **roots** of the equation.

2 **(a)** Now solve these equations. Some have one root, others have two.

(i) $x^3 = 8$ **(ii)** $x^4 = 81$ **(iii)** $x^5 = 32$

(b) Form a theory (a conjecture) about how many roots an equation like this has.

3 How many roots do these equations have?

(a) $x^2 = -4$ **(b)** $x^2 = 0$ **(c)** $x^2 = 3$

Using brackets

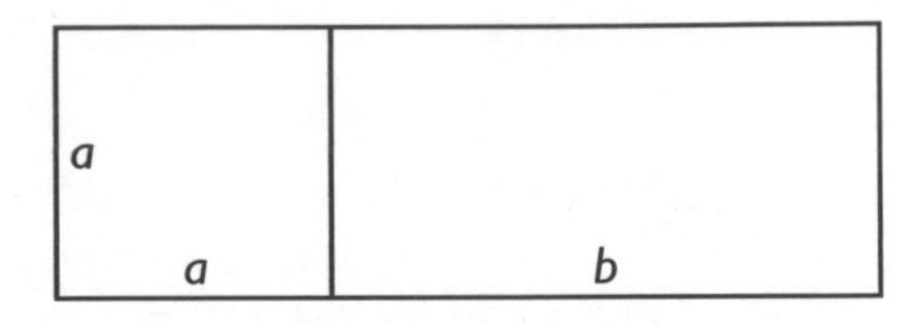

Here are their two answers to the question.

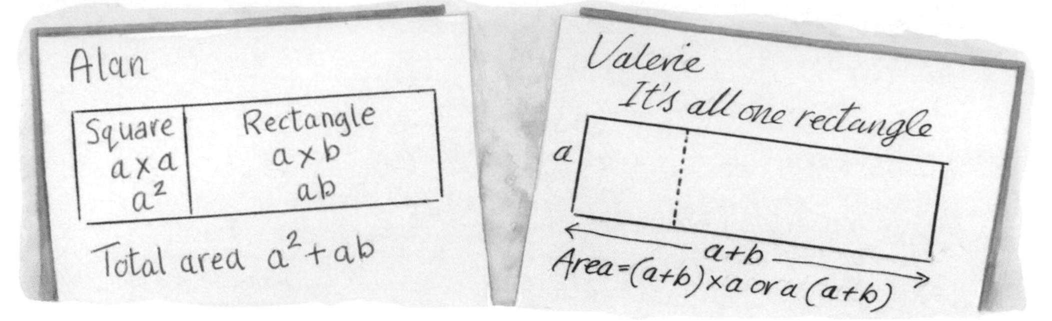

Both of them are right. Alan has given it in **expanded** form, Valerie in **factorised** form.

Work out the answers for particular values of *a* and *b*, like *a* = 5 and *b* = 8. Are they always the same?

Task

Look at the expressions in this list. Some are in factorised form, others in expanded form. Find the matching pairs. Write them in two columns, with the expanded form on the left and the matching factorised form on the right beside it.

$x^2 - xy$	$x(x + 1)$	$xy + y^2$	$x(x - y)$	$x^3 + x^2 + x$	$x^2 + 1$
$x^2 + x$	Cannot be factorised	$x(x - y)$	$x(x^2 + x + 1)$	$y(x + y)$	$x^2 - xy$

Using negative numbers

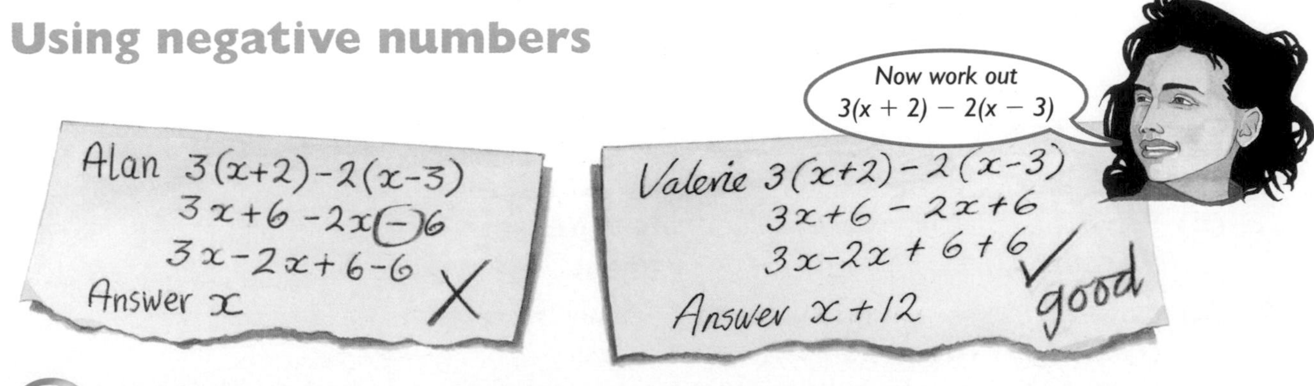

Why is Alan's answer wrong? What mistake has he made?

Always be careful when you see a minus sign outside a bracket. Remember that − × + gives − and that − × − gives +.

Exercise

1 Expand the following expressions.

(a) $3(a - b)$ **(b)** $5(3x + 8y)$ **(c)** $x(x - 1)$
(d) $x(x + 5)$ **(e)** $2x(x + 5)$ **(f)** $2x(3x + 5)$
(g) $x(x + y)$ **(h)** $4x(x - y)$ **(i)** $4x(5x - 7y)$
(j) $x(x^2 - x - 1)$ **(k)** $3x^2(2x + 5)$ **(l)** $3x^2(4x - 11y)$

2 Factorise the following expressions fully.

(a) $5x + 15$ **(b)** $14a + 21b$ **(c)** $xy + xz$
(d) $5xy + 10xz$ **(e)** $3x - ax$ **(f)** $ax^2 + bx^2$
(g) $x^2 - xy$ **(h)** $x^2 - 5xy$ **(i)** $5y + y^2$
(j) $x^3 + x^2 - x$ **(k)** $x^4 + x^2$ **(l)** $x^4 - 9x^2$

3 Expand the brackets in these expressions and then simplify your answers. If you can factorise an answer, do so.

(a) $3(x + 4) + 2(x + 3)$ **(b)** $3(x + 4) + 2(x - 3)$ **(c)** $3(x + 4) - 2(x + 3)$
(d) $3(x + 4) - 2(x - 3)$ **(e)** $3(x - 4) + 2(x + 3)$ **(f)** $3(x - 4) + 2(x - 3)$
(g) $3(x - 4) - 2(x + 3)$ **(h)** $3(x - 4) - 2(x - 3)$ **(i)** $5(2x + 5) - 2(x + 2)$
(j) $4(3x + 2) - 2(x - 6)$ **(k)** $9(3x - 2) - 6(x - 3)$ **(l)** $4(5x - 2) - 2(10x - 7)$

4 Chinua is planning to make this garden pond in Design and Technology. It consists of metal rods joined together. They are then covered in a strong waterproof fabric.

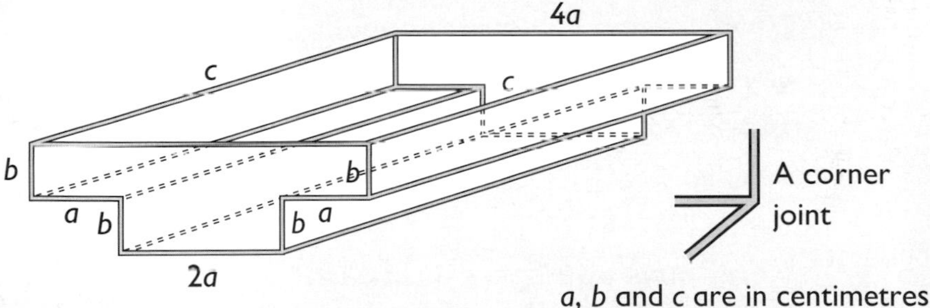

(a) Find expressions for
  (i) the length of metal rod
  (ii) the area of fabric
  (iii) the volume of the pond
  (iv) the number of corner joints.

  Most of your answers will involve the letters a, b and c, and you should give them in factorised form.

(b) Find the values of the four quantities when, $a = 40$, $b = 30$ and $c = 200$. State the units for each answer.

Ruth tells Chinua, 'You should make a and b the same.'

(c) Give the four expressions now, using the letters a and c. They should be in factorised form where possible.

(d) Find their values when $a = 40$ and $c = 150$.

Finishing off

Now that you have finished this chapter you should be able to:

- simplify expressions
- understand the terms: simplify, expand, factorise, power and index
- factorise expressions
- expand brackets
- multiply and divide numbers and expressions in index form.

Review exercise

1 Simplify the following expressions.

(a) $a + a + a + a + a$

(b) $2b + 3b + 4b + 5b$

(c) $5c + 3c - 2c - c$

(d) $2d + 1 + 3d - 4 + 5d$

(e) $2e - 3 + 3e + 5 - e$

(f) $4f + 3g + 2f + 4g + f$

(g) $4h - h + j - 2h + 3j$

(h) $3k - 2m - k - 3m - k + 4m$

(i) $n^2 + n^2 + n^2 + n^2 + n^2$

(j) $4p^2 + 3p^2 + 2p^2 + p^2$

(k) $5q^2 - 3q^2 + 2q^2 + 3q^2$

(l) $7r^2 - 3r^2 - 2r^2 + r^2 - 2r^2$

2 How else can you write the following expressions?

(a) $2 \times a$

(b) $b \times 4$

(c) $c \times c$

(d) $4 \times 3 \times d$

(e) $2e \times 5$

(f) $3f \times 2f$

(g) $5g \times 8g$

(h) $3h \times j \times 3k$

(i) $5m \times 6m \times n$

3 Kerry buys 3 jumpers at £j each and 2 skirts at £s each in the sales.

(a) Write down an expression for what Kerry spends.

(b) Skirts cost £20 each.
Write a new expression for what Kerry spends.

(c) Kerry spends £85 in total.
What is the price of each jumper?

4 Work out the following by using brackets.

(a) 5×34 **(b)** 7×53 **(c)** 9×26

(d) 19×83 **(e)** 81×60 **(f)** 1099×66

5 **(a)** Write each of the following as a single power of 3.

(i) $3 \times 3 \times 3$ **(ii)** 3×3

(iii) $3 \times 3 \times 3 \times 3$ **(iv)** $3 \times 3 \times 3 \times 3 \times 3 \times 3 \times 3$

(b) Work out the answers to (a) by using the index button on your calculator.

6 Write each of the following

(i) in index form

(ii) as an ordinary number.

(a) $3^2 \times 3$ **(b)** $3^2 \times 3^2$ **(c)** $3^5 \times 3^2$

(d) $3^7 \div 3^3$ **(e)** $3^{16} \div 3^{12}$ **(f)** $3^{99} \div 3^{96}$

7 Copy and complete these algebra pyramids.

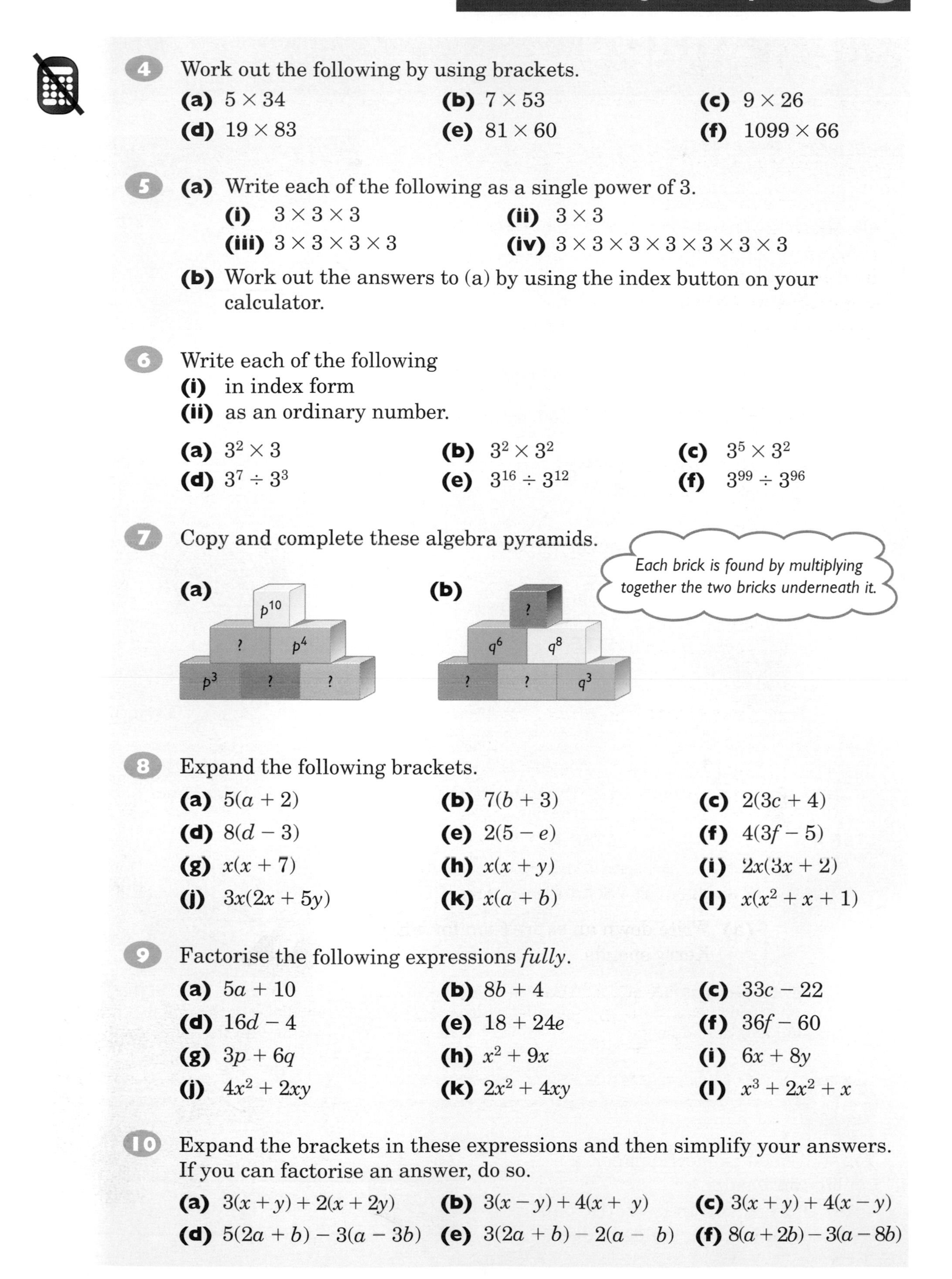

8 Expand the following brackets.

(a) $5(a + 2)$ **(b)** $7(b + 3)$ **(c)** $2(3c + 4)$

(d) $8(d - 3)$ **(e)** $2(5 - e)$ **(f)** $4(3f - 5)$

(g) $x(x + 7)$ **(h)** $x(x + y)$ **(i)** $2x(3x + 2)$

(j) $3x(2x + 5y)$ **(k)** $x(a + b)$ **(l)** $x(x^2 + x + 1)$

9 Factorise the following expressions *fully*.

(a) $5a + 10$ **(b)** $8b + 4$ **(c)** $33c - 22$

(d) $16d - 4$ **(e)** $18 + 24e$ **(f)** $36f - 60$

(g) $3p + 6q$ **(h)** $x^2 + 9x$ **(i)** $6x + 8y$

(j) $4x^2 + 2xy$ **(k)** $2x^2 + 4xy$ **(l)** $x^3 + 2x^2 + x$

10 Expand the brackets in these expressions and then simplify your answers. If you can factorise an answer, do so.

(a) $3(x + y) + 2(x + 2y)$ **(b)** $3(x - y) + 4(x + y)$ **(c)** $3(x + y) + 4(x - y)$

(d) $5(2a + b) - 3(a - 3b)$ **(e)** $3(2a + b) - 2(a - b)$ **(f)** $8(a + 2b) - 3(a - 8b)$

3 Shape

Drawing solids

Heather visits Chicago.

 **Describe the shapes of the buildings.**

Look at this cuboid.
The front face is accurate but the other two are not.

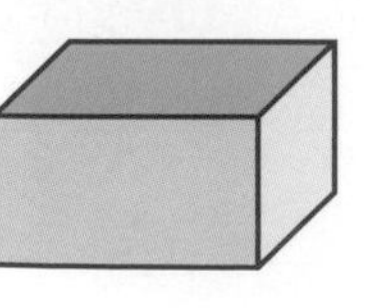

? Which angles in the diagram of the cuboid are really right angles?

You can draw cubes, cuboids and other solid shapes, using isometric paper.

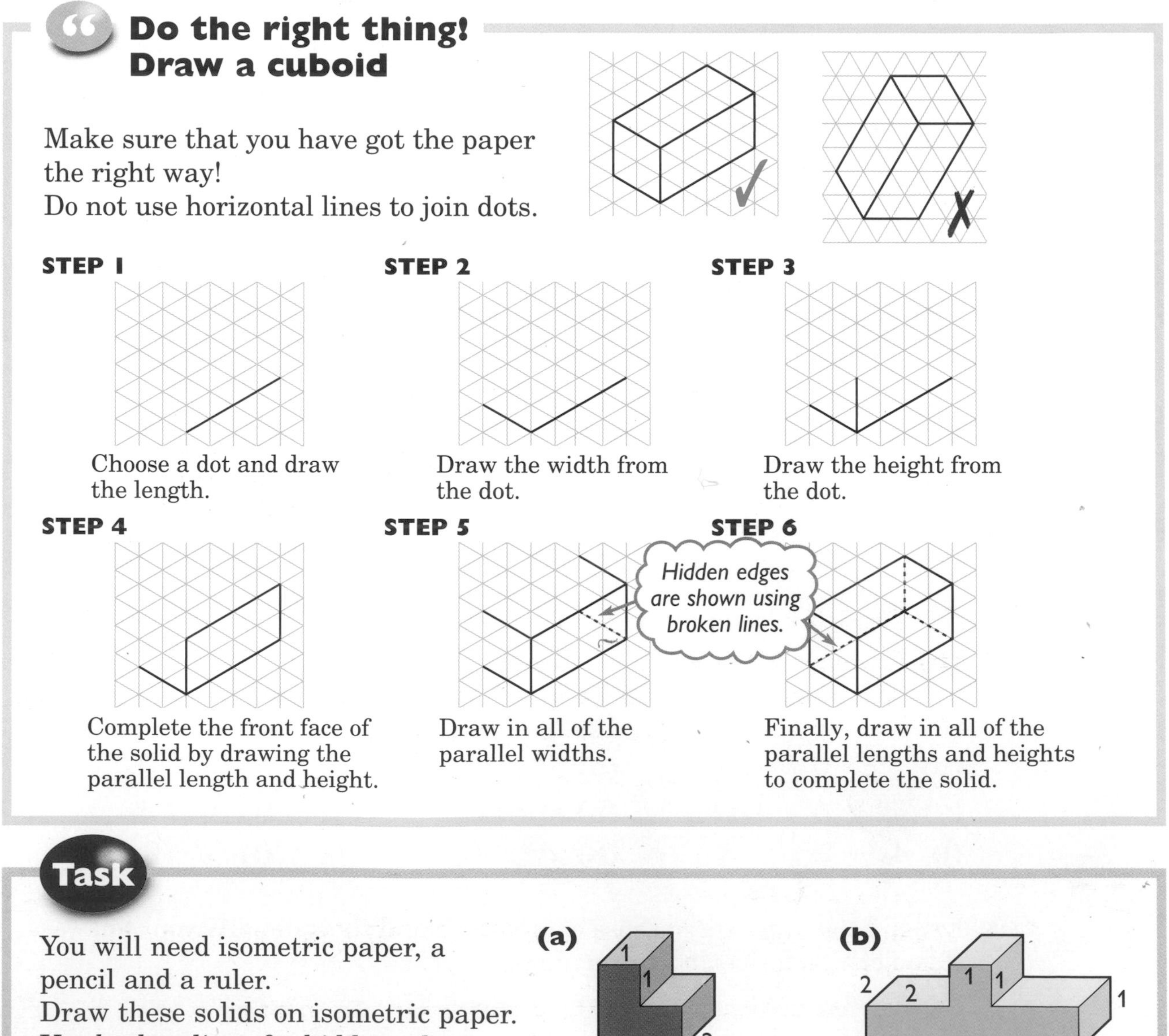

Do the right thing! Draw a cuboid

Make sure that you have got the paper the right way!
Do not use horizontal lines to join dots.

STEP 1
Choose a dot and draw the length.

STEP 2
Draw the width from the dot.

STEP 3
Draw the height from the dot.

STEP 4
Complete the front face of the solid by drawing the parallel length and height.

STEP 5
Draw in all of the parallel widths.

STEP 6
Finally, draw in all of the parallel lengths and heights to complete the solid.

Task

You will need isometric paper, a pencil and a ruler.
Draw these solids on isometric paper.
Use broken lines for hidden edges.

Exercise

1 Draw two **cuboids** of different size on isometric paper.

2 Draw these solids on isometric paper.

(a) **(b)**

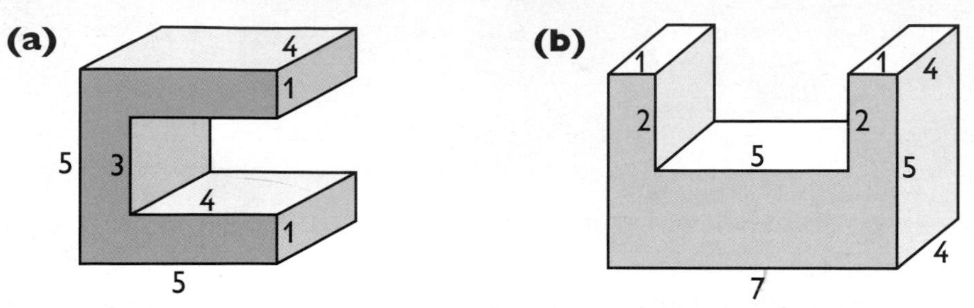

3 Draw the following solids onto isometric paper.

(a)

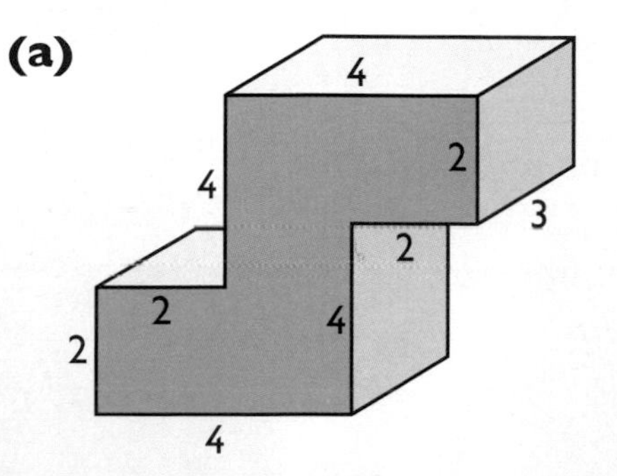

(b)

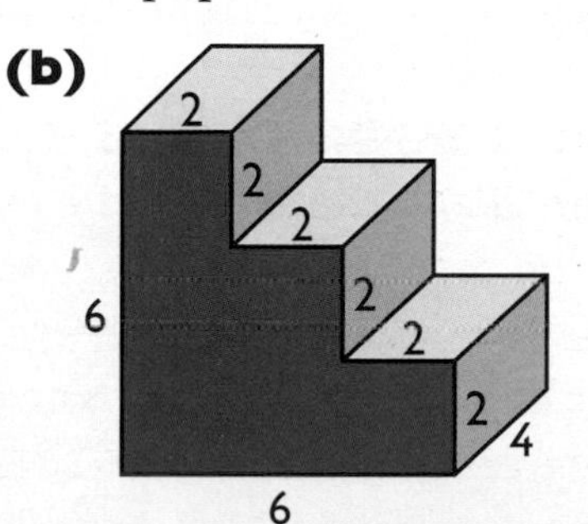

4 **(a)** Carol is playing with toy cubes.
Each cube is 1 cm in length.
She has 9 cubes altogether.
She builds this solid.

Draw this solid on isometric paper.

(b) Carol uses 9 **more** cubes to make another solid.
She joins this solid to her first solid to make a cuboid 3 cm × 3 cm × 2 cm.
Draw Carol's second solid on isometric paper.

Activity Use 1 cm plastic cubes.

Using 4 cubes, make different solids with all of the cubes touching each other, face to face.

Draw your solids on isometric paper and colour them in.

Produce a whole class display.

Plans and elevations

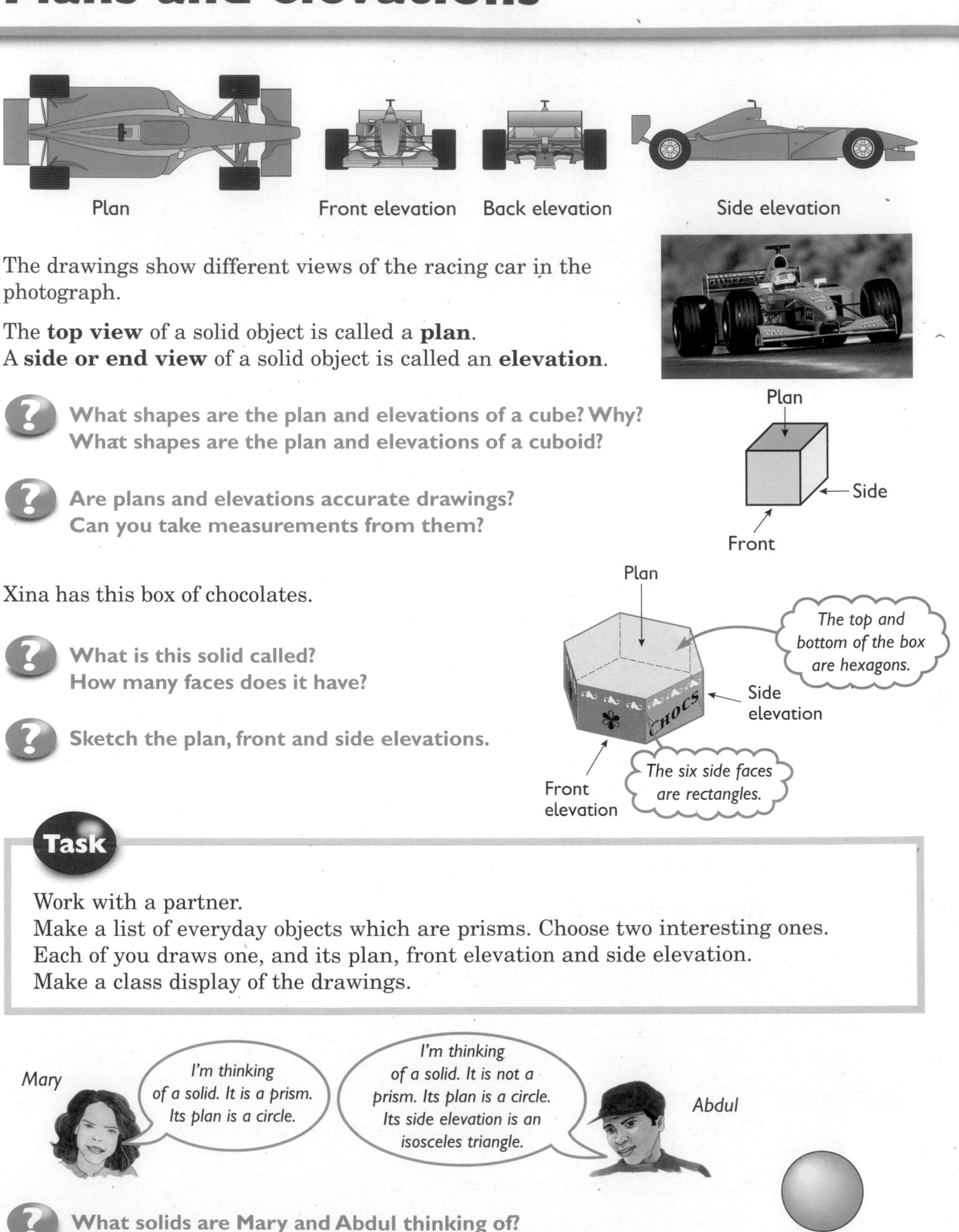

The drawings show different views of the racing car in the photograph.

The **top view** of a solid object is called a **plan**.
A **side or end view** of a solid object is called an **elevation**.

? **What shapes are the plan and elevations of a cube? Why?**
What shapes are the plan and elevations of a cuboid?

? **Are plans and elevations accurate drawings?**
Can you take measurements from them?

Xina has this box of chocolates.

? **What is this solid called?**
How many faces does it have?

? **Sketch the plan, front and side elevations.**

Task

Work with a partner.
Make a list of everyday objects which are prisms. Choose two interesting ones.
Each of you draws one, and its plan, front elevation and side elevation.
Make a class display of the drawings.

? **What solids are Mary and Abdul thinking of?**
What shapes are the plan and elevations of a sphere?

Exercise

1 Look at these solids.
Draw the plan, and front and side elevations of each solid on squared paper.

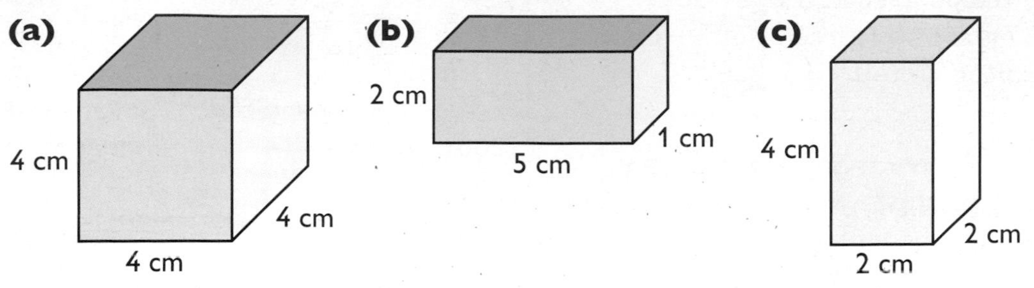

2 Use squared paper to draw the plan, and front and side elevations of this solid.

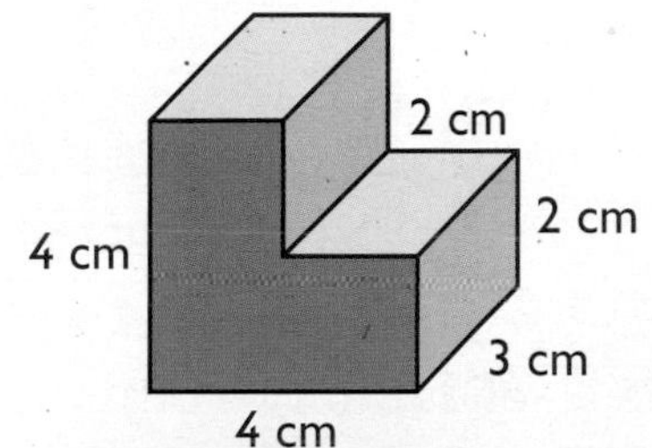

3 A solid object is made from centimetre cubes.
Here are its plan and elevations.

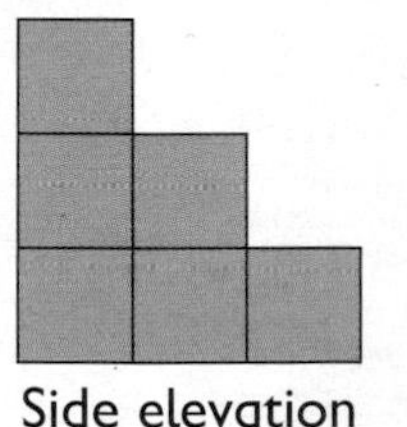
Side elevation

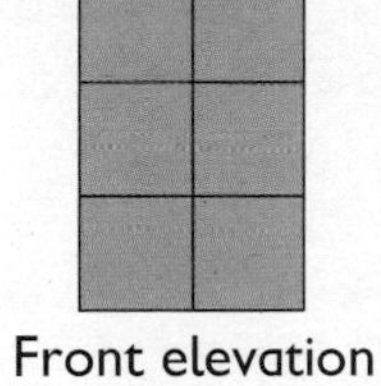
Front elevation

Plan

(a) Work out what the solid looks like.

(b) Draw the solid on isometric paper.

4 Here is a sketch of the plan and elevations of a solid prism.

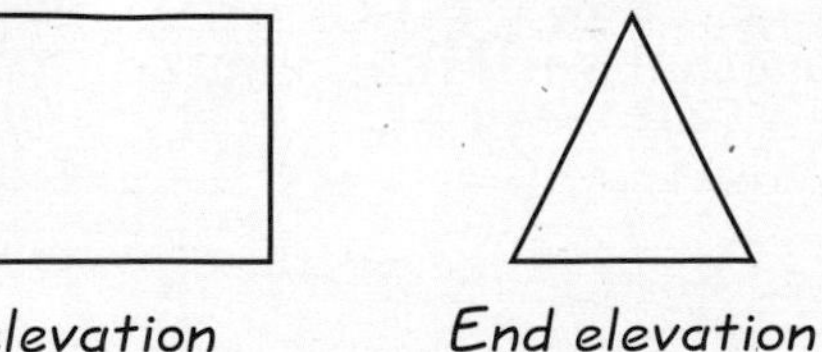
Side elevation *End elevation*

Plan

(a) Copy each sketch.
Work out what the solid looks like.

(b) Name the solid and sketch it.

Planes of symmetry

Aisha cooks a meal for her family. The recipe uses 125 g of butter. She cuts a 250 g rectangular block of butter in half.

Sagalou

125 g butter	1 tablespoon chilli powder
1 large onion	2 green chillies
2 large potatoes, diced	2 tomatoes
200 g spinach	1 teaspoon salt

In how many ways can she do this?

Three ways are shown here.

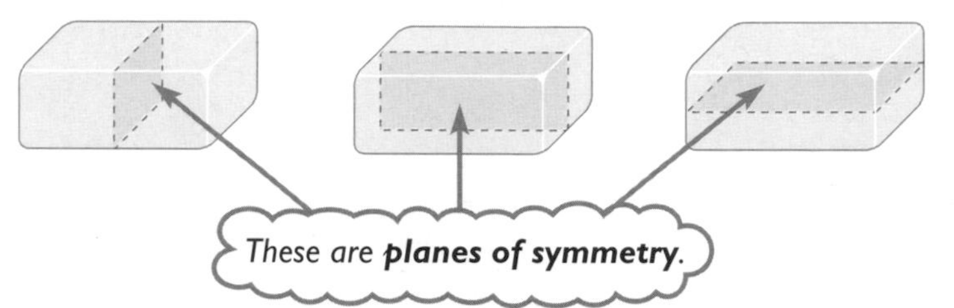

A **plane of symmetry** cuts a solid into two **equal** parts; one part is a **reflection** of the other.
A plane of symmetry is like a flat mirror.

Why does a diagonal cut not give a plane of symmetry?
How many planes of symmetry does a cuboid have?

Here is a net of a cuboid.

1 Copy the net onto centimetre-squared paper, and add flaps.
2 Mark on your net the different lines shown on the diagram, using the colours shown here.
3 Cut it out and stick the cuboid together with the coloured lines on the outside.

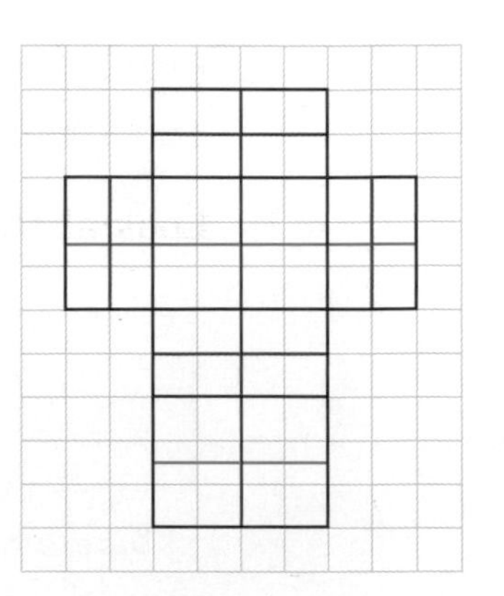

What do the different coloured lines show when the net is folded?
How many planes of symmetry does this cuboid have?

The cross-section of this prism is a regular hexagon.

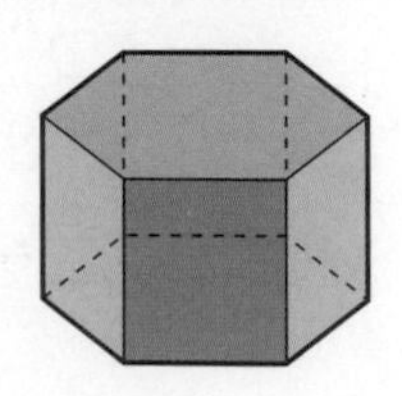

What are the planes of symmetry of this prism?
How many planes of symmetry does this prism have?

Exercise

1 Look at this square-based pyramid.

(a) Using letters, identify the four vertical planes of symmetry (one is shown).

(b) Explain why this solid has no horizontal plane of symmetry.

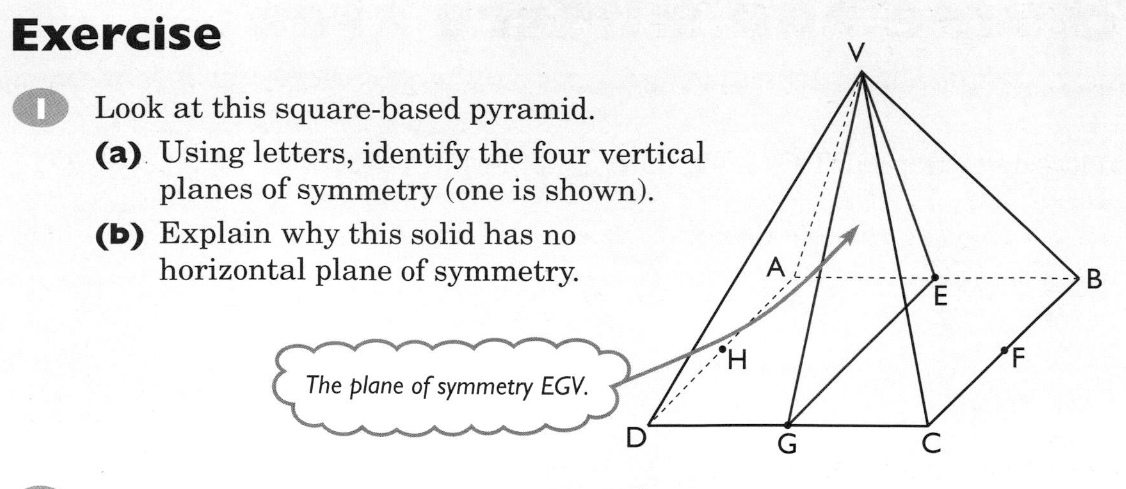

2 Copy this net onto centimetre-squared paper and cut it out.
Stick the cuboid together using flaps.
Label the points marked in the diagram.

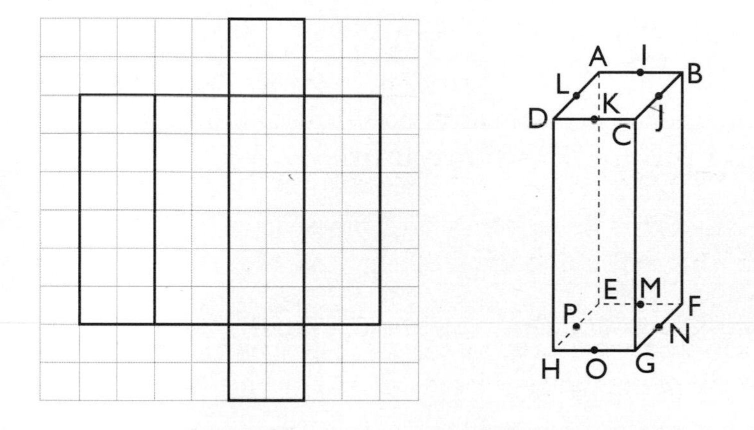

(a) Label the mid-points of edges AE, BF, CG and DH with Q, R, S and T.

(b) Using letters, identify the **horizontal** plane of symmetry in this cuboid.

(c) This cuboid had four vertical planes of symmetry. Using letters, identify them.

3 Look at this triangular prism. Each end is an equilateral triangle.

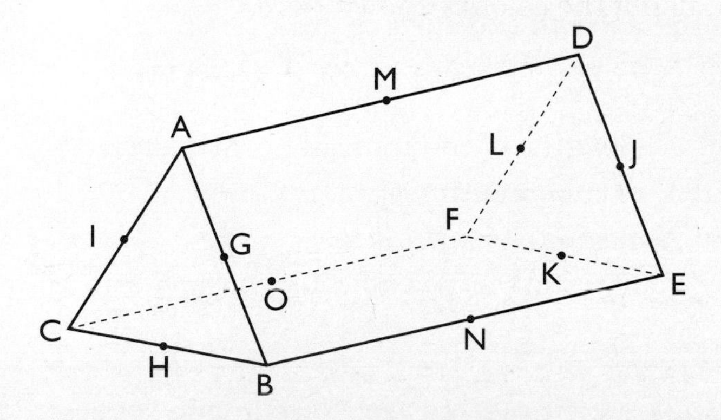

Identify all the planes of symmetry.

Volume and surface area

John is going camping. This is his tent.

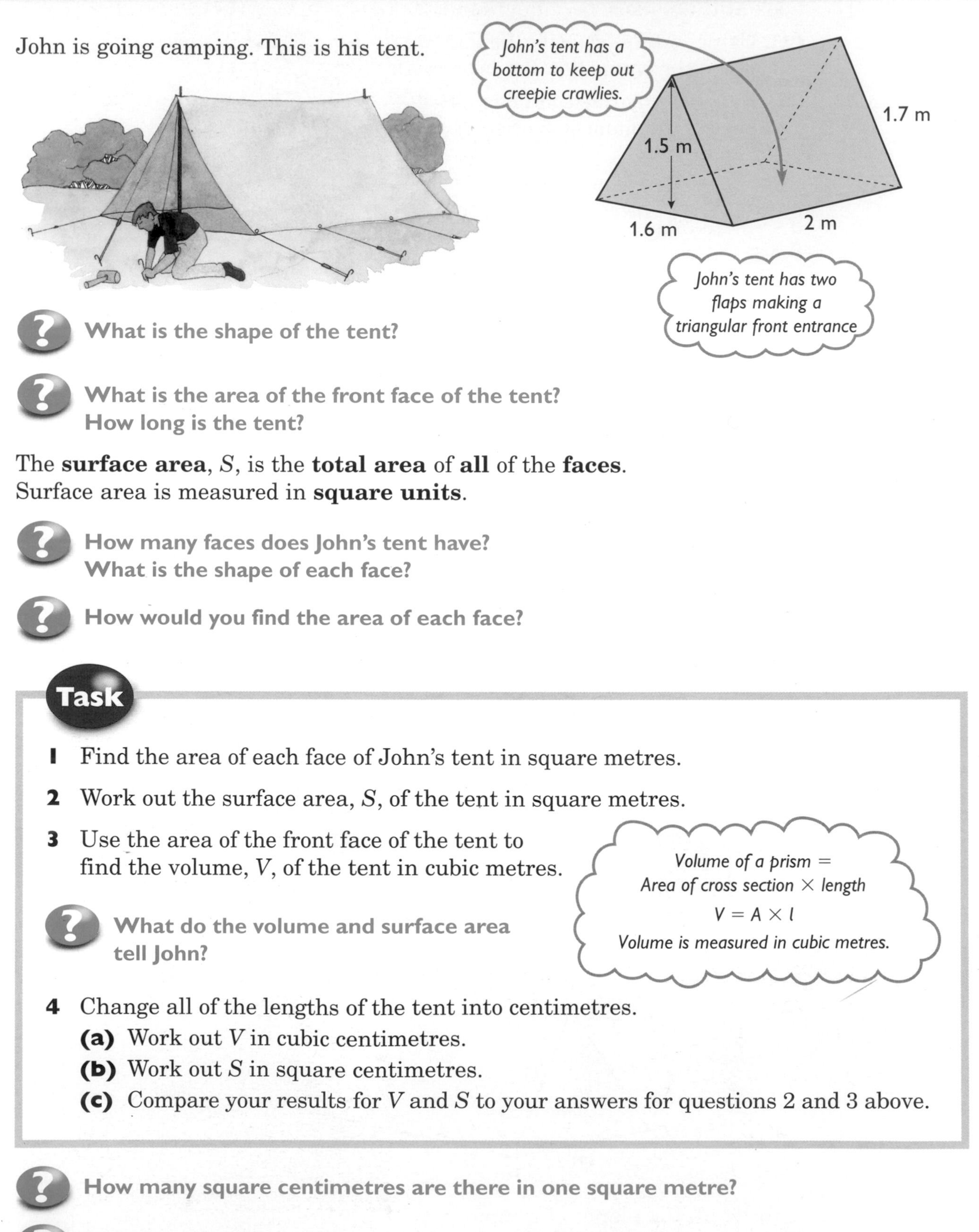

What is the shape of the tent?

What is the area of the front face of the tent?
How long is the tent?

The **surface area**, S, is the **total area** of **all** of the **faces**.
Surface area is measured in **square units**.

How many faces does John's tent have?
What is the shape of each face?

How would you find the area of each face?

Task

1 Find the area of each face of John's tent in square metres.

2 Work out the surface area, S, of the tent in square metres.

3 Use the area of the front face of the tent to find the volume, V, of the tent in cubic metres.

Volume of a prism = Area of cross section × length
$V = A \times l$
Volume is measured in cubic metres.

What do the volume and surface area tell John?

4 Change all of the lengths of the tent into centimetres.
(a) Work out V in cubic centimetres.
(b) Work out S in square centimetres.
(c) Compare your results for V and S to your answers for questions 2 and 3 above.

How many square centimetres are there in one square metre?

How do you change cubic centimetres into cubic metres?

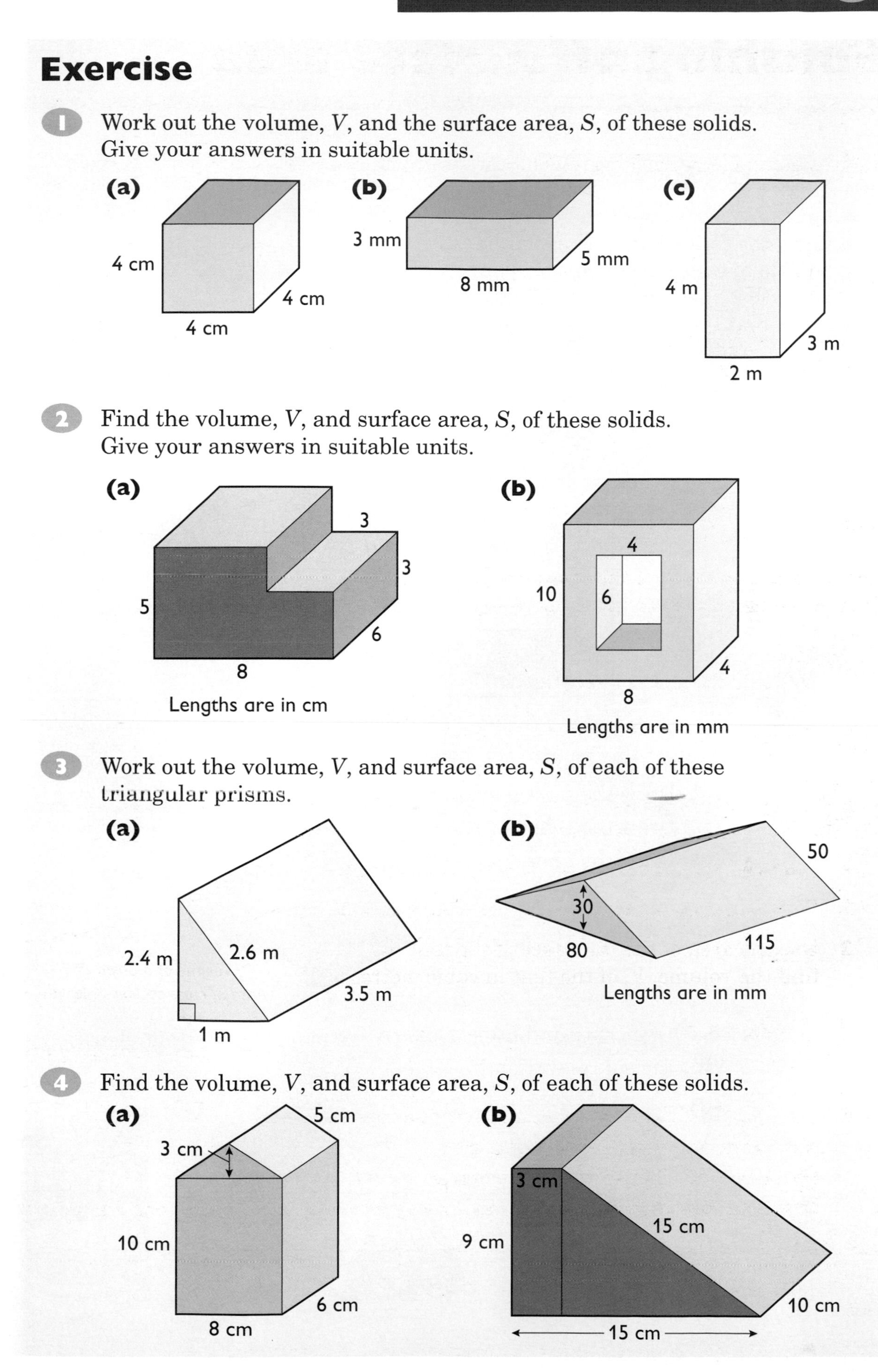

Exercise

1. Work out the volume, V, and the surface area, S, of these solids. Give your answers in suitable units.

2. Find the volume, V, and surface area, S, of these solids. Give your answers in suitable units.

3. Work out the volume, V, and surface area, S, of each of these triangular prisms.

4. Find the volume, V, and surface area, S, of each of these solids.

Finishing off

Now that you have finished this chapter you should:

- be able to draw cubes and cuboids on isometric paper
- know the terms plan and elevation
- be able to recognise planes of symmetry for solids
- know that the surface area of a solid is the total area of all of its faces
- know that the volume of any prism is given by "area of cross-section × length".

Review exercise

1 Draw these solids on isometric paper.

(a)

(b)

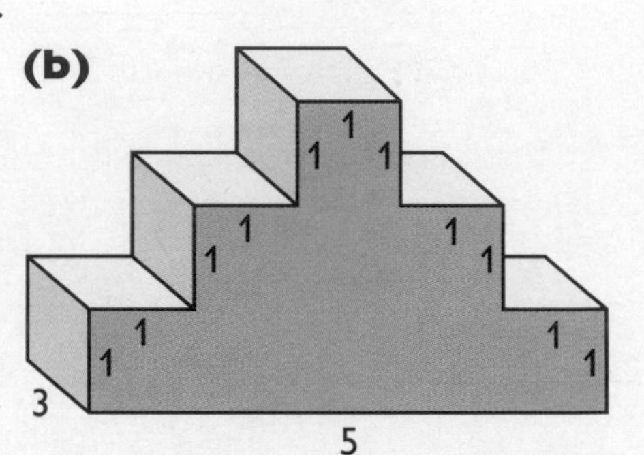

2 Look at this picture of a house.

(a) Draw the front elevation.

(b) Draw the side elevation.

(c) Draw the plan.

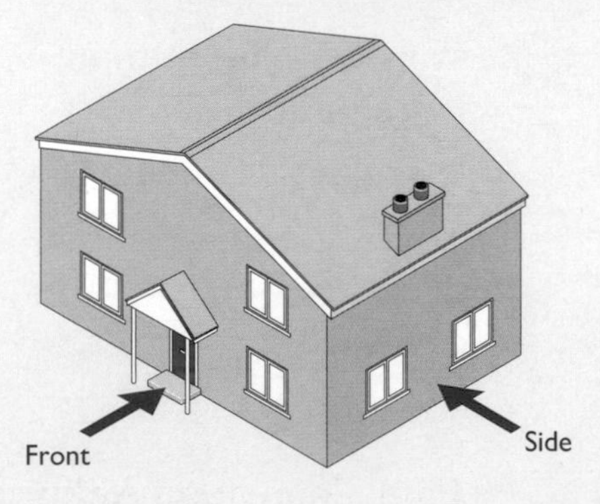

3 For each of these objects, draw **(i)** the plan **(ii)** the front elevation and **(iii)** the side elevation.

(a) **(b)** **(c)**

4 **(a)** What solid is Vashti thinking of?
(b) Sketch the solid.

5 A solid object is made from centimetre cubes.
Its plan and elevations are shown here.

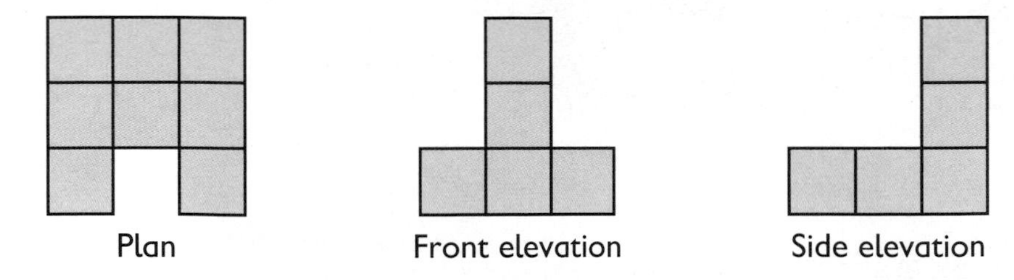

(a) Copy the plan and elevations onto squared paper.
Work out what the solid looks like.
(b) Draw the solid on isometric paper.

6 For each of the following solids
(i) state how many planes of symmetry the solid has
(ii) sketch the solid and mark in the planes of symmetry.

(a) **(b)**

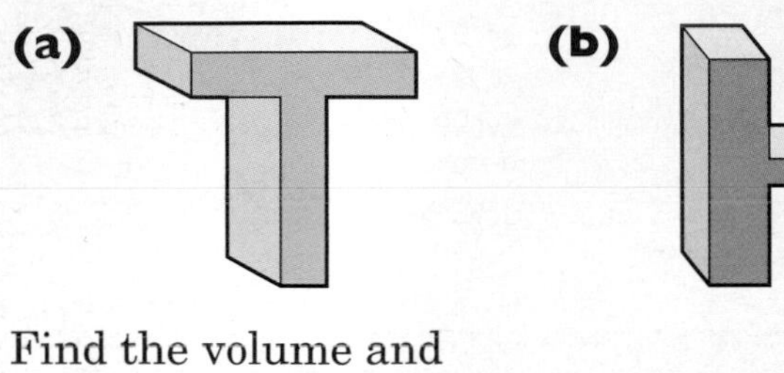

(c)

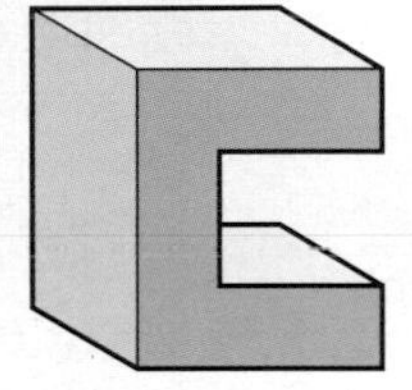
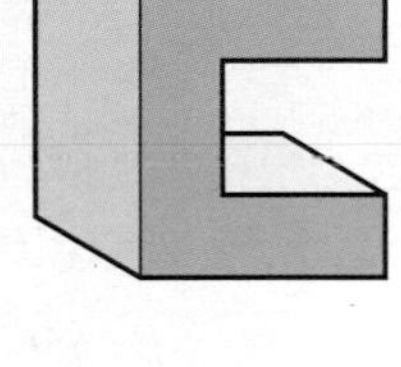

7 Find the volume and surface area of this solid.

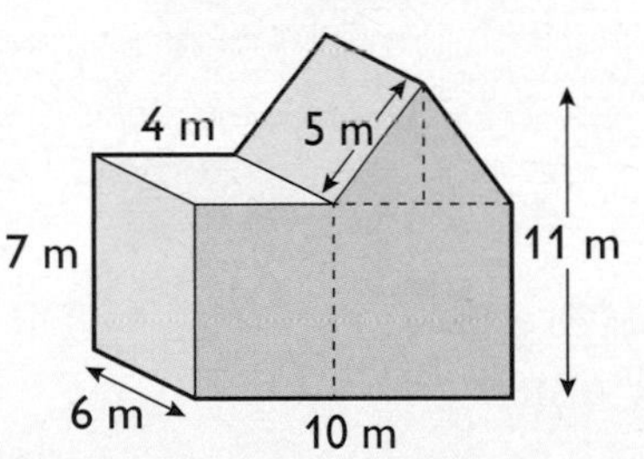

8 Paul, Jenny and Ankur are thinking about a triangular prism.

Who is right?
Give reasons for your answer.

4 Doing a survey

Collecting information

Jack wants to show that his mum is wrong.
He makes a questionnaire.
Here is his first question.

1 Do you watch a lot of TV? Yes ☐ No ☐

Jack's mum says this is not a good question.

What is wrong with this question?

Jack rewrites the question:

1 How many hours of TV do you watch each day?
0–5 hours ☐ 5–10 hours ☐ 10 or more hours ☐

This is still not a good question. What is wrong with it?
How should Jack rewrite the question?

2 I think science fiction is brilliant. What sort of programmes do you like?
Comedy ☐ Science fiction ☐ Soaps ☐

What is wrong with this question?
What is a better question?

Jack tells his friends Molly and Karl about his questionnaire.

Task

Write a short questionnaire that Jack could use.
Include some questions to help Jack find out who is right, Molly or Karl.
Compare your questionnaire with a friend.

Exercise

1 Sally is doing a survey on pocket money.
Here is her questionnaire.

1 How often do you get pocket money?
weekly ☐ monthly ☐

2 What do you spend it on?
clothes ☐ sweets ☐ CDs ☐ magazines ☐ going out ☐

3 How much do you get?
£0–£2 ☐ £2–£5 ☐ £10 ☐ more ☐

4 Do you save any of your pocket money?
yes ☐ no ☐ some of it ☐ all of it ☐ sometimes ☐

5 Everyone should save some of their pocket money, shouldn't they?
yes ☐ no ☐

Sally's questionnaire is not very well designed.

(a) Explain what is wrong with each question.

(b) Write a better questionnaire.

2 Write *one* survey question which you could use to find out about each of the following issues.

Write a sensible set of responses for each question:

(a) how people get to school,

(b) what people think about wearing clothes made of animal fur,

(c) how much exercise people do,

(d) how often people visit the cinema.

3 Design a questionnaire to find out what people think about school uniform. Your questionnaire should cover

- whether people agree with having a school uniform,
- what sort of uniform people would prefer,
- whether there should be rules about jewellery, make-up etc.

Activity Choose a topic that you are interested in.
Design a questionnaire to find out about your chosen topic.

Displaying data

What different ways can you think of to display data?

Here is part of Jack's TV questionnaire.

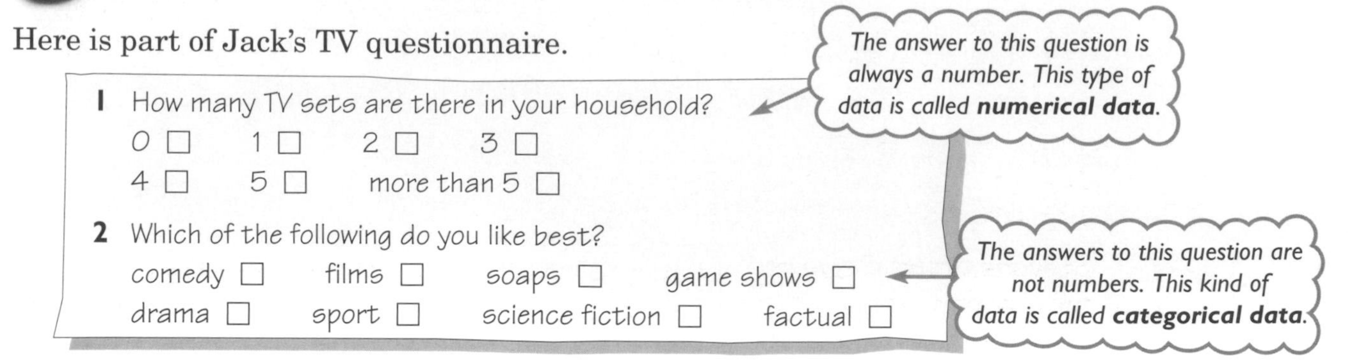

1 How many TV sets are there in your household?
0 ☐ 1 ☐ 2 ☐ 3 ☐
4 ☐ 5 ☐ more than 5 ☐

2 Which of the following do you like best?
comedy ☐ films ☐ soaps ☐ game shows ☐
drama ☐ sport ☐ science fiction ☐ factual ☐

Think of some examples of data to collect for a survey.
Are the data numerical or categorical?

Jack displays his data from Question 2 in his survey.
He makes a tally chart and a bar chart.

Category	Tally	Frequency
Comedy	𝍸	5
Films	|||	3
Soaps	𝍸 |||	8
Game shows			1
Drama	|||	3
Sport	𝍸	5
Science fiction	|||	3
Factual	||	2

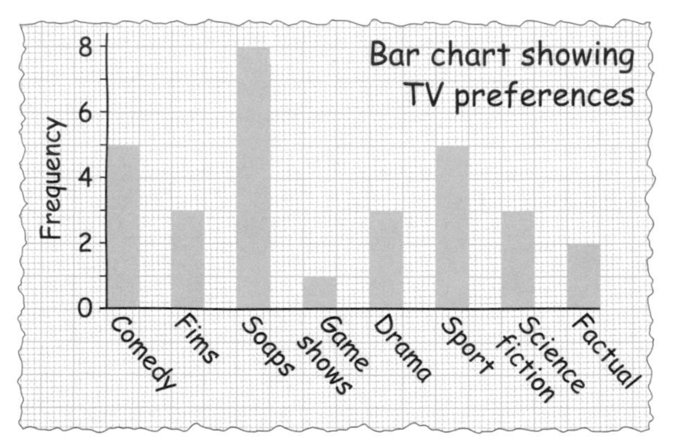

What are the advantages of the bar chart over the tally chart?

Task

Jack makes this two-way table.
It shows the TV preferences of boys and girls separately.

Type	Comedy	Films	Soaps	Game shows	Drama	Sport	Science fiction	Factual
Boys	2	2	1	0	1	5	2	1
Girls	3	1	7	1	2	0	1	1

Display these data in another way.
Your display should make it easy to compare boys' preferences with girls'.

What do these data tell you about the TV preferences of
(a) the boys (b) the girls in Jack's class?

Exercise

1 Mr Berry is a maths teacher.
He wants to know how well his students have done in their Maths GCSE.
Here is a list of his students' grades.

C E B B D E E A F C C C D G
B E C B E F B C A D D E C C
D B C

(a) Make a tally chart to show these results.
(b) How many students are there in Mr Berry's class?
(c) How many students obtained a grade A, B or C?
(d) Draw a bar chart to illustrate these results.

2 Steve is doing research for a travel agent.
He uses this questionnaire.

1 Which of these holiday destinations would you prefer?
Ibiza ☐ Kenya ☐ Florida ☐ Sydney ☐
2 What is your age group?
Under 30 ☐ 30–50 ☐ Over 50 ☐

Steve draws this **compound bar chart** to show his results.

(a) How many people aged 30–50 prefer Ibiza?
Be careful! The answer is *not* 31.

(b) Copy and complete this two-way table.

	Ibiza	Kenya	Florida	Sydney
Under 30	24			
30–50	7			
Over 50	3			
Total	34			

3 Sophie is a keen football supporter.
She writes down the number of goals scored by her team, each match during one season.
Here are her results.

2 1 0 1 1 5 2 3 1 4 3 3 0
1 0 2 2 3 0 1 4 5 2 0 0 2
1 3 2 2 1 0 2 4 1 2 0 3

(a) Make a tally chart to show these results.
(b) How many matches does the team play during the season?
(c) Draw a bar chart to illustrate these results.

Activity Throw two dice 50 times, adding the scores together each time.
1 Record your scores in a tally chart.
2 Draw a bar chart to illustrate your results.
3 Describe the shape of the bar chart. Explain why it is this shape.

Pie charts

Here is another question from Jack's TV questionnaire.

Which of these TV channels do you watch most often?
BBC1 ☐ BBC2 ☐ ITV1 ☐ C4 ☐

Here are the results from this question.

Channel	BBC1	BBC2	ITV1	C4
Frequency	11	6	5	8

Jack wants to draw a pie chart to illustrate these data.

This is how he works out the size of the 'slice' for BBC1:

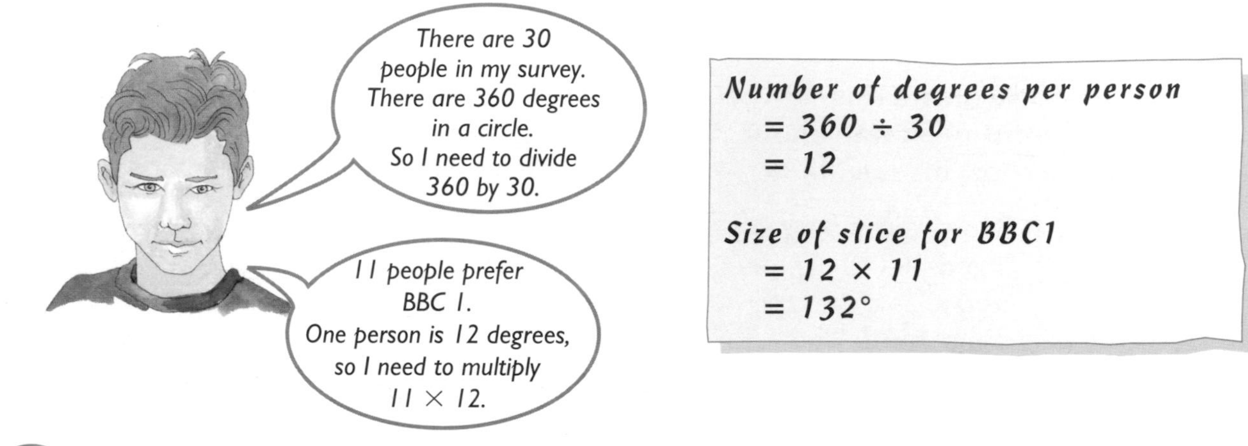

Work out the sizes of the other three 'slices' of the pie chart.

Here is Jack's pie chart.

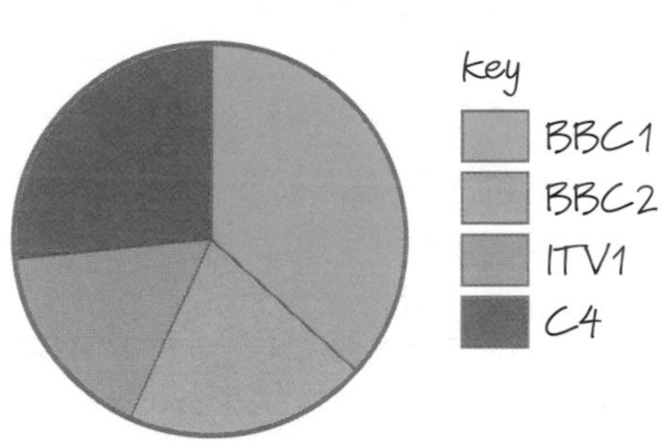

Which is the better way to show categorical data, a bar chart or a pie chart? Which is better for numerical data?

Ask 20 people which TV channel they watch most often.
Draw a pie chart to illustrate your results.

Compare your pie chart with Jack's. What are the differences?

Exercise

1 These two pie charts show the results of two football teams last season.

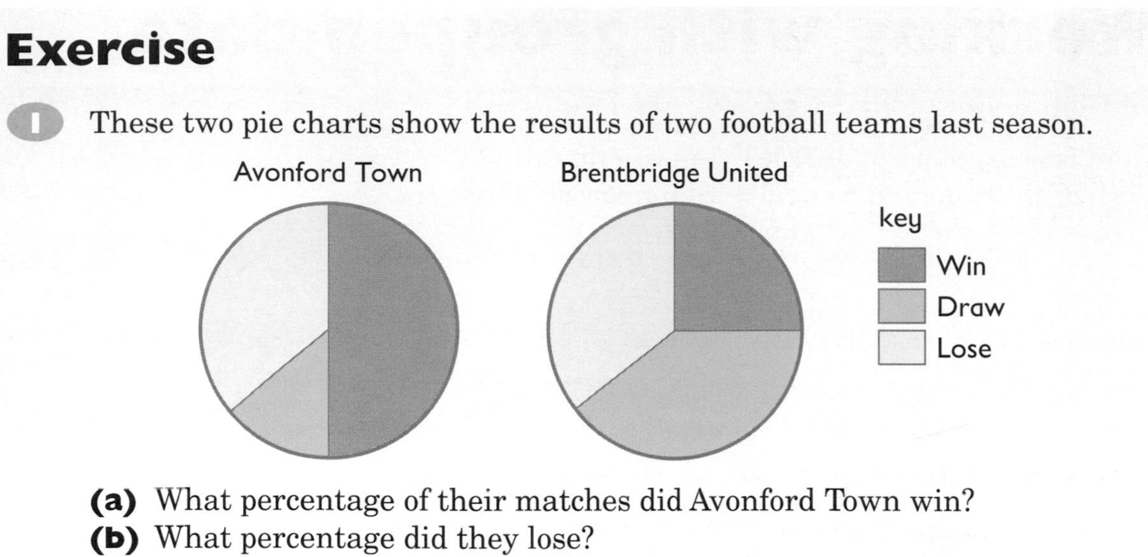

(a) What percentage of their matches did Avonford Town win?
(b) What percentage did they lose?
(c) What percentage of their matches did Brentbridge United draw?
(d) Which team had the better season?

2 Philip wants to draw a pie chart to show how he spends his time one day. He writes down how many hours he spends on each activity.

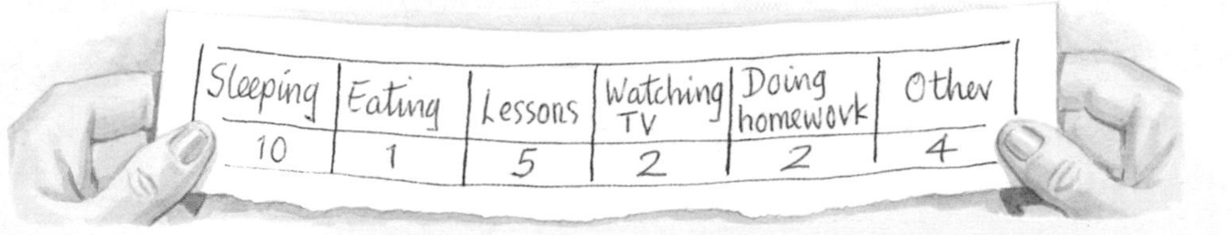

Sleeping	Eating	Lessons	Watching TV	Doing homework	Other
10	1	5	2	2	4

(a) How many degrees should Philip use to represent 1 hour?
(b) Draw Philip's pie chart.

3 Melanie is doing a survey about sport.

Boy ☐ Girl ☐
Which is your favourite sport from the following list?
football ☐ tennis ☐ swimming ☐
hockey ☐ athletics ☐

Here are her results.

	Football	Tennis	Swimming	Hockey	Athletics
Boys	19	5	10	2	4
Girls	3	8	9	11	5

Melanie draws two pie charts to show her results.

One for the boys' results and one for the girls'.

(a) How many boys did Melanie ask?
(b) How many degrees does she use for each boy?
(c) How many girls did she ask?
(d) How many degrees does she use for each girl?
(e) Draw Melanie's two pie charts.

Working with grouped data

Here is some more of Jack's TV questionnaire.
Both of these questions deal with numerical data.

How many TV sets are there in your household?
0 ☐ 1 ☐ 2 ☐ 3 ☐ 4 ☐
5 ☐ More than 5 ☐

How much TV did you watch last week? (t = no. of hours)
$0 \leqslant t < 5$ ☐ $5 \leqslant t < 10$ ☐ $10 \leqslant t < 15$ ☐
$15 \leqslant t < 20$ ☐ $20 \leqslant t < 25$ ☐ $t \geqslant 25$ ☐

*The answer to this question can only take certain values. You cannot have $2\frac{1}{2}$ TV sets! Data like these are called **discrete** data.*

*The answer to this question could take any value. You could have any fraction of an hour. Data like these are called **continuous** data.*

What does $0 \leqslant t < 5$ mean?
Which box should you tick if you watched TV for exactly 10 hours?

Jack draws a bar chart to illustrate the results to the second question.
This kind of bar chart is called a **frequency chart**.

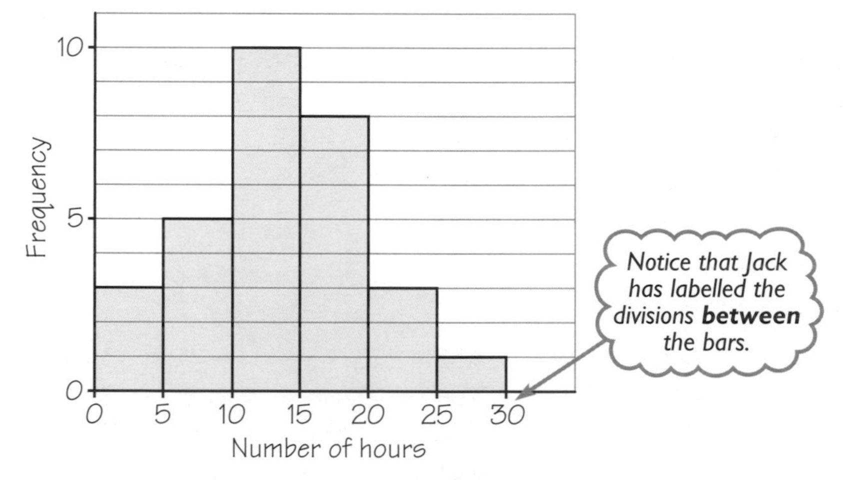

Look back to Jack's conversation with his mum on page 32.
Look at the frequency chart. Who is right, Jack or his mum?

Task

Measure the height of everyone in your maths group.
Choose suitable class intervals and make a **grouped tally chart**.
Draw a frequency chart to illustrate your results.

What does your frequency chart tell you about the heights of people in your maths group?

Exercise

1 State whether each of the following sets of data is categorical, discrete numerical or continuous numerical.

(a) The political parties voted for by people in a village
(b) The heights of a class of schoolchildren
(c) The shoe sizes of a class of schoolchildren
(d) The flavours of crisps preferred by a group of children
(e) The times to run 100 m by a group of athletes.

2 Class 9B have a Science test.
These are their marks, as percentages.
Make a grouped tally chart to show these results.
Use the groups 10–19, 20–29, and so on.

65	72	47	52	55	67
69	58	74	31	66	65
84	22	48	73	82	91
44	50	33	15	65	71
49	52	64	76	88	

3 Ahmed and Sophie both work at a swimming pool.
They each do a survey to find out the ages of people using the pool over a 1-hour period. The charts below show their results.

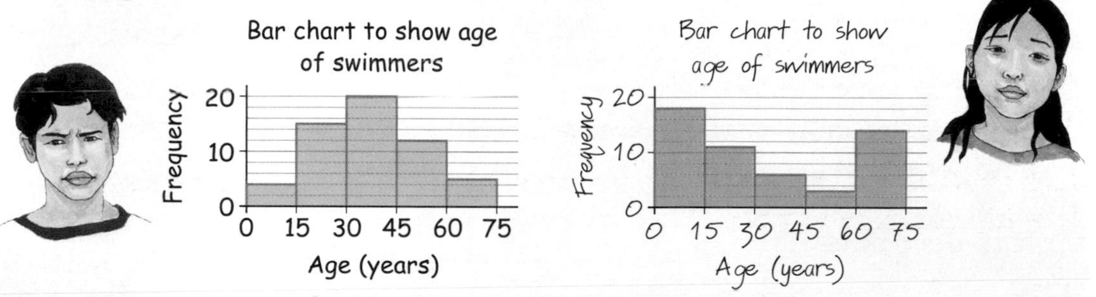

(a) How many people 60 years or over used the pool in each survey group?
(b) How many people under 30 used the pool in each survey group?
(c) One survey was carried out between 7 and 8 pm on a Monday.
The other was carried out between 10 and 11 am on a Tuesday.
Which was which? Why?

4 Liz is a midwife at Avonford Hospital.
She records the birth weights in kilograms for one week.

Boys	3.21	3.64	3.52	2.93	4.13	3.10	3.34	3.81
	3.27	2.75	3.04	3.18	4.09	3.61	3.70	3.15
	2.48	4.46	3.56	3.75	3.22	3.67	2.81	3.94
	3.43	3.87	4.01	2.52				
Girls	2.83	3.42	3.69	3.15	3.04	3.17	2.56	3.21
	3.81	3.60	3.58	2.80	2.95	3.73	3.15	3.31
	2.69	4.12	3.49	3.64	3.52	3.71	3.42	3.83

(a) Draw separate grouped tally charts to show the birth weights of boys and girls.
Use the groups $2.20 \leqslant w < 2.60$, $2.60 \leqslant w < 3.00$, $3.00 \leqslant w < 3.40$, and so on.
(b) Draw a frequency chart to show the boys' weights and another to show the girls' weights.
(c) What differences can you see between the two charts?

Scatter diagrams

Vicki is in the same class as Jack. She is interested in Jack's TV survey.
Vicki thinks that her older sister watches a lot more TV that she does.

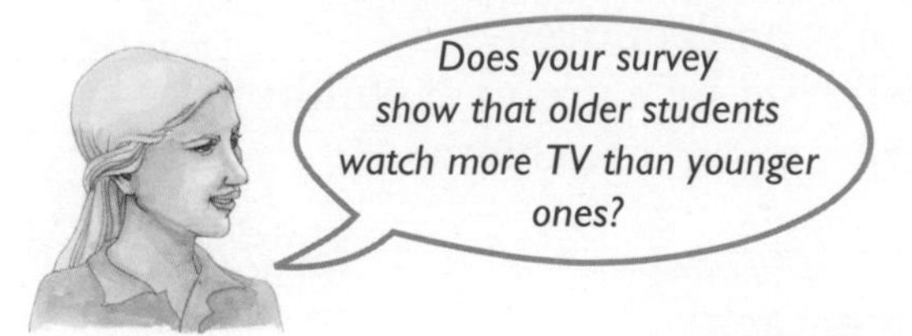

Vicki decides to carry out her own survey.
She chooses 20 people in a variety of year groups.
They each keep a record of how much TV they watch during one week.

Name	Age	Hours of TV	Name	Age	Hours of TV
Amy	11	8	Emily	12	7
Euan	13	9	Matthew	13	16
Hassan	14	23	Vicki	14	19

Vicki draws this **scatter diagram** to show her results.
She also draws a **line of best fit** through the points.

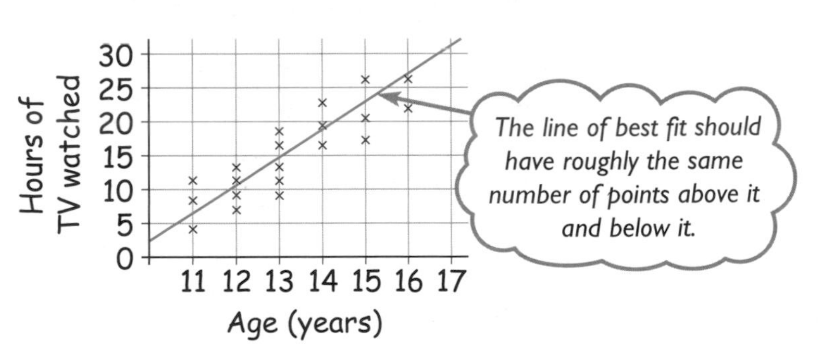

Use the line of best fit to estimate how much TV
(a) a 17-year-old watches
(b) a 10-year-old watches.

Do you think that your estimates are accurate?

1. Measure the height of everyone in your class. Now ask them their shoe size.
2. Draw a scatter diagram to illustrate these data.
3. What does your scatter diagram tell you about your data?

Vicki's scatter diagram shows **positive correlation**.
This means that as one variable increases, so does the other.

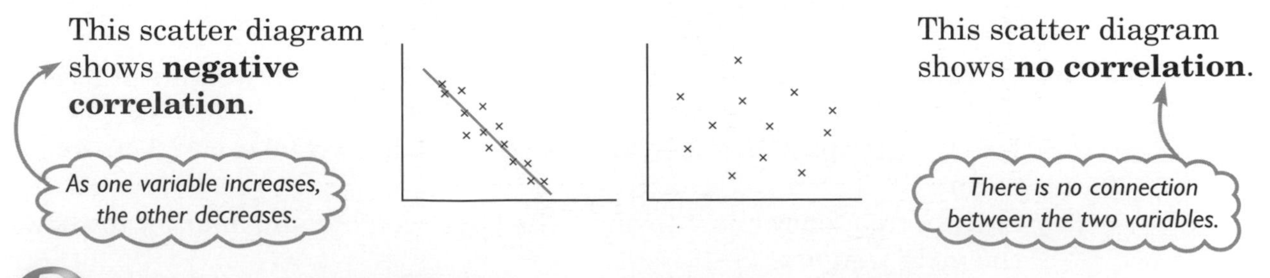

What sets of data might these scatter diagrams represent?

Exercise

1

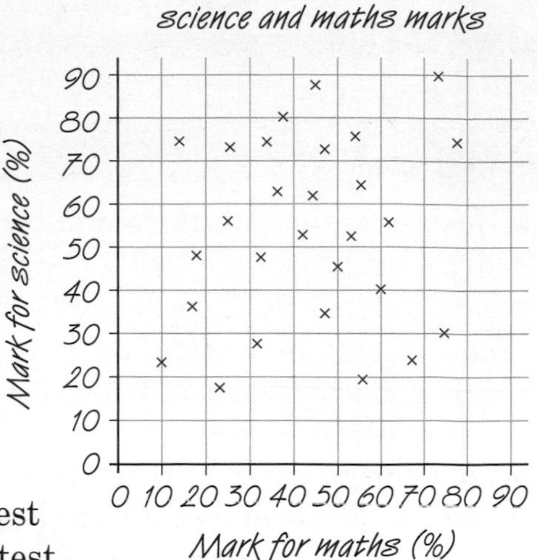

Anita uses the results from Class 9B's maths and science tests to draw this scatter diagram.

(a) Find
  (i) the lowest mark in the maths test
  (ii) the highest mark in the maths test
  (iii) the lowest mark in the science test
  (iv) the highest mark in the science test.

(b) Katie got 40% in the science test. What did she get in the maths test?

(c) James got 61% in the maths test. What did he get in the science test?

(d) Look at the scatter diagram. Is Anita's idea correct?
Explain your answer.

2 The table below shows the maximum temperature and the number of hours of sunshine in 12 British cities on one day in August.

City	Max. temp (°C)	Hours of sunshine	City	Max. temp (°C)	Hours of sunshine
London	24	10	Southampton	25	9
Birmingham	22	9	Norwich	23	11
Manchester	20	8	Liverpool	22	8
Edinburgh	17	9	Exeter	25	10
Glasgow	18	6	Newcastle	20	9
Bristol	23	7	Nottingham	21	8

(a) Draw a scatter diagram to illustrate these data.

(b) Draw a line of best fit on your diagram.

(c) York had a maximum temperature of 19 °C that day.
Estimate the number of hours of sunshine in York that day.

(d) Brighton had 12 hours of sunshine.
Estimate the maximum temperature in Brighton.

3 What do each of the scatter graphs below tell you?

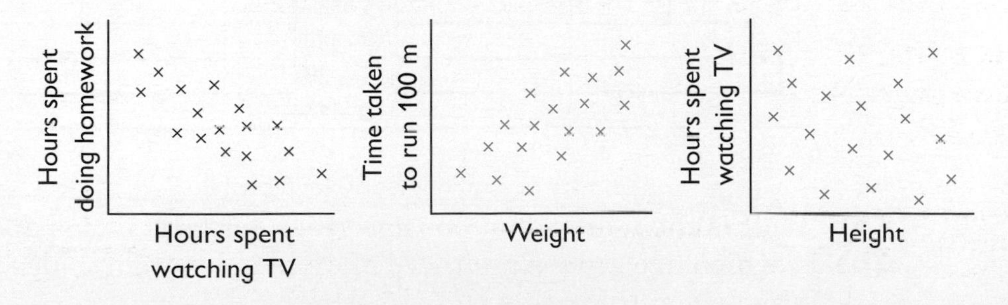

Finishing off

Now that you have finished this chapter you should be able to:

- design a questionnaire and use it to collect data
- draw and interpret tally charts, bar charts, pie charts and two-way tables
- understand the differences between categorical data, discrete numerical data and continuous numerical data
- group data where appropriate
- draw and interpret frequency charts for continuous data
- draw and interpret scatter graphs
- draw the line of best fit on a scatter graph.

Review exercise

1 A council has a budget of £900 million to spend on public services.
It is divided up as shown in the table.

Education services	£580 million
Social services	£200 million
Environmental services	£110 million
Other services	£10 million
Total	£900 million

Jo works for the council.
She wants to draw a pie chart to show these data.

(a) How many degrees should Jo use to represent £1 million?

(b) Draw Jo's pie chart.

2 This is a special kind of bar chart called a **population pyramid**.
This population pyramid shows the population of the UK in 1991.

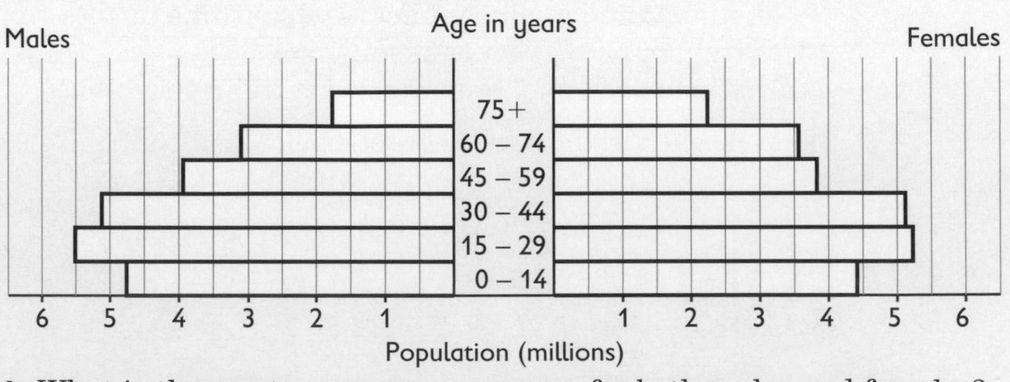

(a) What is the most numerous age group for both males and females?

(b) In which age groups are there more females than males?
Why do you think this is?

3 Stephen and his sister Anna share a computer.

Stephen and Anna both keep a record of the length of time, in minutes, that they use the computer each day for one month.

(a) Make grouped tally charts to show **(i)** Stephen's data **(ii)** Anna's data. Use the groups $0 \leqslant t < 15$, $15 \leqslant t < 30$, etc.

(b) Draw frequency charts to show **(i)** Stephen's data **(ii)** Anna's data.

(c) Do you think that Anna is right? Explain your answer.

4 Shamicka collects some data to find out if she is right.

I think that people who are good at sprinting are also good at long jump.

Time to run 100 m (seconds)	Distance jumped (m)	Time to run 100 m (seconds)	Distance jumped (m)
14.3	1.81	17.3	1.52
15.1	1.76	12.8	2.03
14.4	1.78	16.0	1.68
18.6	1.57	15.3	1.75
13.5	1.95	13.4	1.88
16.1	1.68	13.6	1.92
12.7	1.94	14.7	1.78
13.9	1.84	12.5	2.08
15.4	1.72	16.4	1.58
14.0	1.88	13.7	1.87

(a) Draw a scatter graph to show Shamicka's data.

(b) Describe the relationship between the time taken to run 100 m and the distance jumped.

(c) Draw a line of best fit on your graph.

(d) Shamicka's friend Richard runs 100 m in 14.8 seconds. Use your line of best fit to estimate how far Richard can jump.

5 Ratio

Ratios

How do you simplify (a) 400 g : 2 kg (b) $1\frac{5}{6} : 3\frac{2}{3}$?

Remember, ratios should be in the same units before simplifying.

Alice and Lewis see an eagle at the top of a tree.
They wonder how high up it is.

Why can they not measure the height of the tree directly?

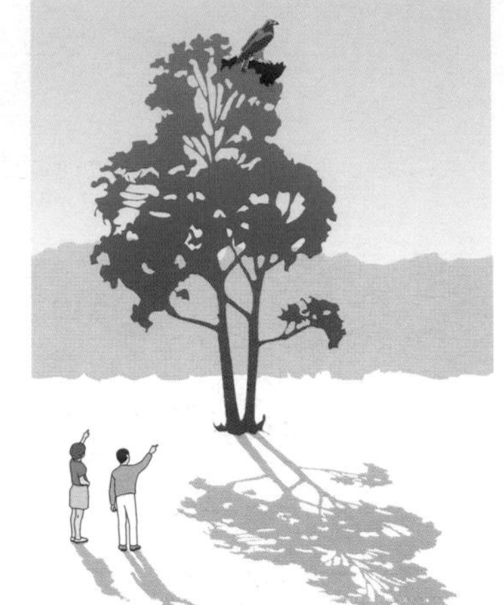

Lewis measures the length of Alice's shadow.
It is 1 m long. He knows Alice is 1.5 m tall.

What is the ratio, Alice's height : Alice's shadow?

Lewis also measures the shadow of the tree. It is 4 m long.

How can they work out the height of the tree?

Task

Which of these might be happening at the same time of day?

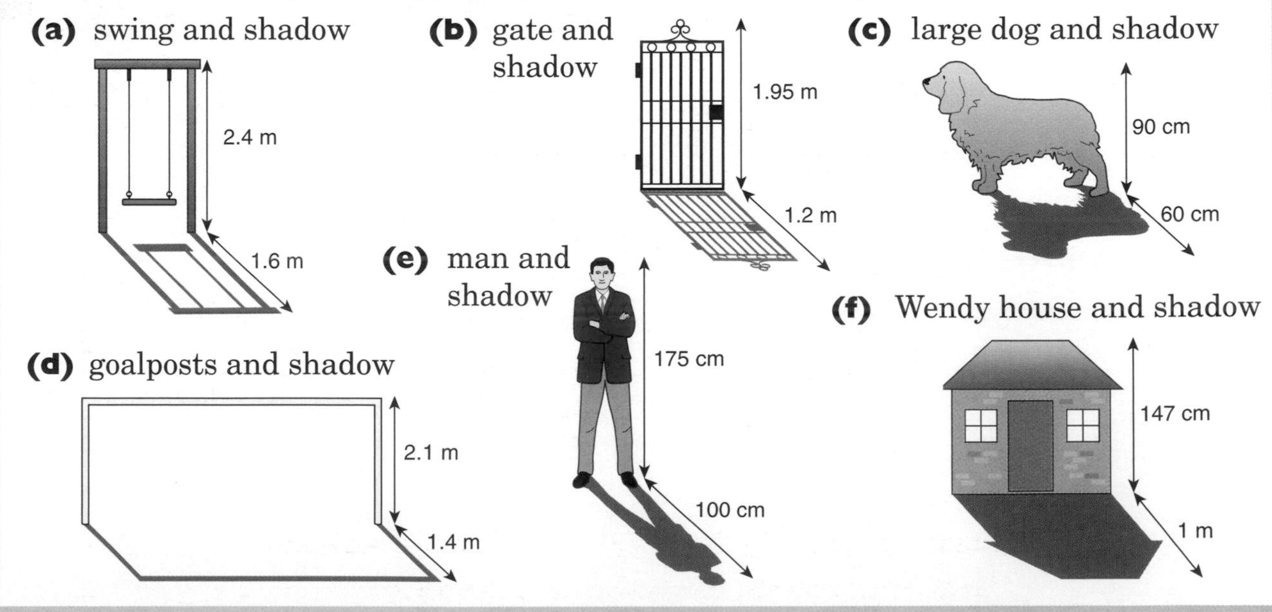

To find a missing quantity in a ratio, it is sometimes useful to use algebra.

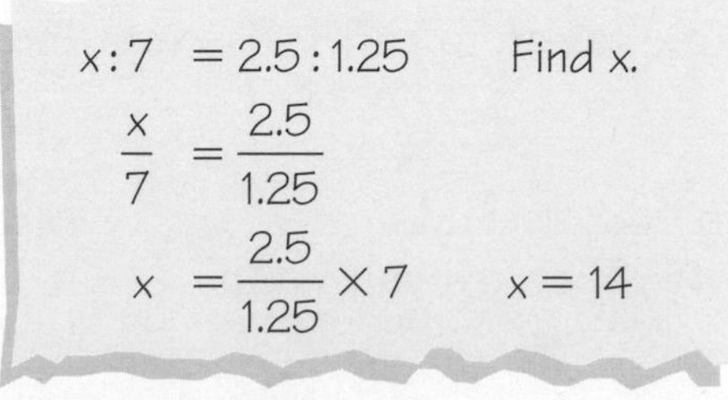

How would you find *x* in $2 : x = 1 : 5$?

Exercise

1 Simplify the following ratios.

(a) $8:12$ **(b)** $24:36$ **(c)** $63:54$ **(d)** $10:2$

(e) $25:35:20$ **(f)** $72:64:80$ **(g)** $56:63:42$ **(h)** $15:5$

(i) $4\frac{1}{2}:5\frac{1}{2}$ **(j)** $3\frac{1}{3}:2\frac{1}{2}$ **(k)** $2\frac{1}{5}:4\frac{2}{5}$ **(l)** $\frac{2}{3}:\frac{5}{6}$

(m) $5:7.5$ **(n)** $2.4:7.2$ **(o)** $6:1.5$ **(p)** $4:2.5$

2 Simplify the following ratios.

(a) £1.50 : £1.20 **(b)** 600 g : 2 kg **(c)** 1.4 kg : 700 g

(d) 50 minutes : 1 hour **(e)** 3 m : 24 cm **(f)** 4.2 m : 56 cm

3 Find the missing quantities in these ratios.

(a) $5:8 = \square:24$ **(b)** $\square:7 = 8:28$

(c) $6:7 = 30:\square$ **(d)** $2\frac{1}{2}:4\frac{1}{2} = \square:9$

(e) $1.5:7.5 = \square:22.5$ **(f)** $1.7:3.2 = 10.2:\square$

(g) $5\frac{1}{3}:\square = 48:17$ **(h)** $4:1 = \square:500$ g

4 **(a)** Draw these triangles *accurately*.

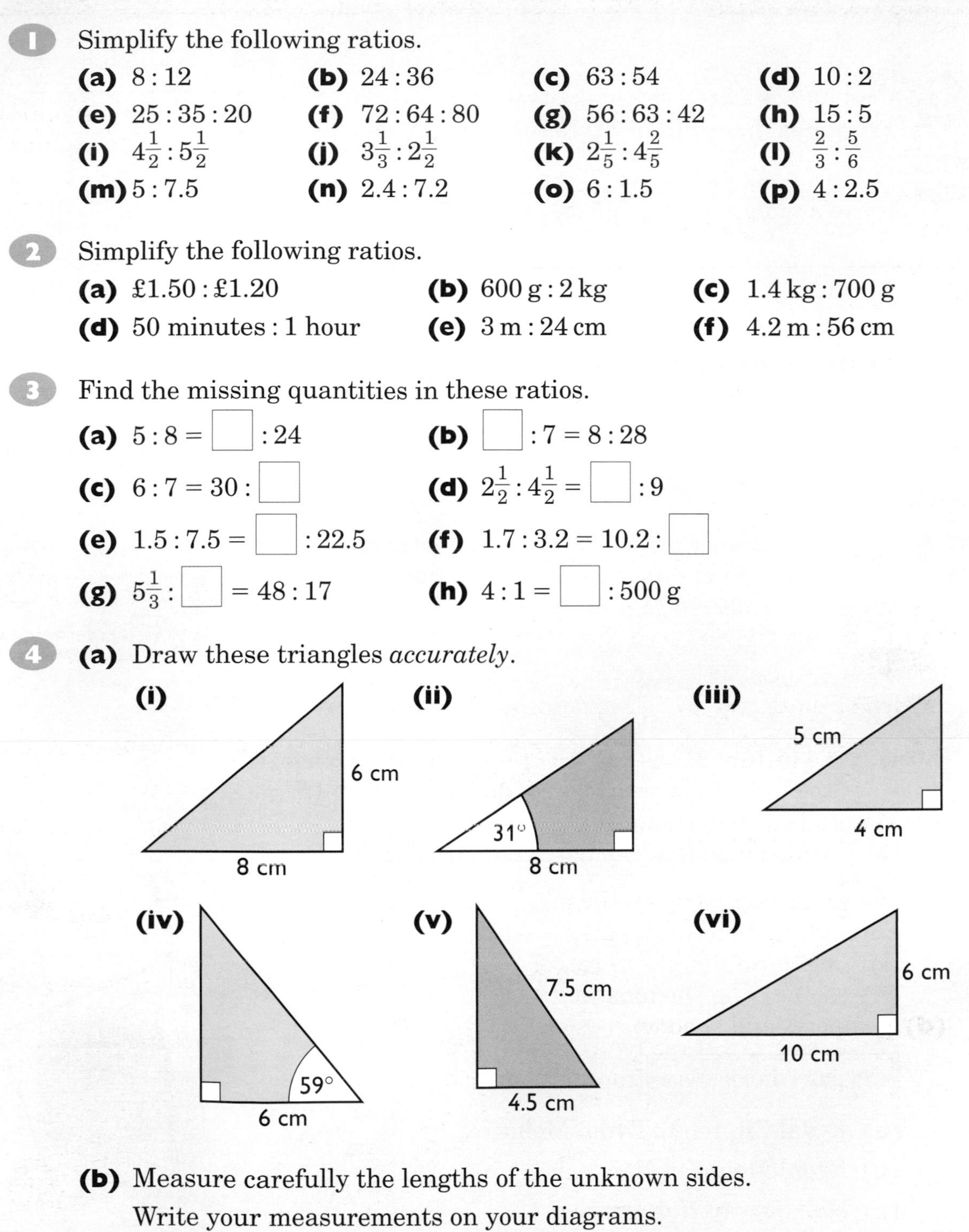

(b) Measure carefully the lengths of the unknown sides.
Write your measurements on your diagrams.

(c) Some of these triangles have their sides in the same ratio as each other.
Sort them into groups having the same ratio.

(d) Are any of these triangles

(i) congruent **(ii)** similar?

Scale

**Paul makes a scale drawing of his room.
Work out the width of Paul's room in real life.**

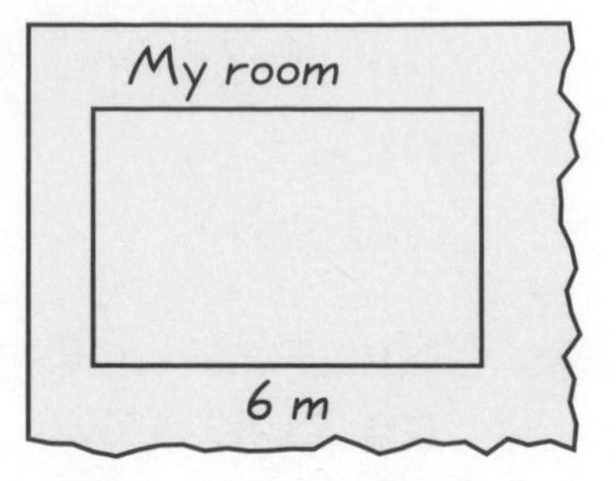

**Using a scale of 1 cm to 50 cm,
make a scale drawing of Paul's room.**

Carina has a model car.
The scale is 1 : 15.

**Carina's model car is 16 cm long.
How long is the real car?**

Using map scales

**Julia is going for a walk. The scale on her map is 1:10 000.
A distance of 1 cm on Julia's map is 10 000 cm in real life.
Convert 10 000 cm to metres.**

Task

Look at the picture at the bottom of the right-hand page.
It shows the view Julia sees looking down over Rydal Water.
Look at Julia's map below.
She has turned it so that the north line matches her view.

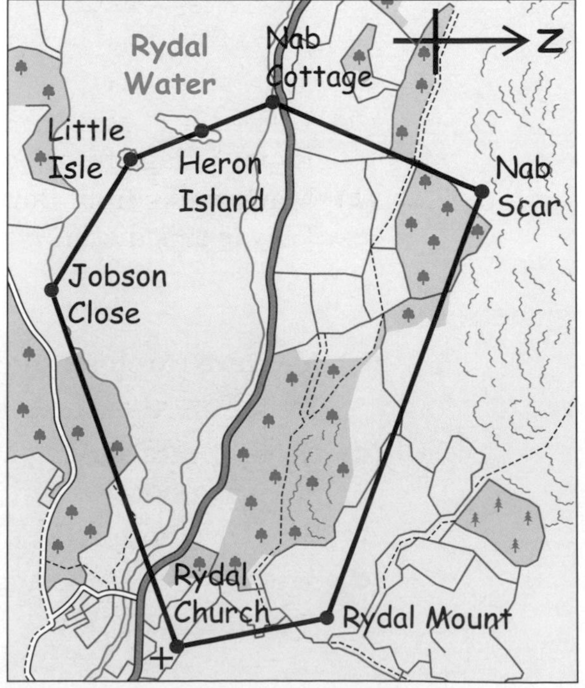

1 Julia marks her trip on the map.
For each section, work out
(i) how far Julia has to travel
(ii) the bearing she must follow.

A bearing is a three figure angle measured clockwise from the north.

(a) Rydal Church to Rydal Mount
(b) Rydal Mount to Nab Scar
(c) Nab Scar to Nab Cottage
(d) Nab Cottage to Heron Island
(e) Heron Island to Little Isle
(f) Little Isle to Jobson Close
(g) Jobson Close to Rydal Church.

2 How far does Julia travel altogether?

Exercise

1 Make a scale drawing of five items on your desk. Use a scale of 1 : 5.
Copy and complete this table to help you:

Item	Actual size	Calculation	Scale size
Ruler			

2 Look at the map on page 46. Jenna starts walking after Julia.
She decides to go directly from Rydal Mount to Nab Cottage.

(a) How far is it from Rydal Mount to Nab Cottage?

(b) What is the bearing of

(i) Nab Cottage from Rydal Mount

(ii) Rydal Mount from Nab Cottage?

3 **(a)** Write down the bearing of

(i) Little Isle from Jobson Close

(ii) Jobson Close from Little Isle.

(b) What is the connection between the answers to part **(a)**?

4 Mark is going walking. Mark's map has a scale of 1 : 50 000.
Copy and complete this table.

Distance:	Bow Fell-Harrison Stickle	Harrison Stickle-Dungeon Ghyll	Dungeon Ghyll-Sargent Man	Sargent Man-Silver Howe
On map		2 cm	5 cm	
In real life	4 km			5 km

5 The picture below is taken looking west from Rydal Village.
Rydal Village to Dungeon Ghyll is 4.5 miles on a bearing of 280°.
Dungeon Ghyll to Bow Fell is 3 miles on a bearing of 200°.

(a) Show this information on a scale diagram. Use a scale of 1 cm to 1 mile.

(b) Using your diagram, work out the distance and bearing of Rydal Village from Bow Fell.

(c) Mark walks from Bow Fell to Rydal Village. How far does he walk? Why is this distance greater than your answer to part **(b)**?

Ratios and similar figures

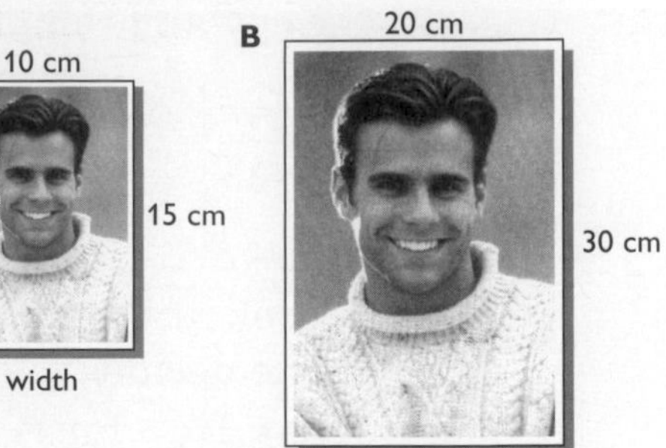

Rory is a pop star.
He is choosing some photographs to send to fans.
Photograph B is an enlargement of photograph A.

What is the ratio of the widths, smaller : larger?
What is the ratio of the heights, smaller : larger?

What do you notice?

These rectangles are **similar** because their widths and heights are in the same ratio.

What is the ratio of width : height for each photograph?

What do you notice?

The ratios can be used either way:

width of A : height of A = width of B : height of B

or width of A : width of B = height of A : height of B

Take care to keep the same order (width : height or A : B).

Task

A photograph is 10 cm by 15 cm.

1 The diagrams show some possible enlargements.
Find the missing measurements.

(a) 30 cm, ?

(b)

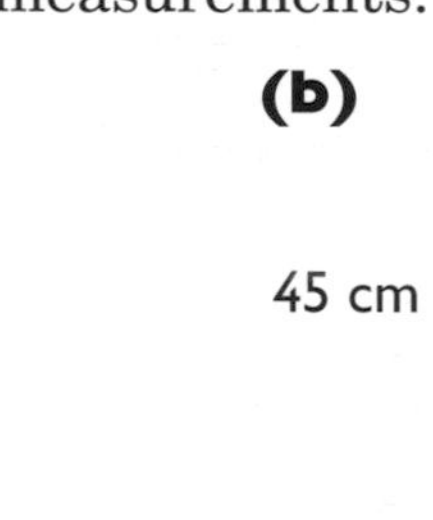

(c)

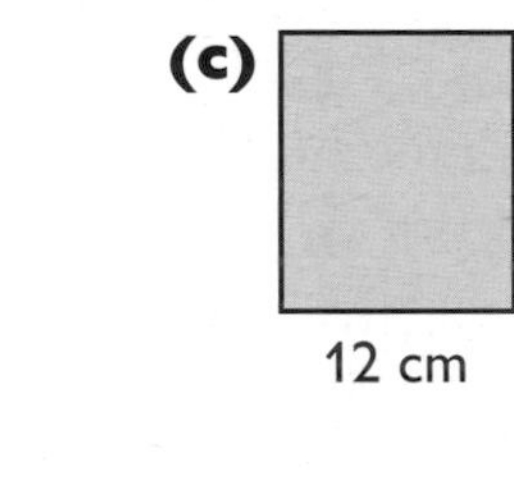

2 Look back at the first photographs A and B again.
(a) What is the ratio of the widths?
(b) Work out the area of each photograph.
(c) What is the ratio of area of photograph A : area of photograph B?
(d) What do you notice?

3 Rory can also have posters made.
Posters are in the ratio
height : width = 7 : 5.
(a) Copy and complete the table opposite.
(b) What is the ratio of the width of poster X : width of poster Y?
(c) What is the ratio of the area of poster X : area of poster Y?

Poster	Height	Width	Area
X		50 cm	
Y	42 cm		

Exercise

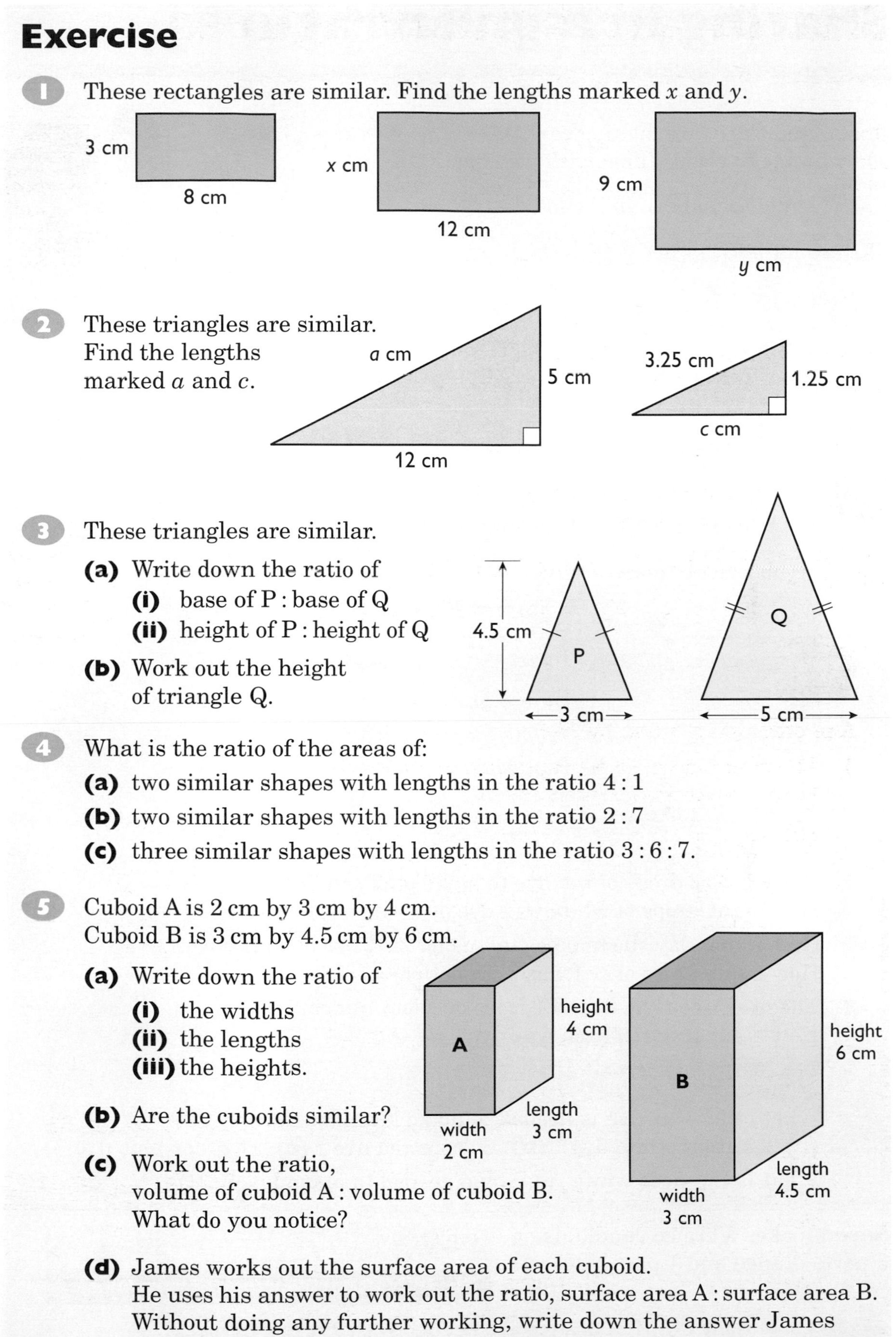

1 These rectangles are similar. Find the lengths marked x and y.

2 These triangles are similar. Find the lengths marked a and c.

3 These triangles are similar.

(a) Write down the ratio of

(i) base of P : base of Q

(ii) height of P : height of Q

(b) Work out the height of triangle Q.

4 What is the ratio of the areas of:

(a) two similar shapes with lengths in the ratio 4 : 1

(b) two similar shapes with lengths in the ratio 2 : 7

(c) three similar shapes with lengths in the ratio 3 : 6 : 7.

5 Cuboid A is 2 cm by 3 cm by 4 cm.
Cuboid B is 3 cm by 4.5 cm by 6 cm.

(a) Write down the ratio of

(i) the widths

(ii) the lengths

(iii) the heights.

(b) Are the cuboids similar?

(c) Work out the ratio,
volume of cuboid A : volume of cuboid B.
What do you notice?

(d) James works out the surface area of each cuboid.
He uses his answer to work out the ratio, surface area A : surface area B.
Without doing any further working, write down the answer James should find.

Sharing in a given ratio

Simon and Carly are making candles.
They decide to colour them with wax dye.

Carly wants to make a pink candle.

PINK: 6 parts white + 1 part red

So white : red = 6 : 1.

Carly puts in 3 drops of red wax dye.
How many drops of white dye does she need?

Simon decides to make a moss green candle.

MOSS GREEN: blue : yellow : red = 3 : 3 : 2

Simon puts in 24 drops of wax dye altogether.
How many drops of each colour does he use?

Here are his calculations:

$3 + 3 + 2 = 8$ parts altogether
8 parts = 24 drops
1 part = $\frac{24}{8} = 3$ drops
Blue, 3 parts = 3×3 drops = 9 drops
Yellow, 3 parts = 3×3 drops = 9 drops
Red, 2 parts = 2×3 drops = 6 drops

Task

1. Carly uses 2 drops of red dye to make pink candles.
   How many drops of white wax dye does she need?
2. On another day she makes pink using 24 drops of white dye.
   How many drops of red wax dye does she need?
3. The next week she uses 42 drops of colour altogether.
   How many drops of each wax dye does she use?
4. LAVENDER: 2 parts pink + 1 part blue

   What other wax dye is needed to make lavender if Carly uses:
   (a) 24 drops white dye **(b)** 6 drops red dye **(c)** 21 drops blue dye
   What is the ratio white : red : blue needed to make lavender?

Simon makes a brown candle using 5 parts orange and 1 part black.

ORANGE: 1 part red : 1 part yellow

BROWN: 5 parts orange : 1 part black

How many drops of orange are needed with 2 drops of black?
What is the ratio of red : yellow : black needed for his brown candle?

Exercise

1 In a netball match the ratio of goals scored to goals missed is 3 : 1.
60 shots are attempted altogether.
How many goals are scored?

2 In a football season the ratio of penalties scored to penalties saved is 14 : 3.
51 penalties were awarded.

(a) How many penalty goals were scored?

(b) How many penalty saves were made?

3 A play has animal characters in it.

(a) The mask for the rabbit's head is 35 cm tall including the ears.
The ratio of ears : face = 3 : 4.

(i) How long are the ears?

(ii) How big is the face?

(b) The mask for the mouse's head is 24 cm wide.
The ratio of ear : face : ear = 1 : 2 : 1.

(i) How wide is each ear?

(ii) How wide is the face?

4 In shortbread biscuits the ratio of butter : sugar : flour is 2 : 1 : 4.

(a) 120 g of butter is used.
How much **(i)** sugar and **(ii)** flour are needed?

(b) 160 g of flour is used.
How much **(i)** butter and **(ii)** sugar are needed?

(c) The total mass of the ingredients is 350 g.
How much of each ingredient is used?

5 The table below refers to goals scored in a local football league. It shows the total number of goals scored by each team and the ratio of how the goals were scored.

Work out how many of each type were scored in each case.

	left-footed : right-footed : headed	total
Shellbury	4 : 7 : 4	45
Plystar	3 : 5 : 1	36
Swynton	2 : 6 : 1	27
Avonford	4 : 6 : 1	55

Finishing off

Now that you have finished this chapter you should be able to:

- simplify a ratio
- find missing quantities in a given ratio
- use scale
- understand similarity
- find the ratio of areas and volumes of similar shapes
- share quantities in a given ratio.

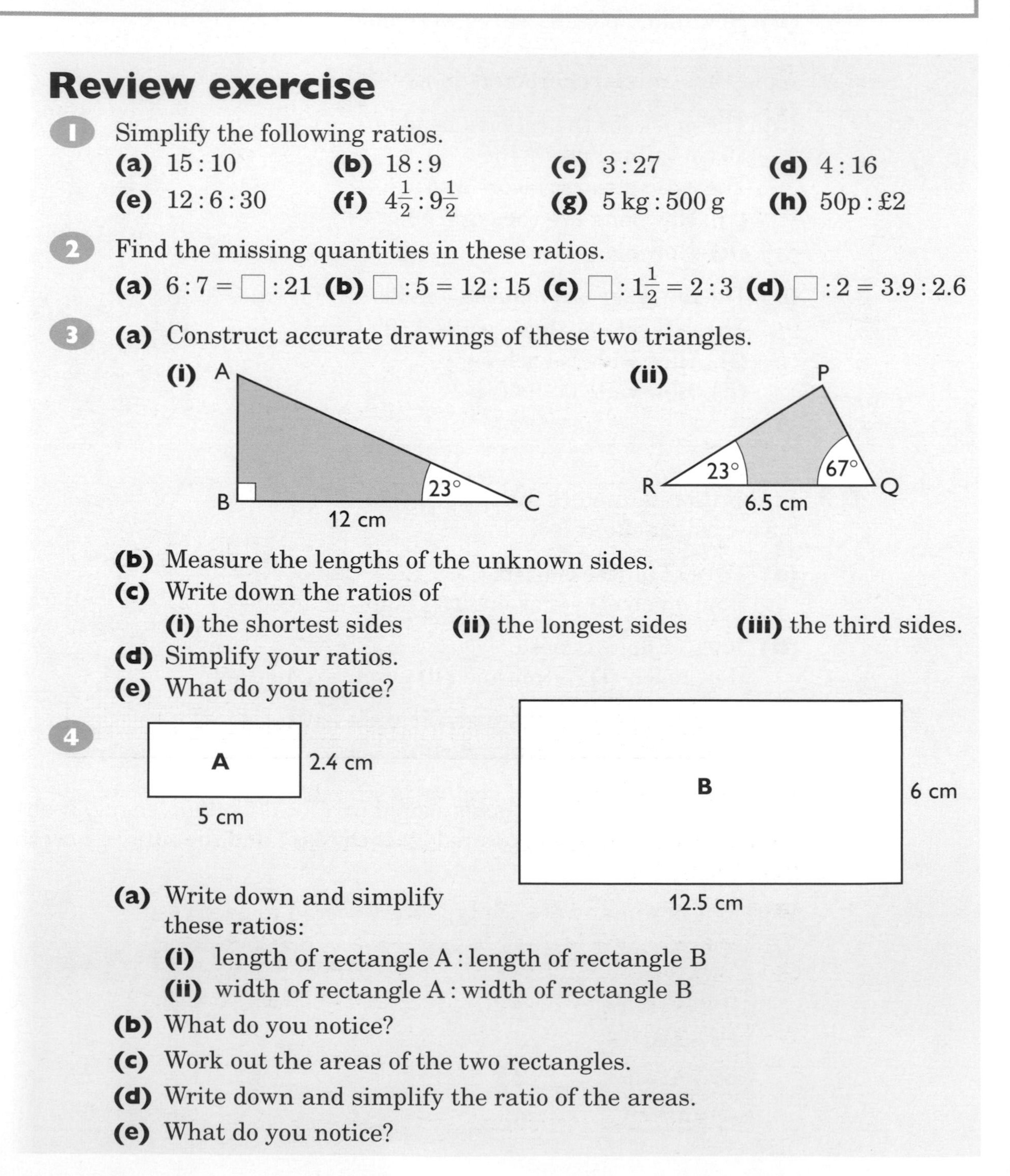

Review exercise

1 Simplify the following ratios.

(a) 15 : 10 **(b)** 18 : 9 **(c)** 3 : 27 **(d)** 4 : 16

(e) 12 : 6 : 30 **(f)** $4\frac{1}{2} : 9\frac{1}{2}$ **(g)** 5 kg : 500 g **(h)** 50p : £2

2 Find the missing quantities in these ratios.

(a) 6 : 7 = □ : 21 **(b)** □ : 5 = 12 : 15 **(c)** □ : $1\frac{1}{2}$ = 2 : 3 **(d)** □ : 2 = 3.9 : 2.6

3 **(a)** Construct accurate drawings of these two triangles.

(i)

(ii)

(b) Measure the lengths of the unknown sides.

(c) Write down the ratios of

(i) the shortest sides **(ii)** the longest sides **(iii)** the third sides.

(d) Simplify your ratios.

(e) What do you notice?

4

(a) Write down and simplify these ratios:

(i) length of rectangle A : length of rectangle B

(ii) width of rectangle A : width of rectangle B

(b) What do you notice?

(c) Work out the areas of the two rectangles.

(d) Write down and simplify the ratio of the areas.

(e) What do you notice?

5 These two cubes have edges of 2 cm and 3 cm.

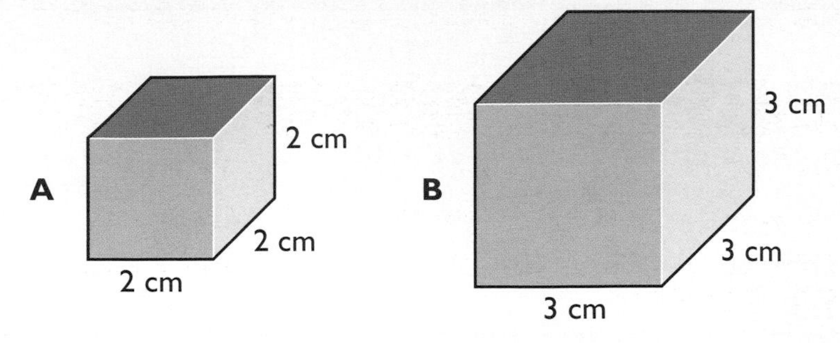

(a) Find the ratio of the area of a face of A to the area of a face of B.

(b) Find the ratio, volume A : volume B.

(c) What is the relationship between these answers and the ratio, an edge of A : an edge of B?

6 A map is drawn to a scale of 1 : 50 000.
What distances are represented by

(a) 1 cm on the map **(b)** 3 cm on the map?

7 A map has a scale of 1 : 200 000.
The distance on the map between Avonford and Banton is 3.2 cm.
How far apart are they in real life, in kilometres?

8 Divide £15 000 profit between Mrs. Shah and Miss Kachinska in the ratio 3 : 7.

9

How many of each type of goal were scored?

10 A recipe for Luncheon Cake includes flour, rice flour, and fat in the ratio 6 : 2 : 3.

(a) John uses 120 g of flour.
How much of the other two ingredients does he need?

(b) Myra uses 150 g of fat.
How much of the other two ingredients does she need?

(c) The total mass of a cake is 1050 g.
It uses flour, rice flour, fat, sugar, currants and candied peel in the ratio 12 : 4 : 6 : 8 : 4 : 1.
How much of each ingredient is needed?

6 Equations

Equations review

Jim is solving the equation $6x + 4 = 4x + 10$.

	$6x + 4 = 4x + 10$
Subtract 4 from both sides	$6x + 4 - 4 = 4x + 10 - 4$
Tidy up	$6x = 4x + 6$
Subtract $4x$ from both sides	$6x - 4x = 4x + 6 - 4x$
Tidy up	$2x = 6$
Divide both sides by 2	$x = 3$

? **Look at Jim's problem again.**
This time divide both sides by 2 first.
What answer do you get?

Task

Work with a friend.
Each of you takes one of these equations, **1(a)** or **1(b)**.
Solve your equation, step by step.
Then pass your work to your friend.
Your friend describes your steps in words, for example, 'Add 3 to both sides'.
You do the same for your friend's equation.
Then do the same for the other sets of equations.

1 **(a)** $5x - 3 = 2x - 12$
(b) $7x + 8 = 3x + 28$

2 **(a)** $3(x + 1) + 2 = x + 15$
(b) $2(x - 5) + 3x = 35$

3 **(a)** $2(x + 1) - 3(7 - x) = 16$
(b) $4(x + 1) - 3(x - 1) = 8$

? **How do you check the answers to the equations in the task?**

? **Look at this equation involving fractions.**
Explain each step in the solution.

$$\frac{3}{4}x = 9$$
$$4 \times \frac{3}{4}x = 4 \times 9$$
$$3x = 36$$
$$3x \div 3 = 36 \div 3$$
$$x = 12$$

Exercise

In this exercise make certain that you set out your working step by step.

1 Solve the following equations and check your answers.

(a) $3x + 1 = 10$ **(b)** $5x - 6 = 9$

(c) $4x + 2 = 18$ **(d)** $6x - 15 = 15$

(e) $3x - 7 = x + 13$ **(f)** $8x + 4 = 5x - 5$

(g) $7x + 9 = 5x + 9$ **(h)** $12x + 4 = 6x - 8$

(i) $5x - 5 = 4x - 9$ **(j)** $3x + 9 = x - 5$

(k) $2.1x + 0.4 = 1.1x + 2.4$ **(l)** $2.1x + 0.4 = 1.6x + 2.4$

(m) $2.4x + 5.9 = 14.3$ **(n)** $2.1x - 0.5 = 0.9x + 3.1$

2 For the following **(i)** write down an equation for the balance
(ii) solve the equation.

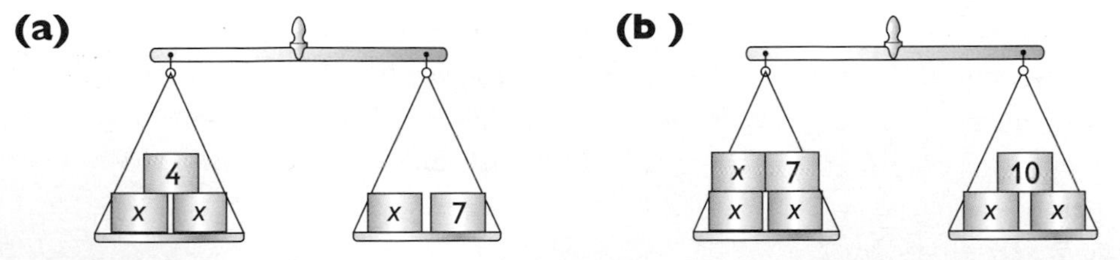

3 Alan and Ruth are making up number puzzles for each other.

I think of a number.
I multiply it by 3 and add 3.
My answer is 4 times
my first number.

I think of a number.
I double it and subtract 5.
My answer is the same as
my first number.

(a) (i) Write down an equation for Alan's number.
(ii) What number was Alan thinking of?

(b) (i) Write down an equation for Ruth's number.
(ii) What number was Ruth thinking of?

4 Solve each of these equations showing all your working.

(a) $5(x + 1) - 1 = 9$ **(b)** $3(x + 1) + 1 = 10$

(c) $3(x + 2) + x = 22$ **(d)** $5(x - 1) + 2 = 4x$

(e) $3(x - 11) + x = 15$ **(f)** $2(x + 1) + 3(x + 2) = 2(x + 6) - 1$

(g) $5(x + 1) - 4(x - 1) = 0$ **(h)** $3(x - 3) - 2(x - 4) - 3 = 0$

Check your answers in the original equations.

5 Solve each of these equations showing all your working.

(a) $\frac{x}{5} = 2$ **(b)** $\frac{x}{3} = 20$ **(c)** $\frac{x}{2} = \frac{5}{2}$

(d) $\frac{1}{2}x = 5$ **(e)** $\frac{1}{4}x = 3$ **(f)** $\frac{3}{4}x = 100$

Solving equations

Ahmed has an equation to solve: $\mathbf{12 - 2x = 10}$

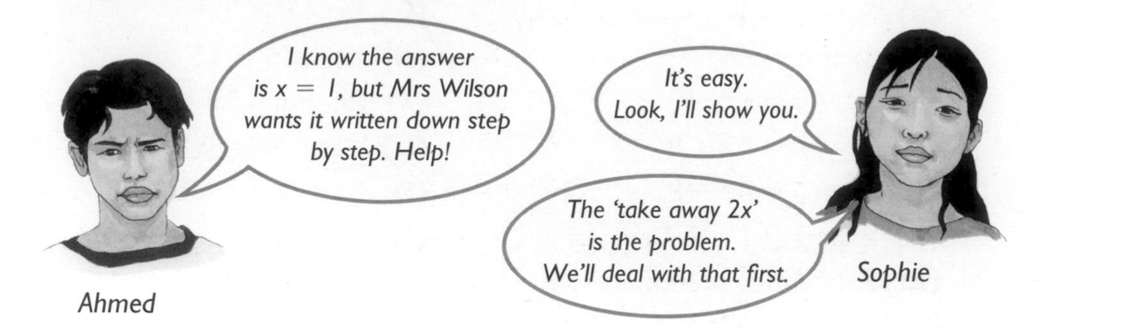

Sophie writes

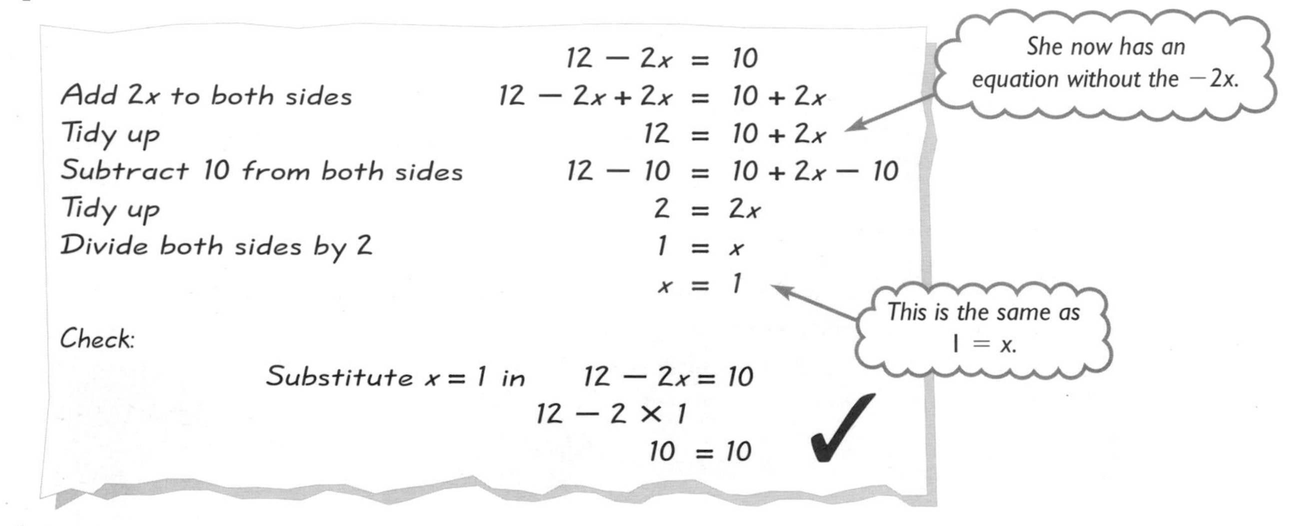

	$12 - 2x = 10$
Add $2x$ to both sides	$12 - 2x + 2x = 10 + 2x$
Tidy up	$12 = 10 + 2x$
Subtract 10 from both sides	$12 - 10 = 10 + 2x - 10$
Tidy up	$2 = 2x$
Divide both sides by 2	$1 = x$
	$x = 1$

Check:

Substitute $x = 1$ in $12 - 2x = 10$

$12 - 2 \times 1$

$10 = 10$ ✓

Always deal with the '**subtract *x***' bit of the equation first.

Task

Make up an equation with a 'take away x' bit in it.
Write down the answer secretly.
Ask your partner to solve the equation in writing.
Do you both get the same answer?

Vicki writes

I can solve this a different way. — Vicki

	$12 - 2x = 10$
Subtract 12 from both sides	$-2x = -2$
Divide both sides by -2	$x = 1$

Remember that $-2 \div -2 = 1$.

Which method do you think is better? Why?

Exercise

On the opposite page you met two ways of solving these equations.
Do some of this exercise one way and the rest the other.

1 Solve these equations showing all your working.
Check your answers in the original equation.

(a) $10 - 2x = 4$ **(b)** $13 - 3x = 1$ **(c)** $15 - 4x = 7$
(d) $11 - 3x = 2$ **(e)** $20 - 2r = 15$ **(f)** $29 - 5w = 14$
(g) $12 - 4y = 10$ **(h)** $55 - 5t = 0$ **(i)** $17 - 3x = 2$

2 Write an equation for each of the following situations.
Let x stand for the unknown value of the coin or note.
Solve each of the equations.

(a) Suzy has £11 in her pocket.
She drops 3 coins, all of the same value.
When she checks, she has £8 left.

(b) Natalie goes shopping with £250.
When she checks her purse she finds 3 notes (all of the same value) are missing. She has £190 left.

3 Solve these equations.

Be careful, some answers are negative and some not integers (whole numbers). Use substitution to check your answers.

(a) $10 - 4x = 2$ **(b)** $12 - 2x = 18$ **(c)** $17 - 4x = 21$
(d) $25 - 3x = 4$ **(e)** $13 - 8x = 7$ **(f)** $32 - 5x = 42$
(g) $2(3 - x) + 1 = 17$ **(h)** $3(x - 4) - 4(x - 1) + 12 = 0$

Investigation

Match each of these equations with its solution.

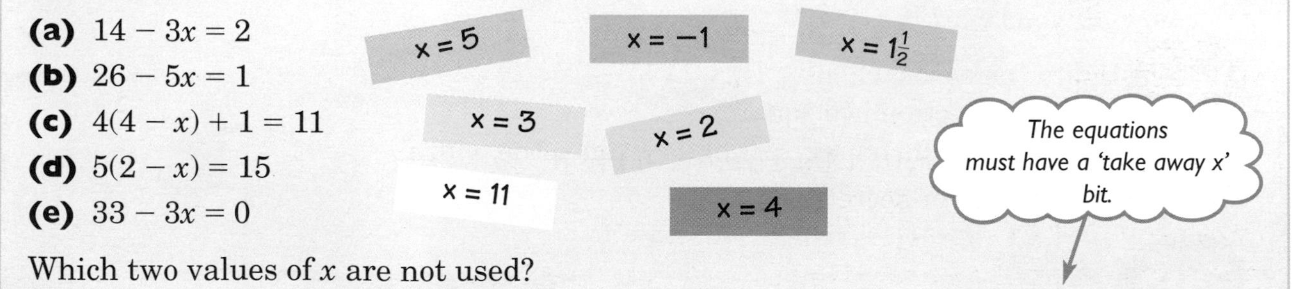

Which two values of x are not used?
Make up similar equations which have these two values of x as their solutions.

Investigation

(a) Here is a solution of the equation $\frac{4}{x-3} = 2$

Copy this solution and write in the steps involved.

$$\frac{4}{x-3} = 2$$
$$4 = 2(x-3)$$
$$4 = 2x - 6$$
$$4 + 6 = 2x$$
$$10 = 2x$$
$$x = 5$$

(b) Solve these equations.

(i) $\frac{2}{x+1} = 1$ **(ii)** $\frac{8}{2x+3} = 2$ **(iii)** $\frac{3}{x+1} = \frac{6}{x+6}$

Forming equations

Len the park keeper fences a rectangular section of Avonford Park.

He has 140 metres of fence. He makes the fenced section 30 metres wide.

He writes

30m ?

Perimeter = $x + 30 + x + 30$
Perimeter = $2x + 60$
$140 = 2x + 60$
$80 = 2x$
$40 = x$ so $x = 40$

Len has 140 m of fencing.

The fenced section will be 40 metres long.

I'll let x stand for the length.

Look at Len's solution and say what he has done at each line.
How can you check Len's answer?

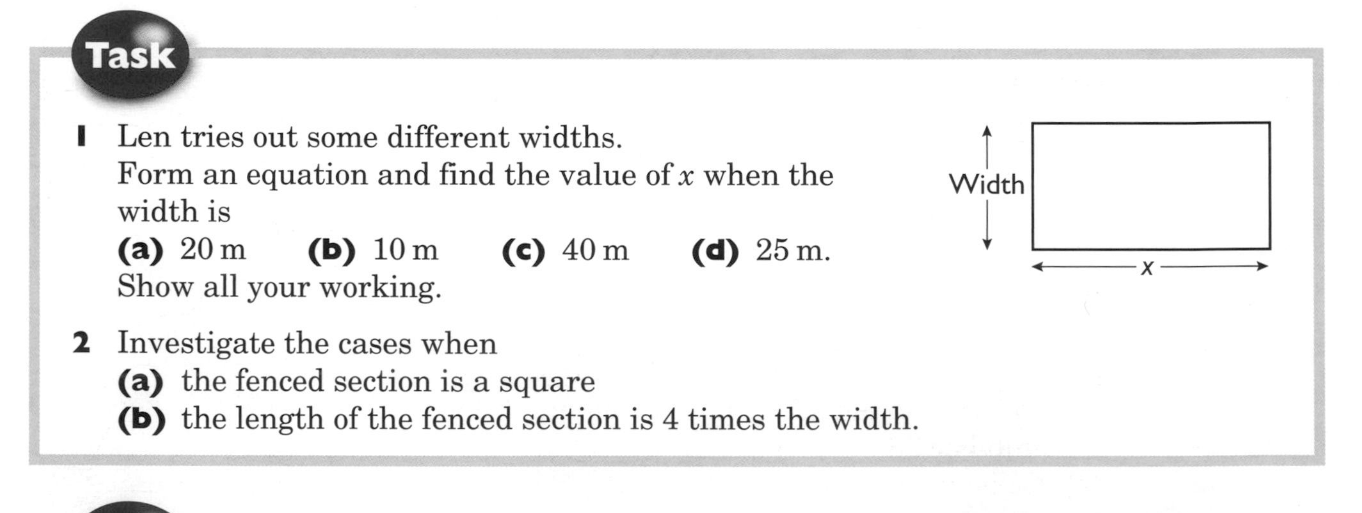

Task

1 Len tries out some different widths.
Form an equation and find the value of x when the width is
(a) 20 m **(b)** 10 m **(c)** 40 m **(d)** 25 m.
Show all your working.

Width
x

2 Investigate the cases when
(a) the fenced section is a square
(b) the length of the fenced section is 4 times the width.

Task

Look at this Arithmagon. The number in each square is the sum of the numbers in the circles on each side of it.

Form an equation for x and solve it.

Now form an equation for y and solve it.

What is the value of z?

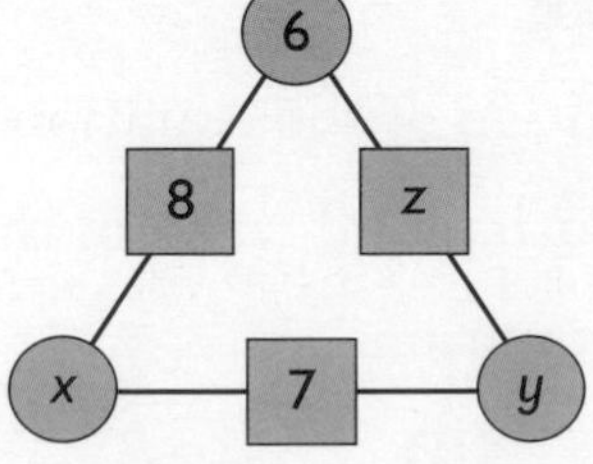

Exercise

1 **(a)** 1 packet of sweets costs p pence.
Tim buys 5 packets and has 60p change from £2.00.
Write down an equation for p and solve it. Check your answer.

(b) Cinema tickets cost t pounds.
Pat buys five tickets. She has £2.50 change from £25.
Write down an equation for t and solve it. Check your answer.

(c) Ruth cuts 8 ribbon lengths from a piece 15 metres long.
The remaining piece is 1 metre. r is the length in metres of each piece.
Write down an equation for r and solve it. Check your answer.

(d) Rachel buys 8 tins of baby food.
She has 8p change from £4.00. 1 tin of food costs t pence.
Write an equation for t and solve it. Check your answer.

2 Bill needs to build a run for his rabbit.
He builds it against his garden wall.
He has 60 metres of wire netting.

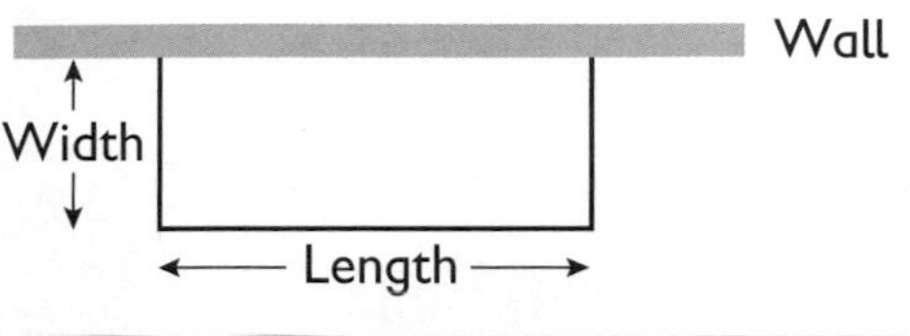

(a) He makes the run 10 metres wide.
Form an equation to work out the length of the run.
Let x stand for the length.
Solve your equation.
Set out your working clearly.

(b) Bill wants to make the area of the run as big as possible.
He tries some different widths.
Copy and complete Bill's table of values to find out what width will give the maximum area.

Width	Length	Area
10		
8		
6		
4		
12		

3 **(a)** Copy this Arithmagon.
Fill in the bottom circles in terms of a.

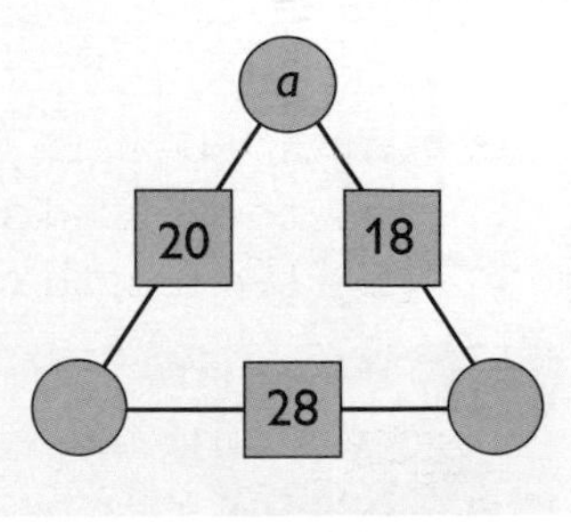

(b) Use your answers to the bottom two circles to write an equation in terms of a and the number 28.

(c) Solve your equation to find a.
Check your value fits back into the Arithmagon.

Investigation

Find the values of $2(x + 2) + 1$ and $3x + 5 - x$ when
(a) $x = 2$ **(b)** $x = 5$ **(c)** $x = 100$. What do you notice?
Try to solve $2(x + 2) + 1 = 3x + 5 - x$. What happens?

? **How many more values of x can you find that make the two expressions equal?**

A statement that is true for **all** values is called an **identity**. The symbol $\equiv$ is used.
For example, $3(a - 2) \equiv 3a + 6$. Write three identities of your own.

Trial and improvement

Sam designs a box to hold 200 cm^3 of bubble bath.

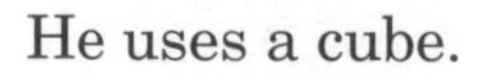

He uses a cube.

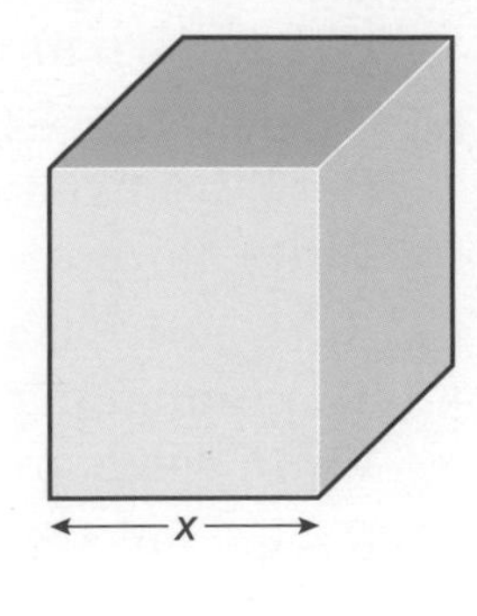

The volume of the cube $= x \times x \times x$
$= x^3$

Sam writes

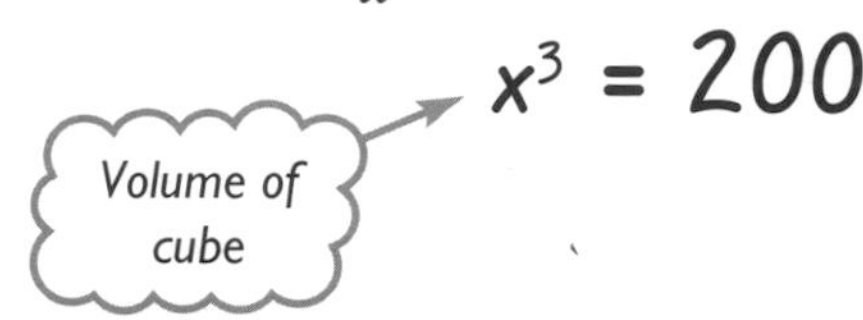

Sam draws this table.

x cm	x^3	small/large
4	4 × 4 × 4 = 64	too small
6	6 × 6 × 6 = 216	too large
5	5 × 5 × 5 = 125	too small
5.5	5.5 × 5.5 × 5.5 = 166.375	too small
5.8		

How does Sam choose his values of *x*?

Task

Copy and continue Sam's table.
Keep going until your answer is correct to the nearest mm.

What size cube will Sam make for the new economy size?
Set out your working like Sam.

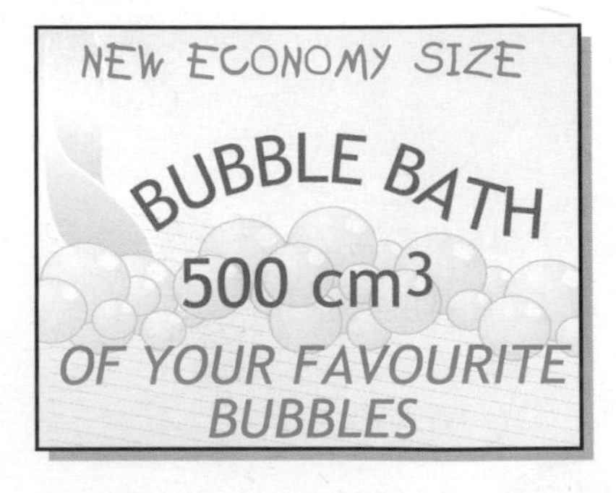

Sam changes the shape of the economy size box.
The length of the box is now twice the size of the square end.

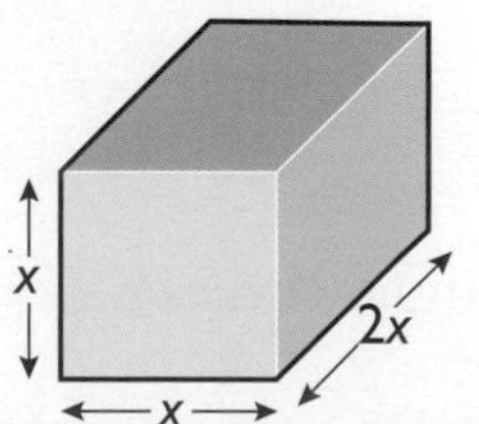

Write the formula for the volume of the box.
Use a table of values to work out *x* to 2 decimal places.

Exercise

1 Copy and complete the table to solve the equation $x^3 + x = 100$.
Give your answer to 2 decimal places.

x	x^3	$x^3 + x$	small/large

2 Make your own table to solve the equation $x^2 - x = 50$.
Give your answer to 2 decimal places.

3 Janet is making a tray.
She uses a 20 cm square piece of card.
She cuts a square out of each corner
and folds the card into a tray.

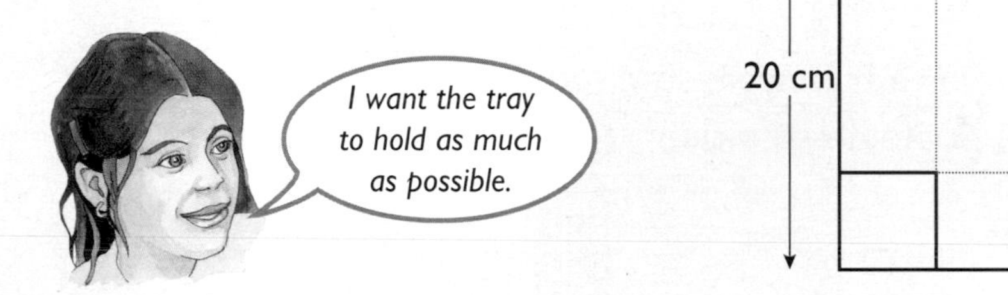

(a) Copy and complete Janet's table of values.
Find the value of x that gives the largest volume.
Give your answer to 2 decimal places.

x cm	Volume (cm³)
2	$16 \times 16 \times 2 = 512$
3	$14 \times 14 \times 3 = 588$

(b) Why must the value of x always be less than 10 cm?

4 Make up tables to solve the following equations.
Give your answers to 2 decimal places.

(a) $x + x^2 = 150$

(b) $2x^2 + x = 75$

(c) $x(1 + x) = 150$

(d) Explain your answers to parts (a) and (c).

Finishing off

Now that you have finished this chapter you should be able to:

- solve equations
- form your own equations
- use 'trial and improvement' to solve equations.

Review exercise

1 Solve the following equations.

(a) $3x - 4 = 5$ **(b)** $5x - 7 = 13$ **(c)** $4t + 3 = 15$

(d) $9x + 3 = 30$ **(e)** $12 = 7x - 9$ **(f)** $3s + 7 = 28$

(g) $6x + 7 = 37$ **(h)** $29 = 9x - 7$ **(i)** $8p - 5 = 59$

(j) $6x + 8 = 8$ **(k)** $23 - 2x = 15$ **(l)** $16 - 5x = 6$

2 For each balance below.

(i) write an equation **(ii)** solve your equation **(iii)** check your answer.

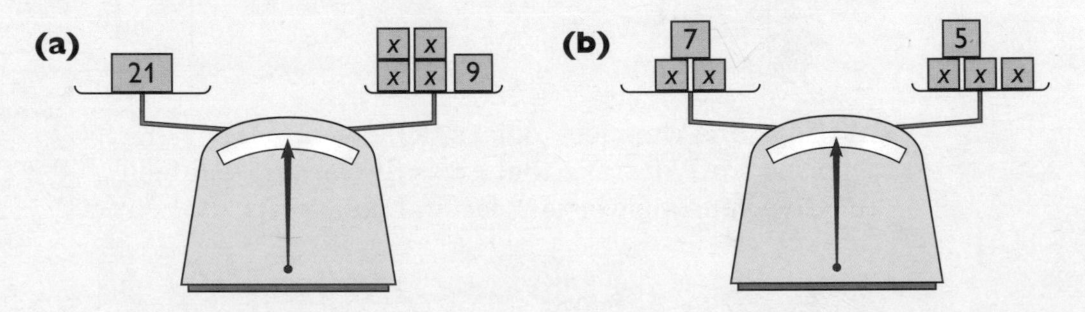

3 Solve the following equations and check your answers.

(a) $3x - 2 = 4x - 5$ **(b)** $2x + 9 = 3x - 5$ **(c)** $4x + 6 = 8 + 3x$

(d) $7x - 8 = 4x + 4$ **(e)** $5x - 5 = 2x + 4$ **(f)** $7 + 3x = 4x + 2$

(g) $3 + 7x = 7 + 3x$ **(h)** $5x + 1 = 6x - 1$ **(i)** $3x - 2 = 2x + 6$

(j) $4x + 1 = 7 + 2x$ **(k)** $5(x + 2) + 2 = 3x + 16$ **(l)** $4(x - 1) + x = 1$

(m) $2(x + 3) + 5 = x$ **(n)** $3(x - 4) + 2x = 8$ **(o)** $7(2x - 3) - 10x = x$

4 Solve these equations and check your answers.

(a) $\frac{2x}{3} = 6$ **(b)** $\frac{x}{8} = \frac{3}{4}$ **(c)** $\frac{x}{5} = \frac{1}{2}$

(d) $\frac{x}{6} = 2$ **(e)** $\frac{2x}{9} = \frac{2}{3}$ **(f)** $\frac{2x}{3} = \frac{7}{3}$

5 Tim and Pat are testing each other with number problems.

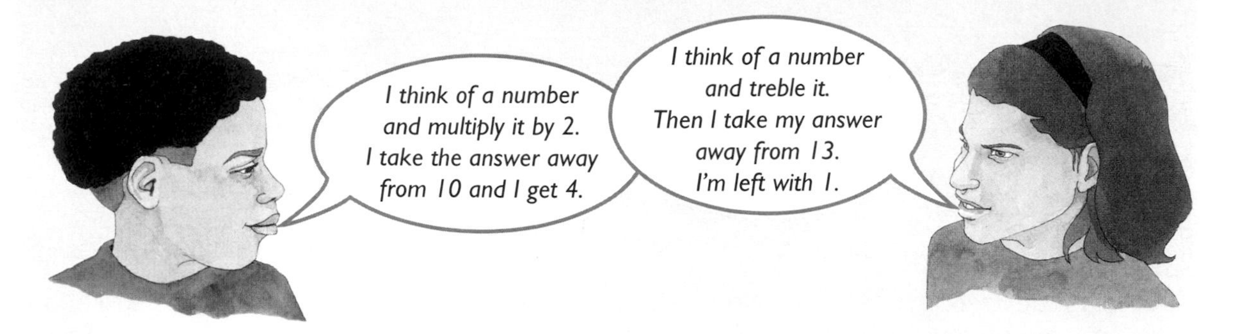

(a) (i) Write down an equation for Tim's starting number.
(ii) Solve the equation.

(b) (i) Write down an equation for Pat's starting number.
(ii) What number was Pat thinking of?

6 Mike has a set of 4 steps.
The tread is twice the riser.
He has a 108 cm length of carpet.
This covers the steps exactly.

(a) Write down an equation to show this information.

(b) Solve the equation to find
(i) the height of the riser
(ii) the width of the tread.

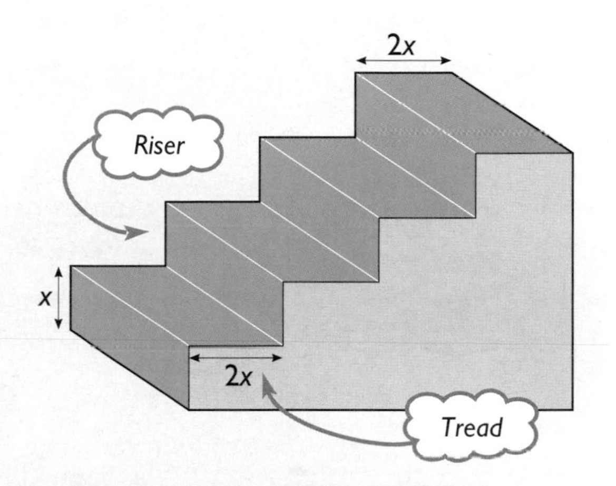

7 Copy the table and continue it to solve the equation $x^3 + 2x = 100$.
Find the value of x to 2 decimal places.

x	x^3	$2x$	$x^3 + 2x$	small/large
4	64	8	$64 + 8 = 72$	too small
5	125	10	$125 + 10 = 135$	too large
4.5				

8 Make your own tables to solve these equations.
In each case find the value of x to 1 decimal place.

(a) $x^3 - x^2 = 100$ **(b)** $x^3 + x = 50$

7 Angles and polygons

Angle conventions

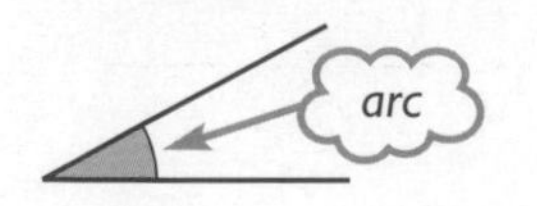

An angle is the amount of turning between two lines. An angle is indicated by an arc.

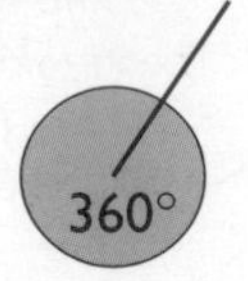

A whole turn is 360°.

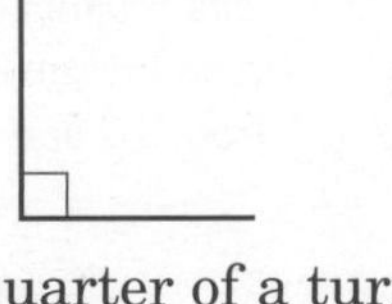

A quarter of a turn (90°) is called a right angle. A right angle is indicated by a small square.

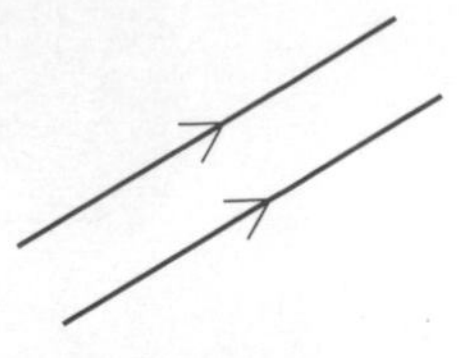

Parallel lines are shown by arrows.

Angle facts

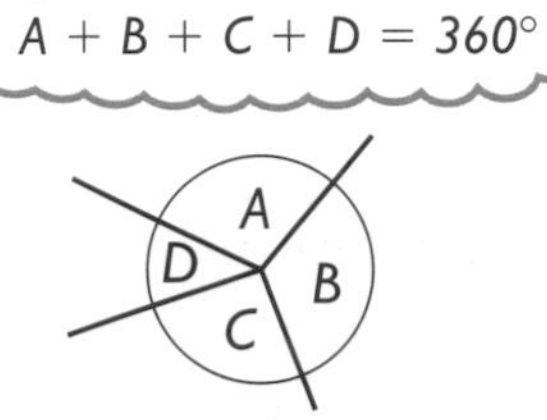

Angles **at a point** add up to 360°.

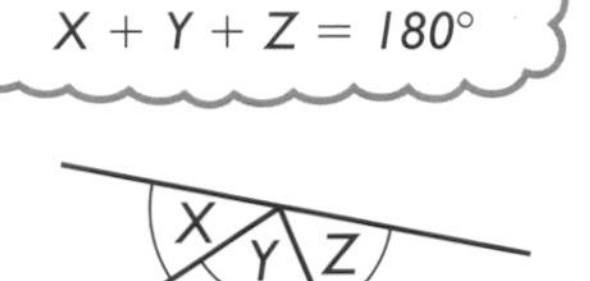

Angles **on a straight line** add up to 180°.

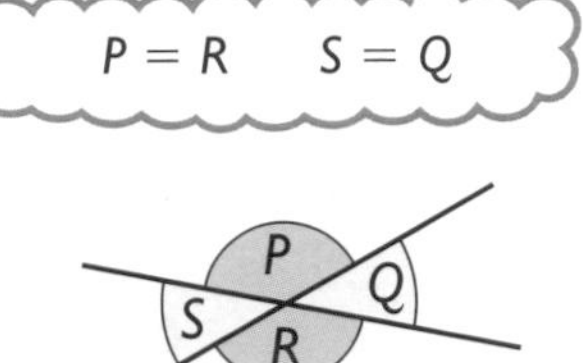

Vertically opposite angles are equal.

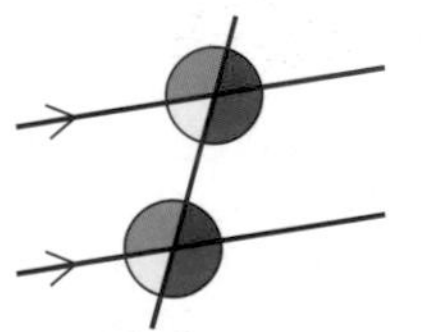

Corresponding angles are equal.

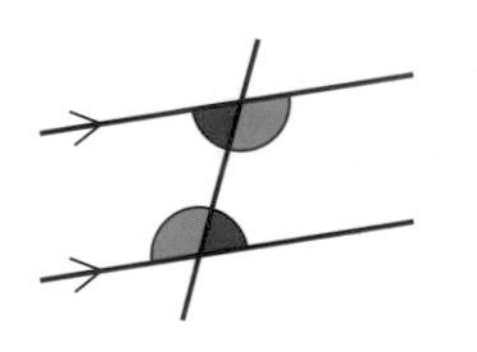

Alternate angles are equal.

What does 'corresponding' mean?

What does 'alternate' mean?

What is the difference between a convention and a fact?

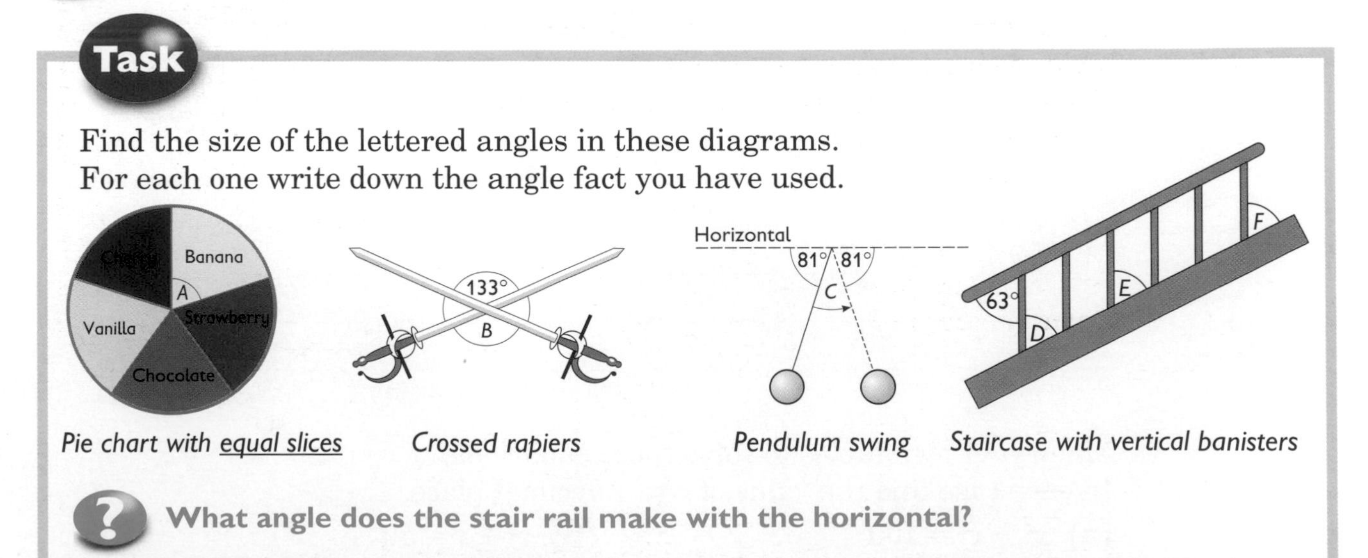

Task

Find the size of the lettered angles in these diagrams.
For each one write down the angle fact you have used.

Pie chart with equal slices — *Crossed rapiers* — *Pendulum swing* — *Staircase with vertical banisters*

What angle does the stair rail make with the horizontal?

Angle facts are used to solve angle problems.

Exercise

1 For each diagram **(i)** calculate the value of the lettered angle(s).
(ii) write down the angle fact you use for each angle.

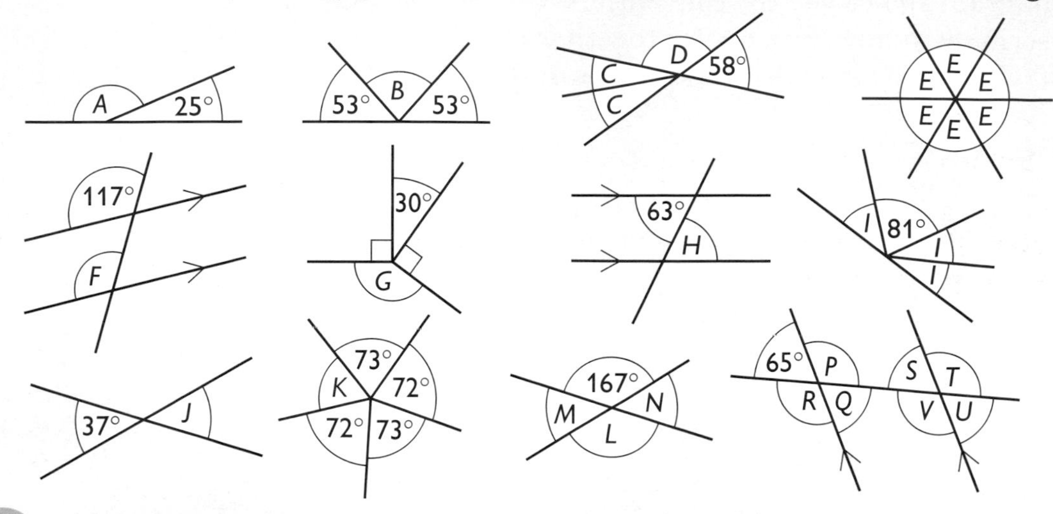

2 A metronome produces a steady beat for musical practice.
These diagrams show the *symmetrical* swings for different tempos.

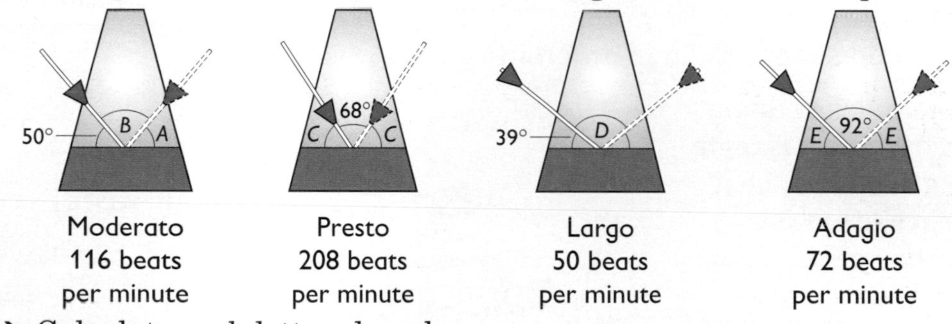

(a) Calculate each lettered angle.
(b) Write down the tempos in order of speed, slowest to fastest.
(c) Look carefully at the diagrams. How is the metronome adjusted?

3 A dealer sells cars in three colours.
The pie chart shows the cars sold at one showroom.
Twice as many green cars are sold as blue ones.
The number of red cars sold is three times the number of blue ones.
Calculate each lettered angle in the pie chart.

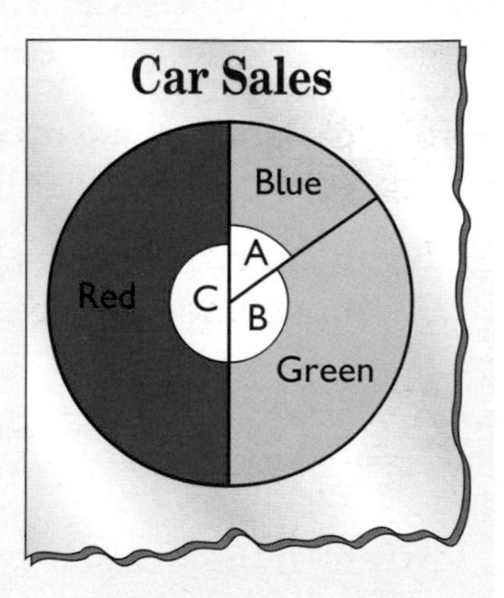

4 This fan has rotational symmetry order 8.
Calculate angle X.

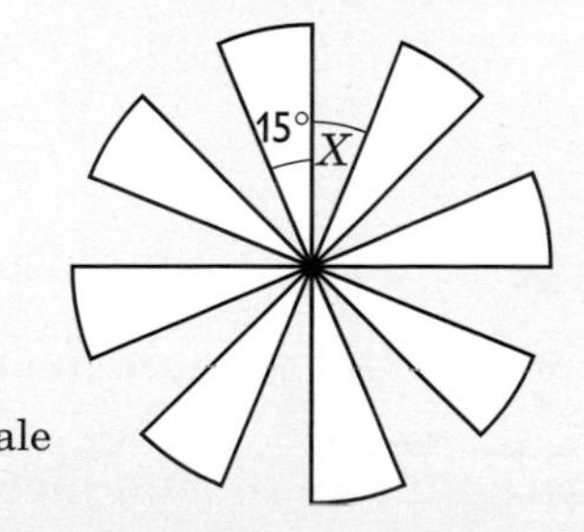

Not to scale

5 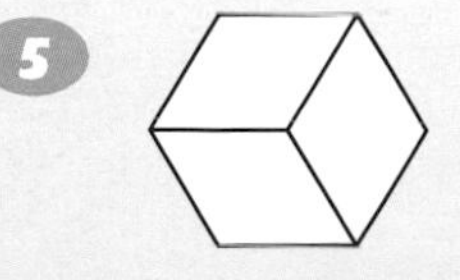

The three shapes in this pattern are congruent rhombi.
What can you say about the hexagon they form?
Explain your answer.

Angles in triangles and quadrilaterals

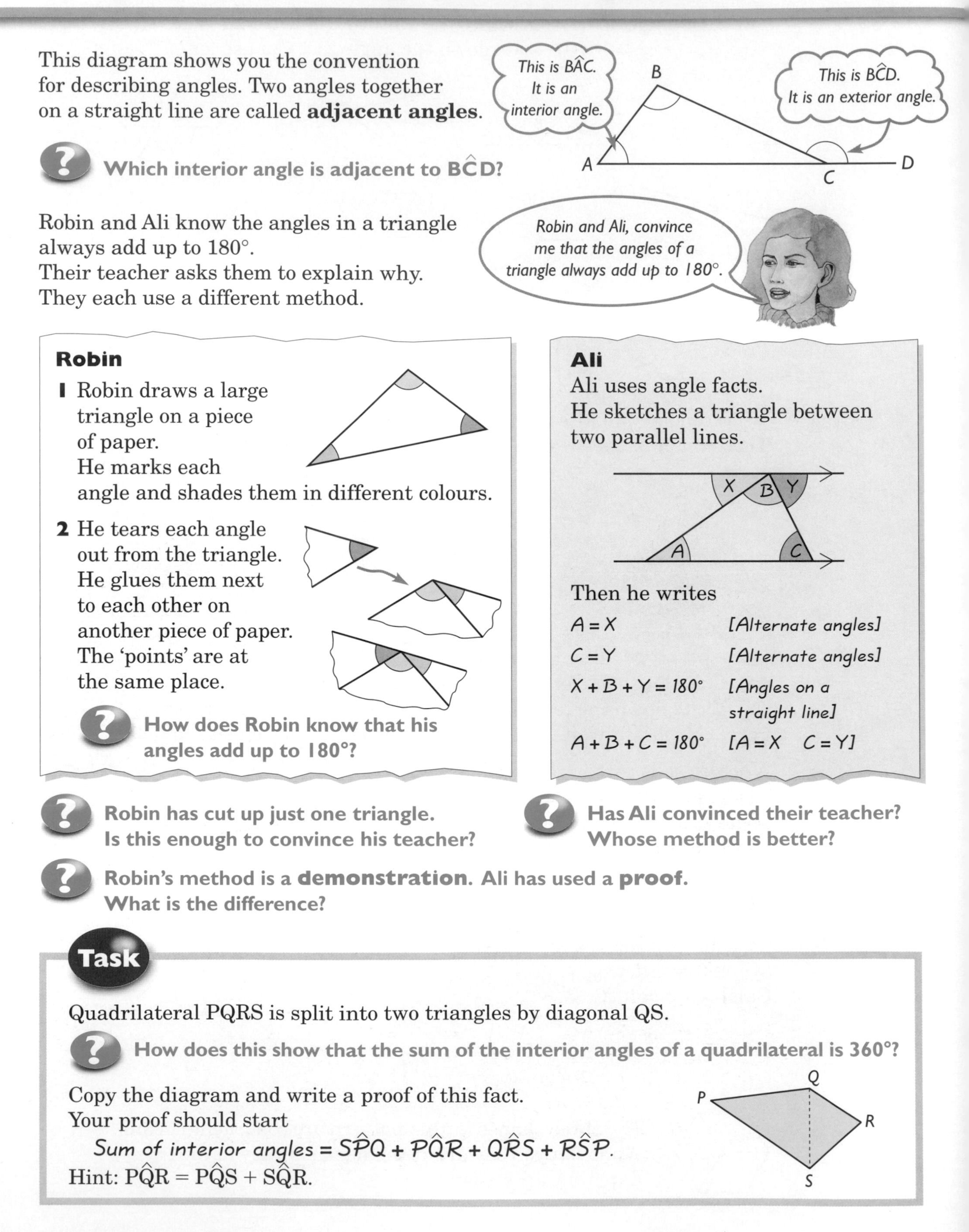

This diagram shows you the convention for describing angles. Two angles together on a straight line are called **adjacent angles**.

? Which interior angle is adjacent to $B\hat{C}D$?

Robin and Ali know the angles in a triangle always add up to 180°.
Their teacher asks them to explain why.
They each use a different method.

Robin

1. Robin draws a large triangle on a piece of paper. He marks each angle and shades them in different colours.
2. He tears each angle out from the triangle. He glues them next to each other on another piece of paper. The 'points' are at the same place.

? How does Robin know that his angles add up to 180°?

Ali

Ali uses angle facts.
He sketches a triangle between two parallel lines.

Then he writes

$A = X$	[Alternate angles]
$C = Y$	[Alternate angles]
$X + B + Y = 180°$	[Angles on a straight line]
$A + B + C = 180°$	[$A = X \quad C = Y$]

? Robin has cut up just one triangle. Is this enough to convince his teacher?

? Has Ali convinced their teacher? Whose method is better?

? Robin's method is a demonstration. Ali has used a proof. What is the difference?

Task

Quadrilateral PQRS is split into two triangles by diagonal QS.

? How does this show that the sum of the interior angles of a quadrilateral is 360°?

Copy the diagram and write a proof of this fact.
Your proof should start

Sum of interior angles $= S\hat{P}Q + P\hat{Q}R + Q\hat{R}S + R\hat{S}P$.

Hint: $P\hat{Q}R = P\hat{Q}S + S\hat{Q}R$.

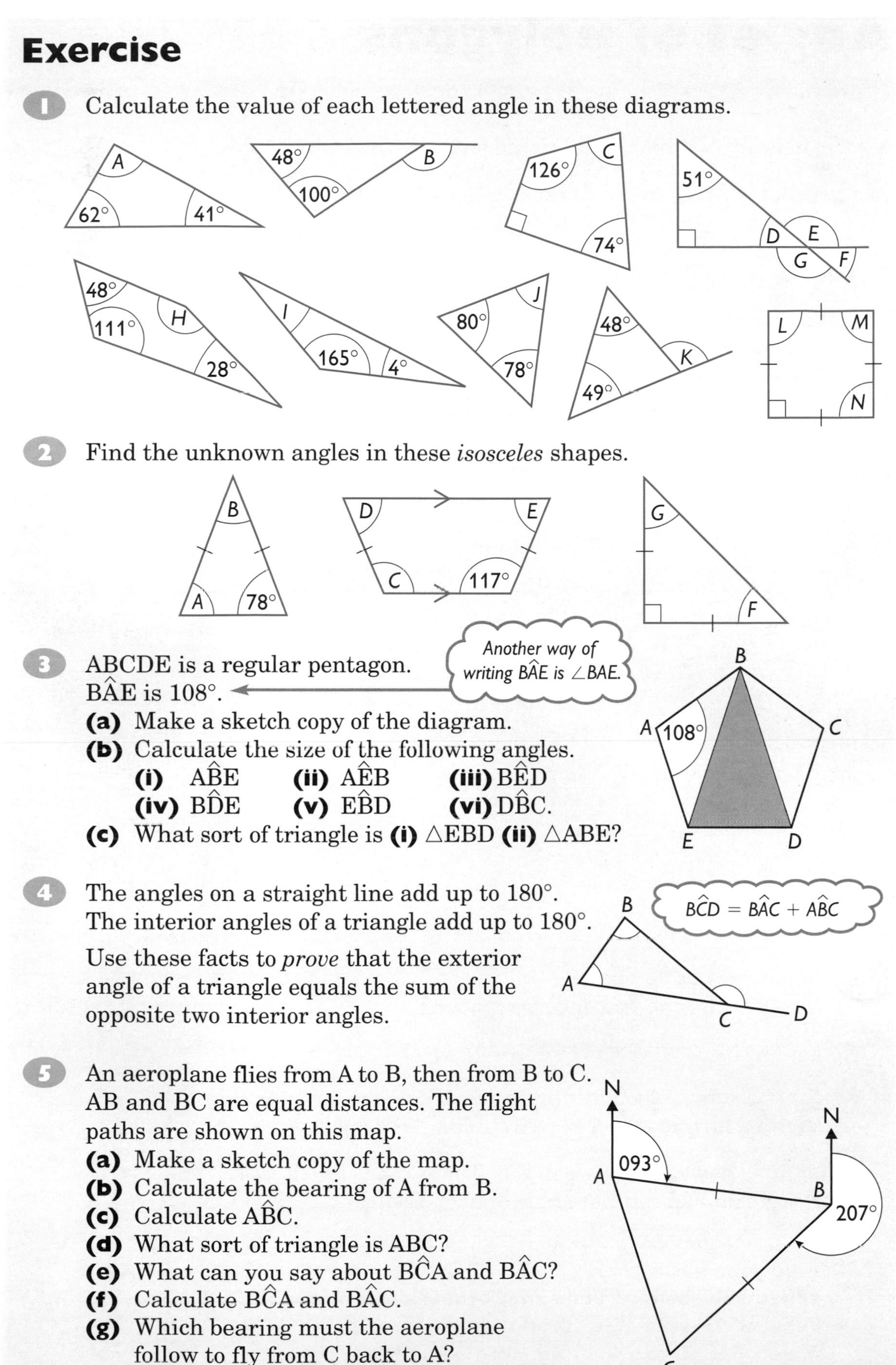

Exercise

1 Calculate the value of each lettered angle in these diagrams.

2 Find the unknown angles in these *isosceles* shapes.

3 ABCDE is a regular pentagon.
$\mathrm{B\hat{A}E}$ is 108°.

Another way of writing $\mathrm{B\hat{A}E}$ is $\angle\mathrm{BAE}$.

(a) Make a sketch copy of the diagram.
(b) Calculate the size of the following angles.
(i) $\mathrm{A\hat{B}E}$ **(ii)** $\mathrm{A\hat{E}B}$ **(iii)** $\mathrm{B\hat{E}D}$
(iv) $\mathrm{B\hat{D}E}$ **(v)** $\mathrm{E\hat{B}D}$ **(vi)** $\mathrm{D\hat{B}C}$.
(c) What sort of triangle is **(i)** $\triangle\mathrm{EBD}$ **(ii)** $\triangle\mathrm{ABE}$?

4 The angles on a straight line add up to 180°.
The interior angles of a triangle add up to 180°.

$\mathrm{B\hat{C}D} = \mathrm{B\hat{A}C} + \mathrm{A\hat{B}C}$

Use these facts to *prove* that the exterior angle of a triangle equals the sum of the opposite two interior angles.

5 An aeroplane flies from A to B, then from B to C. AB and BC are equal distances. The flight paths are shown on this map.
(a) Make a sketch copy of the map.
(b) Calculate the bearing of A from B.
(c) Calculate $\mathrm{A\hat{B}C}$.
(d) What sort of triangle is ABC?
(e) What can you say about $\mathrm{B\hat{C}A}$ and $\mathrm{B\hat{A}C}$?
(f) Calculate $\mathrm{B\hat{C}A}$ and $\mathrm{B\hat{A}C}$.
(g) Which bearing must the aeroplane follow to fly from C back to A?

Angles of polygons

A polygon is a two-dimensional closed shape with straight sides.

What is (a) the smallest number of sides a polygon can have?
(b) the largest number?

Task

1 Look at this heptagon.
Four diagonals have been drawn from one **vertex**.
The diagonals split the heptagon into five triangles.

What is the sum of all the interior angles of the five triangles?
What is the sum of the interior angles of the heptagon?

2 Copy and complete this table.

Polygon	Sides	Diagonals	Triangles	Sum of interior angles
	3	0	1	$1 \times 180° = 180°$
	4	1	2	$2 \times 180° = 360°$
	5	2		
	6			
	7	4	5	
	19			
	n			

S = the sum of the interior angles of a polygon with n sides.
Write a formula for S in terms of n.

This diagram shows part of a polygon. There is an exterior angle and an interior angle at each vertex.

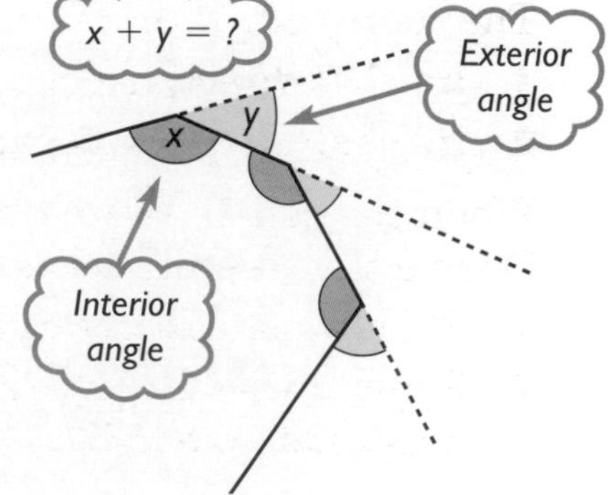

A polygon has n sides.
What is the sum of all its interior and exterior angles?
What is the sum of its interior angles?

What is the sum of the exterior angles of a polygon?

Exercise

1 Calculate the sum of the interior angles of

(a) a hexagon
(b) an octagon
(c) a polygon with 23 sides
(d) a square
(e) a triangle
(f) a polygon with 501 sides.

2 A *regular* polygon has equal sides and equal angles.
Calculate the size of the interior angle for

(a) a regular hexagon
(b) a regular octagon
(c) a regular triangle
(d) a regular quadrilateral.

What are the regular polygons (c) and (d) usually called?

3 The tessellation of a regular polygon is called a *regular* tessellation.

(a)

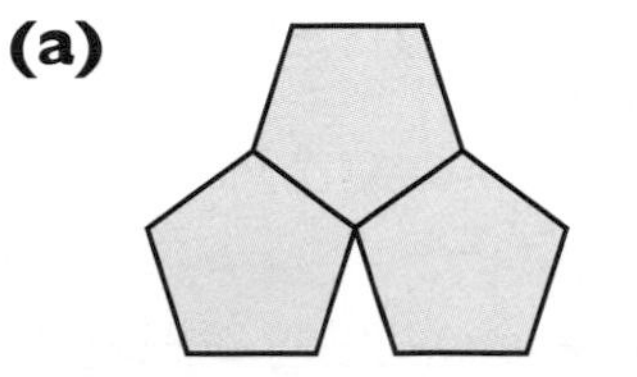

Explain why regular pentagons *cannot* form a regular tessellation.

(b)

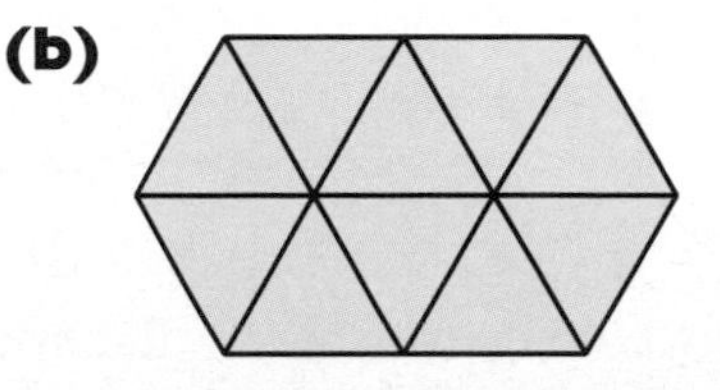

Explain why equilateral triangles form a regular tessellation.

(c) Find the two other regular polygons that form regular tessellations. Draw sketches to show how they tessellate.

(d) Explain why there are only three regular tessellations.

4 This diagram shows a pentagon split into five triangles.

(a) What is the sum of

(i) all angles in all the triangles?
(ii) the angles at the centre?
(iii) the interior angles of the pentagon?

(b) Use the diagram to explain why the sum of the interior angles of a polygon with n sides is $n \times 180° - 360°$.

(c) Show that the formula in part (b) is the same as $(n - 2) \times 180°$.

Investigation

Naseema runs a cross country race.
The course starts and finishes at A.

1 In what direction does she run at the beginning of the race?
2 In what direction is she running at the end of the race?

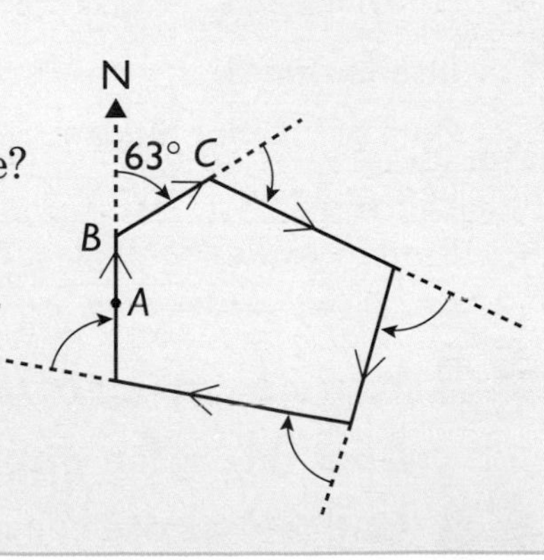

When she gets to B she turns through 63° and runs along BC.
During the race she turns through five angles in all.

3 What must the angles add up to? Explain your answer.
4 Explain why the interior angles of the pentagon must add up to 540°. (Use your answer to part 3.)

Regular polygons

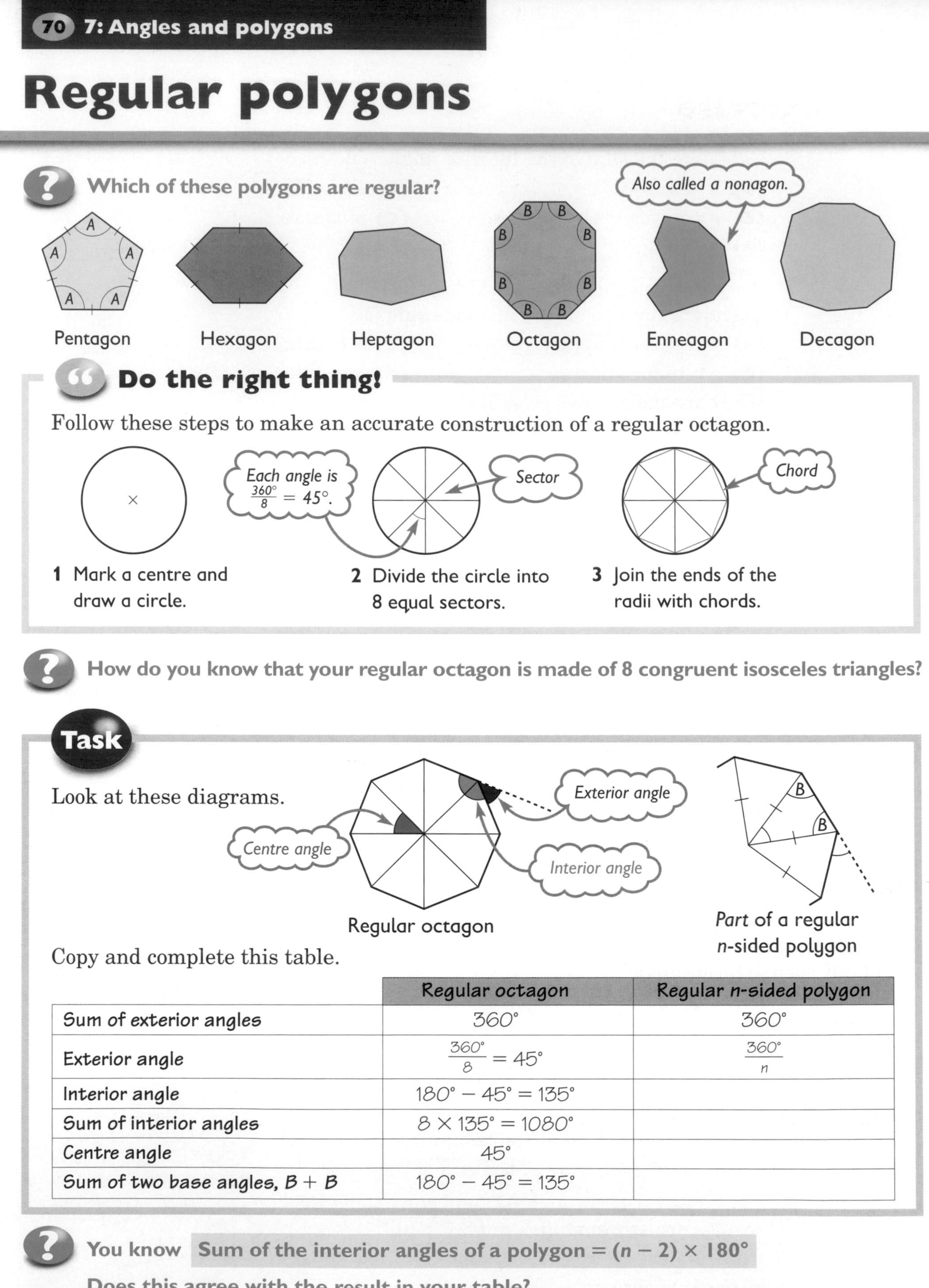

Which of these polygons are regular?

Pentagon · Hexagon · Heptagon · Octagon · Enneagon · Decagon

Do the right thing!

Follow these steps to make an accurate construction of a regular octagon.

1 Mark a centre and draw a circle.

2 Divide the circle into 8 equal sectors.

3 Join the ends of the radii with chords.

How do you know that your regular octagon is made of 8 congruent isosceles triangles?

Task

Look at these diagrams.

Regular octagon

Part of a regular n-sided polygon

Copy and complete this table.

	Regular octagon	Regular n-sided polygon
Sum of exterior angles	360°	360°
Exterior angle	$\frac{360°}{8} = 45°$	$\frac{360°}{n}$
Interior angle	$180° - 45° = 135°$	
Sum of interior angles	$8 \times 135° = 1080°$	
Centre angle	45°	
Sum of two base angles, $B + B$	$180° - 45° = 135°$	

You know Sum of the interior angles of a polygon = $(n - 2) \times 180°$

Does this agree with the result in your table?

Notice that the centre angle of a regular polygon is the same as its exterior angle. Explain why this is so.

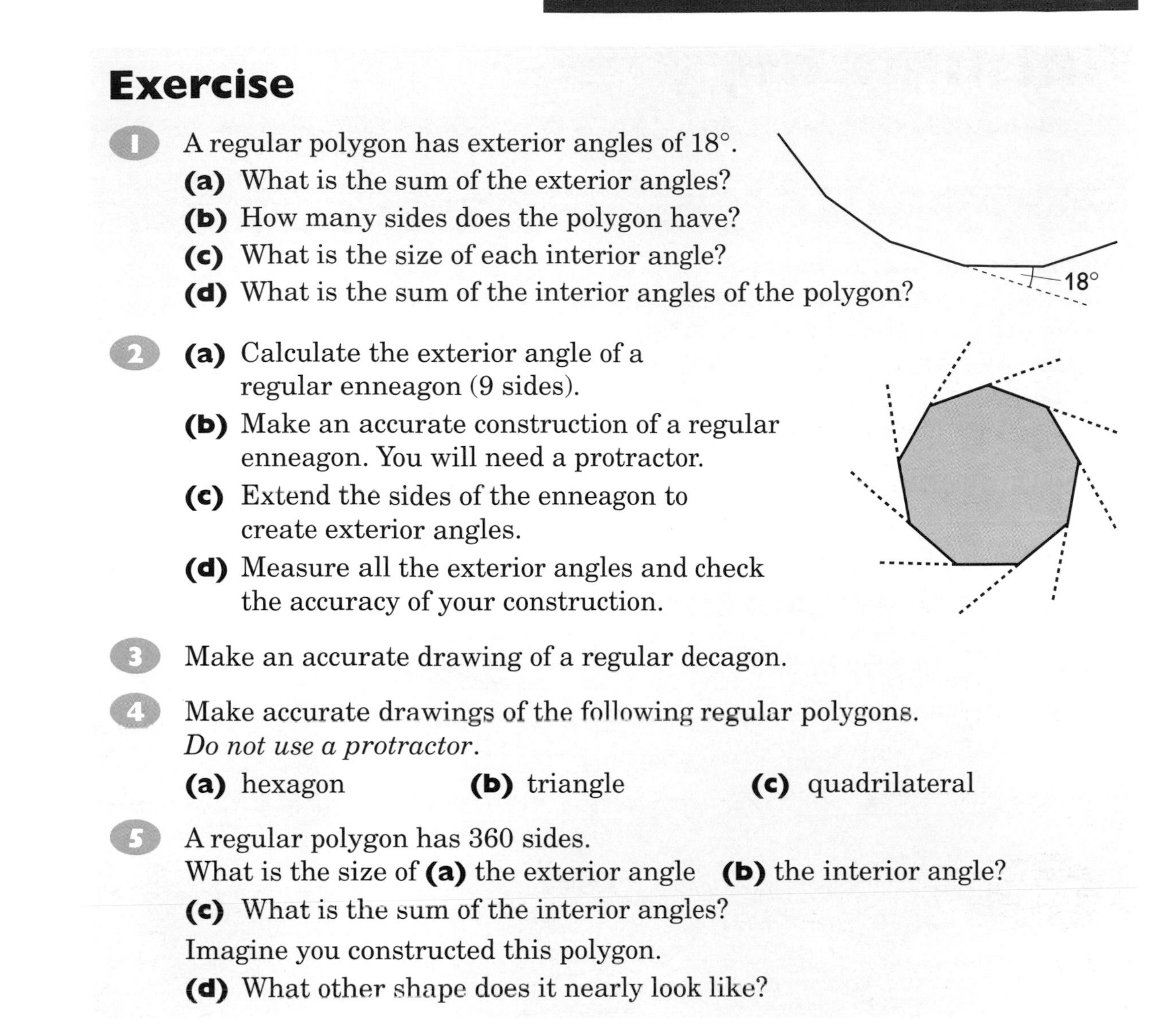

Exercise

1. A regular polygon has exterior angles of 18°.
   (a) What is the sum of the exterior angles?
   (b) How many sides does the polygon have?
   (c) What is the size of each interior angle?
   (d) What is the sum of the interior angles of the polygon?

2. **(a)** Calculate the exterior angle of a regular enneagon (9 sides).
   (b) Make an accurate construction of a regular enneagon. You will need a protractor.
   (c) Extend the sides of the enneagon to create exterior angles.
   (d) Measure all the exterior angles and check the accuracy of your construction.

3. Make an accurate drawing of a regular decagon.

4. Make accurate drawings of the following regular polygons.
   Do not use a protractor.
   (a) hexagon **(b)** triangle **(c)** quadrilateral

5. A regular polygon has 360 sides.
   What is the size of **(a)** the exterior angle **(b)** the interior angle?
   (c) What is the sum of the interior angles?
   Imagine you constructed this polygon.
   (d) What other shape does it nearly look like?

Investigation

Make an accurate drawing of a regular pentagon.
Extend the sides to form exterior angles.
Cut out each exterior angle.
Glue the exterior angles next to each other on another piece of paper with the points at the same place.

Do the exterior angles fit together as you expected?

Repeat using an *irregular* pentagon.

What can you say about the exterior angles of your irregular pentagon?

Now try it with two more irregular polygons which have different numbers of sides.

What does this investigation show you about the exterior angles of polygons? Is it a demonstration or a proof?

Finishing off

Now that you have finished this chapter you should know:

- angle conventions and facts
- how to prove that the angles in a triangle add up to 180°
- how to prove that the angles in a quadrilateral add up to 360°
- that the sum of the interior angles of a polygon with n sides is $(n - 2) \times 180°$
- the size of the interior angles of a regular (a) pentagon (b) hexagon
- that the sum of the exterior angles of a polygon is 360°
- how to construct a regular polygon
- that a chord is a line across a circle, from one point on the circumference to another.

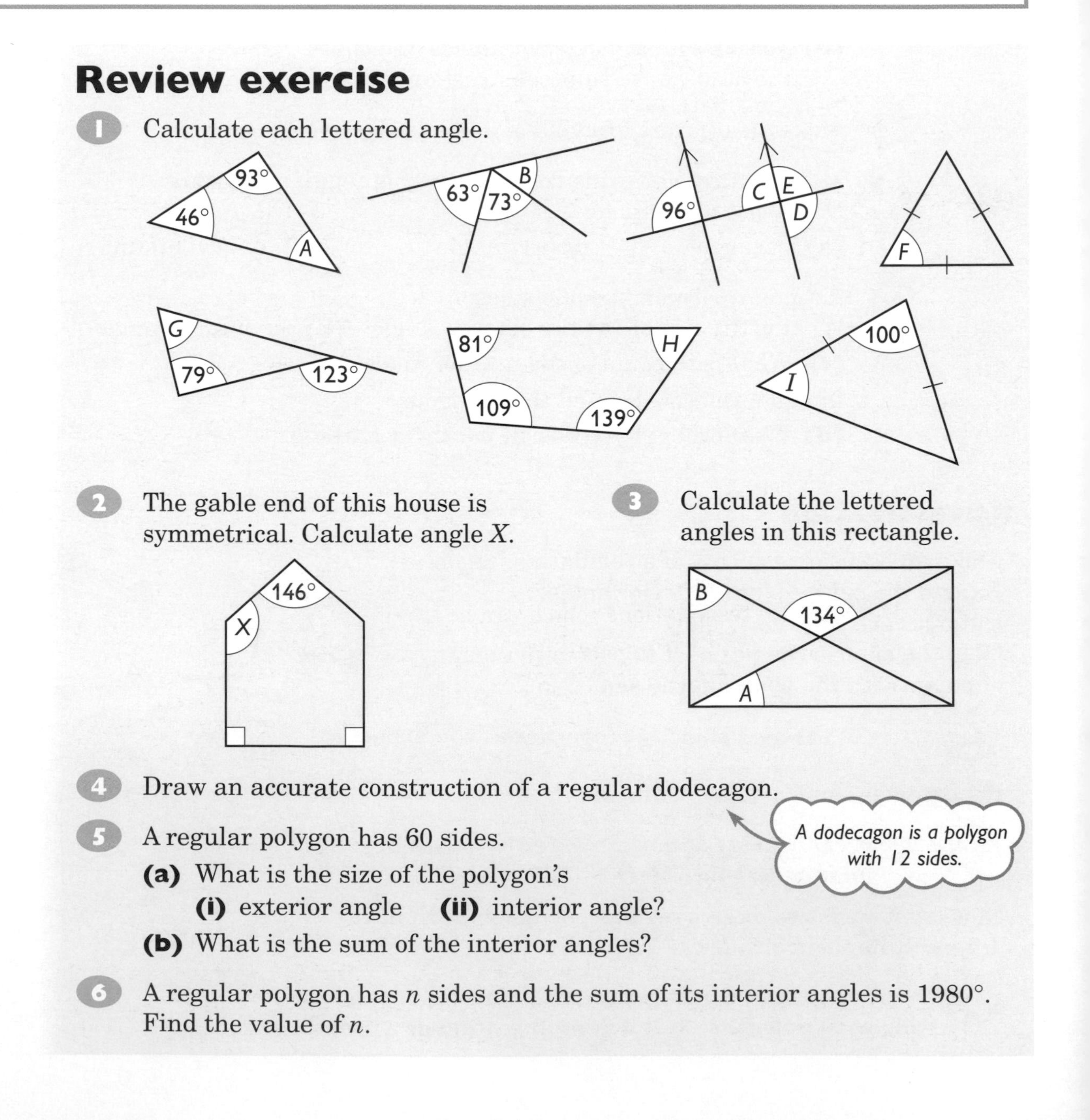

Review exercise

1 Calculate each lettered angle.

2 The gable end of this house is symmetrical. Calculate angle X.

3 Calculate the lettered angles in this rectangle.

4 Draw an accurate construction of a regular dodecagon.

5 A regular polygon has 60 sides.

(a) What is the size of the polygon's

(i) exterior angle **(ii)** interior angle?

(b) What is the sum of the interior angles?

6 A regular polygon has n sides and the sum of its interior angles is 1980°. Find the value of n.

7 Look at the diagram of a heptagon. Some of the sides are parallel.

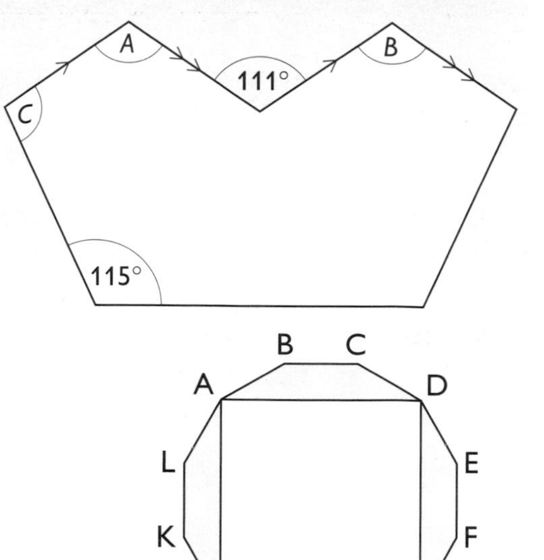

(a) Write down the size of angle A.

(b) Explain why $A = B$.

(c) Calculate the sum of the interior angles of the heptagon.

The heptagon is symmetrical.

(d) Use this information to calculate angle C.

8 The diagram shows a quadrilateral inside a regular dodecagon.

(a) Calculate $\mathrm{A\hat{B}C}$.

(b) Calculate $\mathrm{B\hat{A}D}$.

(c) Write down $\mathrm{L\hat{A}J}$.

(d) Explain why quadrilateral ADGJ must be a square.

Investigation

The tiled floor below is a tessellation made from octagons and squares. Any tessellation made from two or more *regular* polygons is called a **semi-regular** tessellation.

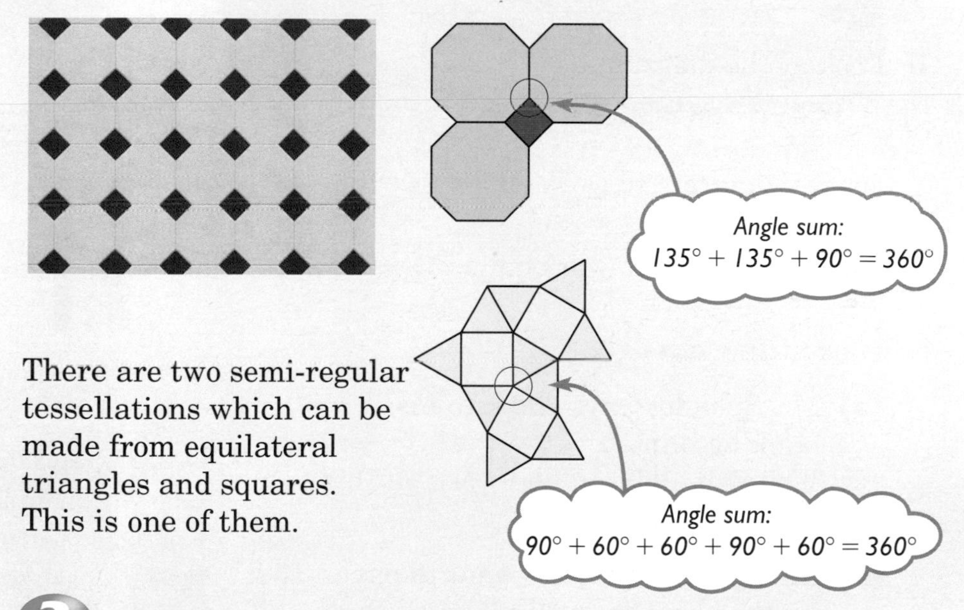

There are two semi-regular tessellations which can be made from equilateral triangles and squares. This is one of them.

How does the angle sum show you that the triangles and squares may tessellate?

Another way to write the angle sum is $90° + 90° + 60° + 60° + 60° = 360°$. How does this describe a different tessellation?

1 Sketch the other semi-regular tessellation of equilateral triangles and squares.

2 There are two semi-regular tessellations which can be made from hexagons and triangles. Use isometric paper to draw them.

8 Directed numbers

Adding negative numbers

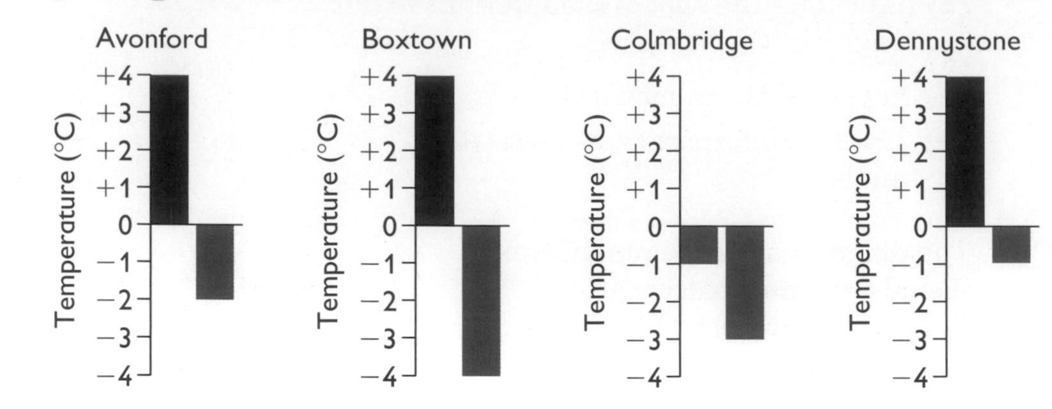

These graphs show temperatures at noon and 6 pm on a cold day. Work out the mean temperature for each place.

Work out the missing number in these equations:

(a) $(-3) + \square = 0$ **(b)** $(-3) + 4 = \square$ **(c)** $(-3) + 2 = \square$

Task

1 Look at the diagrams.
Write a few sentences to explain them.

2 Draw a diagram to work out $(-3) + 5$.

3 Draw a diagram to work out $(-3) + (-1)$.

4 Look at the letter.

(a) Mr. Spender pays £30 into his bank account.
Write an addition and work out his new balance.

(b) Mr. Spender does not want to pay in £30.
Instead he wants to write a cheque for £30.
Write an addition to show what his new balance would be.

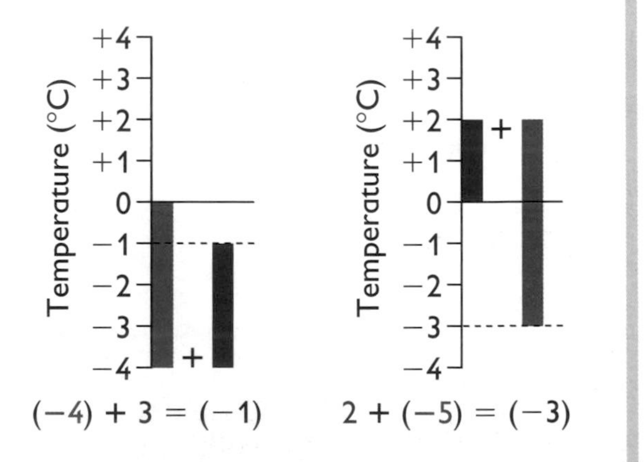

$(-4) + 3 = (-1)$ $2 + (-5) = (-3)$

Dear Mr. Spender,
I am writing to let you know that your bank account is £20 overdrawn.
Please repay the money at your earliest convenience.
Yours faithfully,

M. O'Ney

Ms. M. O'Ney (Manager)

Fill in the boxes in this statement with six different pairs of numbers.

$3 + 4 + \square + \square = 9$

Exercise

1 Draw diagrams to show these additions.

(a) $(-4)+(-2)$ **(b)** $(-4)+6$ **(c)** $(-4)+2$

2 Work out these additions.

(a) $(-5)+5$ **(b)** $(-5)+4$ **(c)** $(-5)+6$

(d) $(-6)+8$ **(e)** $(-6)+10$ **(f)** $(-6)+2$

(g) $5+(-5)$ **(h)** $5+(-10)$ **(i)** $5+(-2)$

3 Copy this number wall and find the missing numbers.
The number in each brick is the sum of the two bricks beneath it.

−2

7 −3

5 2 −5 2 −9

4 Peter's bank account is £50 overdrawn. He pays in £80.
What is his new balance?

5 Sarah's bank account is £10 overdrawn. She takes out £15.
What is her new balance?

6 Paul puts £20 into his bank account. His account is now £5 in credit.
What was his bank balance to begin with?

7 The answer to an addition is (-5). Write down four possible questions.

8 Work out these additions.

(a) $(-4)+4+6$ **(b)** $(-4)+10$ **(c)** $(-5)+5+4$ **(d)** $(-5)+9$

Activity

'End Target'

START

−8	−7	−6	−5	−4	−3	−2	−1	0	1	2	3	4	5	6	7	8

1 Petra's die always counts as positive.
Nicola's die always counts as negative.
They place a counter at zero and take it in turns to throw their die.

Petra throws a 3 and moves the counter to (3).
Nicola throws (-5) and moves the counter to (-2).
Petra throws a 6 and moves to (4).

(a) Write down three additions to explain what is happening.

Petra aims for 8 or higher. She throws first.
Petra wins in two goes.

(b) Write down different additions to show how this can happen.

Nicola aims for (-8) or lower. She throws first.
Nicola wins in two goes.

(c) Write down different additions to show how this can happen.

2 Play a game of 'End Target'.

Subtracting negative numbers

Ruth is on floor 3. Nina is on floor (-2).

How many floors is Ruth above Nina?
What about Nina above Ruth?

How can you show that one floor is lower than another?

Subtraction compares numbers:

$(-4) - (-5) = 1$ tells us that
(-4) is one floor higher than (-5).

Work out $(-3) - (-2)$ and $(-3) - (2)$.
Explain how you get your answers.

1 Matt is on floor 5. He is thirsty.
There are drinks machines on floors 8 and (-1).
Which is closer to Matt?

2 John is on floor 4. He must visit Rachel on (-2).
He also wants to visit Sandra on (-1) and Roger on 6.
How does he plan for the least time in the lift?

3 Tim is on floor 1 and Peter is on (-3).
They need to meet.
Which floor is half way between them?

4 The answer to a subtraction is (-2).
Write down five possible questions.

5 Copy these equations and fill in the missing numbers.

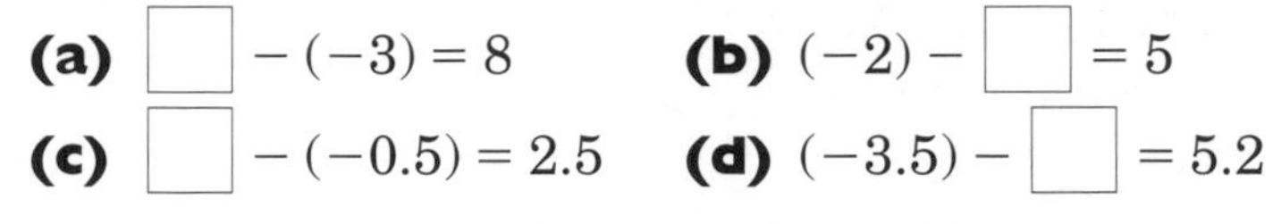

(a) $\square - (-3) = 8$ **(b)** $(-2) - \square = 5$

(c) $\square - (-0.5) = 2.5$ **(d)** $(-3.5) - \square = 5.2$

During the 1900s, the highest temperature in Britain was 37.1 °C (recorded in Cheltenham on 3/8/90).
The lowest temperature was −27.2 °C (in the Scottish Highlands on 30/12/95).
Work out the range in temperatures.

Find groups of three numbers to go in the boxes in this equation.

$\square - \square - \square = -2$

Exercise

1. Sam is on floor 20. Mike is on floor (−4).
   Work out **(a)** 20 − (−4) **(b)** (−4) + 20

2. Charlene is on floor (4).
   She needs to do some photocopying.
   There are photocopiers on floors 8 and (−1).
   Which is closer to Charlene?

3. Work out these subtractions.
   (a) (−5) − (−10) **(b)** (−10) − (−5) **(c)** 5 − (−10)
   (d) 0 − (−30) **(e)** 0 − (30) **(f)** (−30) − 0
   (g) 6 − (−2) **(h)** (−2) − 6 **(i)** (−2) − (−6)

4. Copy and complete this number wall.
   The number in each brick is the sum of those in the two bricks beneath it.

2
−1
−6
4

5. **(a)** Lucy is altering a photograph on her computer.
   She alters the brightness.
   Work out 60 − (−50).

   (b) Lucy sets the brightness half way between (60) and (−50).
   What brightness does she use?

60

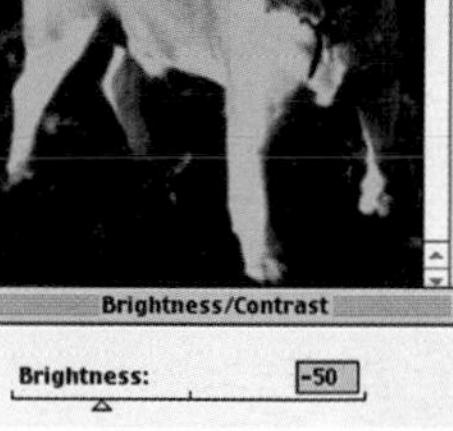

(−50)

6. The temperature at midday is 10 °C.
   The temperature at midnight is (−5) °C.
   Find the range in temperature.

7. The highest temperature is 8 °C.
   The range in temperature that day is 10 °C.
   What is the lowest temperature that day?

8. Copy and complete this table.

Maximum temperature (°C)	15	−2	10		20
Minimum temperature (°C)	−2	−15		−5	−6
Range in temperature (°C)			15	10	

Activity Draw a subtraction number wall with five layers.
Write (−4) as the top number. Fill in the other numbers.
There is more than one answer to this question.

Multiplying and dividing negative numbers

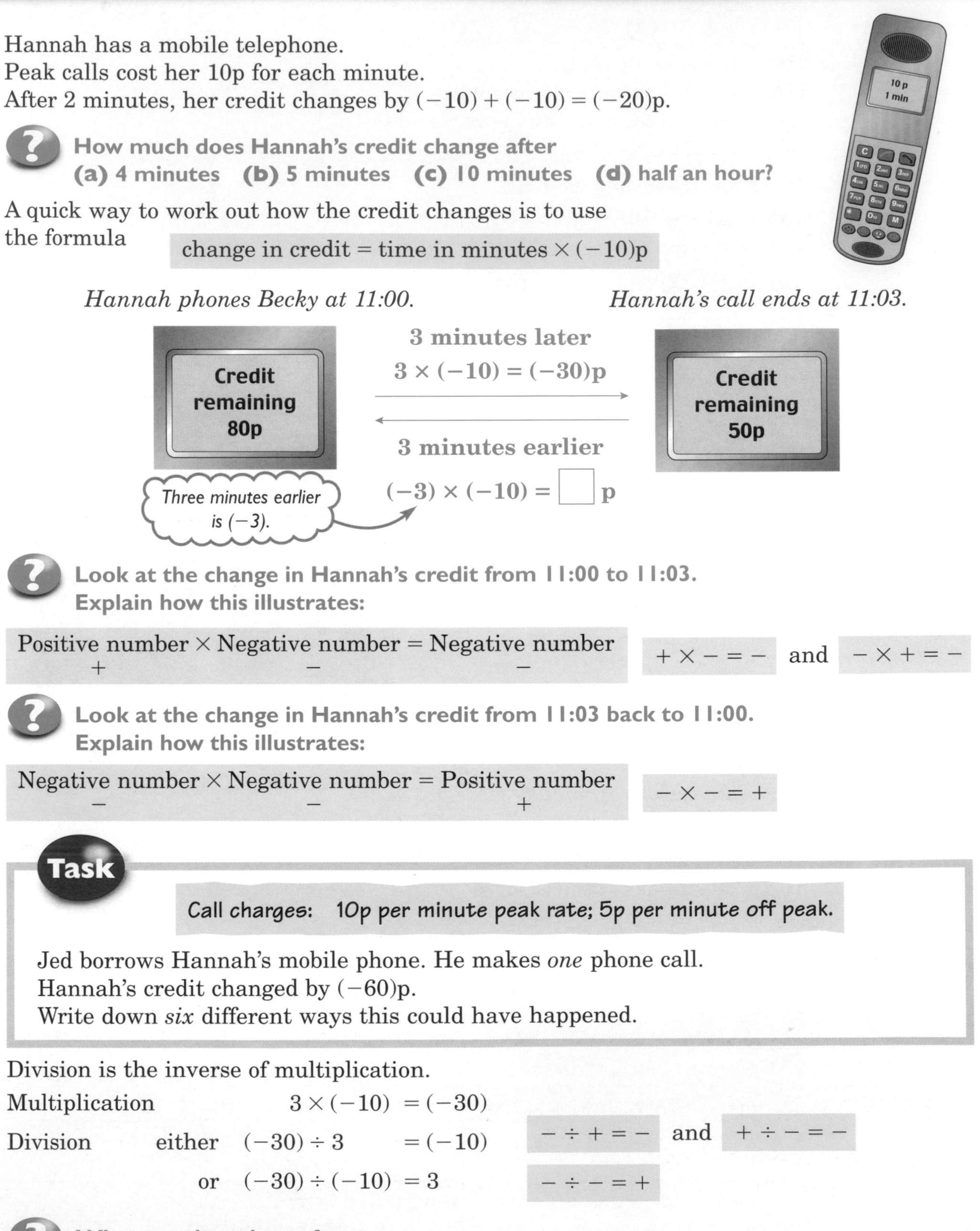

Hannah has a mobile telephone.
Peak calls cost her 10p for each minute.
After 2 minutes, her credit changes by $(-10) + (-10) = (-20)$p.

? How much does Hannah's credit change after
(a) 4 minutes (b) 5 minutes (c) 10 minutes (d) half an hour?

A quick way to work out how the credit changes is to use the formula

change in credit = time in minutes × (-10)p

Hannah phones Becky at 11:00. *Hannah's call ends at 11:03.*

Credit remaining 80p

3 minutes later
$3 \times (-10) = (-30)$p

3 minutes earlier
$(-3) \times (-10) = \square$ p

Credit remaining 50p

Three minutes earlier is (−3).

? Look at the change in Hannah's credit from 11:00 to 11:03.
Explain how this illustrates:

Positive number × Negative number = Negative number
(+ × − = −)

$+ \times - = -$ and $- \times + = -$

? Look at the change in Hannah's credit from 11:03 back to 11:00.
Explain how this illustrates:

Negative number × Negative number = Positive number
(− × − = +)

$- \times - = +$

Task

Call charges: 10p per minute peak rate; 5p per minute off peak.

Jed borrows Hannah's mobile phone. He makes *one* phone call.
Hannah's credit changed by (-60)p.
Write down *six* different ways this could have happened.

Division is the inverse of multiplication.

Multiplication $3 \times (-10) = (-30)$

Division either $(-30) \div 3 = (-10)$ $- \div + = -$ and $+ \div - = -$

or $(-30) \div (-10) = 3$ $- \div - = +$

? What are the values of
(a) $50 \div 5$ (b) $50 \div (-5)$ (c) $(-50) \div 5$ (d) $(-50) \div (-5)$?

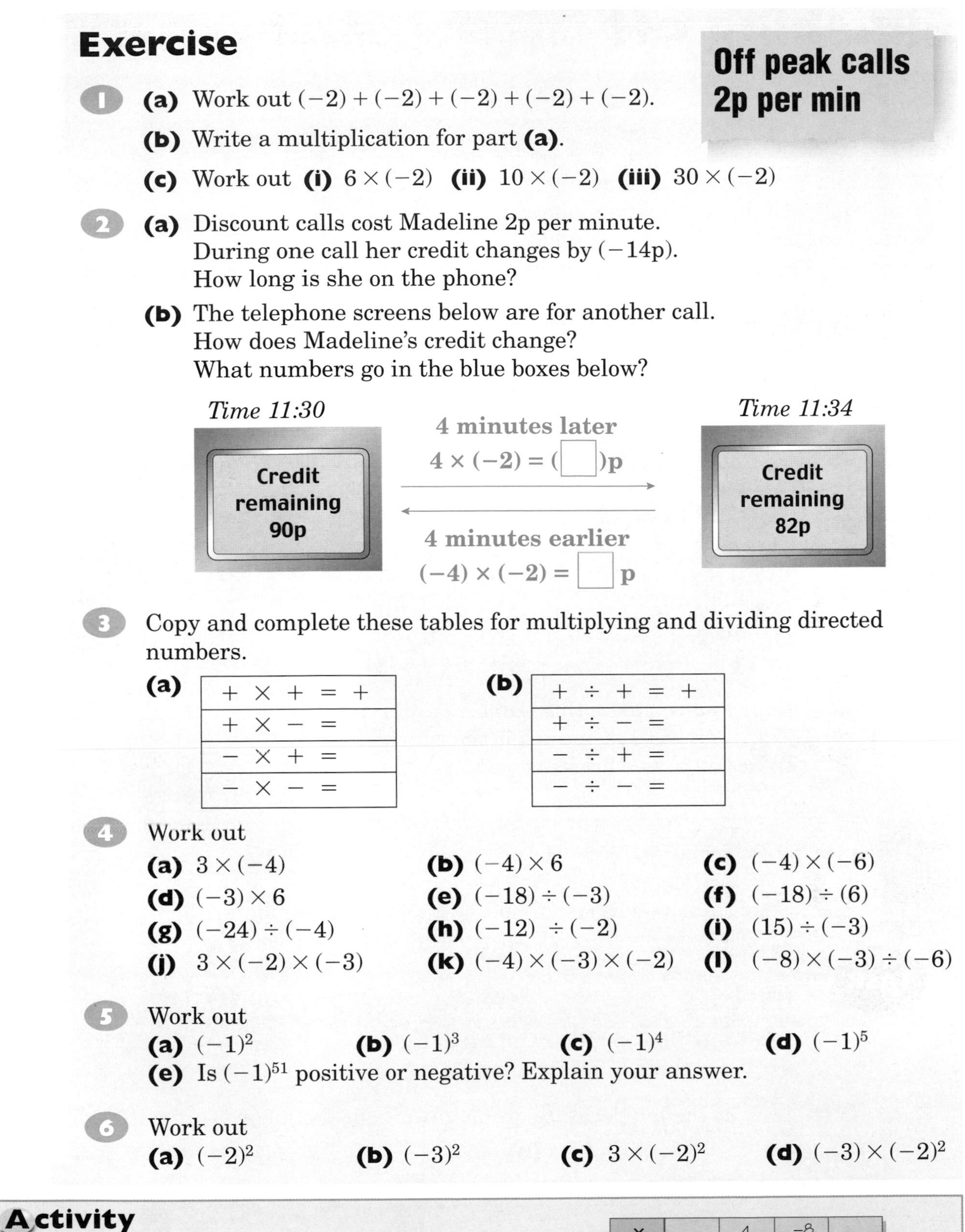

Exercise

Off peak calls 2p per min

1 **(a)** Work out $(-2)+(-2)+(-2)+(-2)+(-2)$.

(b) Write a multiplication for part **(a)**.

(c) Work out **(i)** $6 \times (-2)$ **(ii)** $10 \times (-2)$ **(iii)** $30 \times (-2)$

2 **(a)** Discount calls cost Madeline 2p per minute.
During one call her credit changes by (−14p).
How long is she on the phone?

(b) The telephone screens below are for another call.
How does Madeline's credit change?
What numbers go in the blue boxes below?

Time 11:30

Credit remaining 90p

4 minutes later

$4 \times (-2) = (\square)$p

4 minutes earlier

$(-4) \times (-2) = \square$ p

Time 11:34

Credit remaining 82p

3 Copy and complete these tables for multiplying and dividing directed numbers.

(a)

+ × + = +
+ × − =
− × + =
− × − =

(b)

+ ÷ + = +
+ ÷ − =
− ÷ + =
− ÷ − =

4 Work out

(a) $3 \times (-4)$ **(b)** $(-4) \times 6$ **(c)** $(-4) \times (-6)$

(d) $(-3) \times 6$ **(e)** $(-18) \div (-3)$ **(f)** $(-18) \div (6)$

(g) $(-24) \div (-4)$ **(h)** $(-12) \div (-2)$ **(i)** $(15) \div (-3)$

(j) $3 \times (-2) \times (-3)$ **(k)** $(-4) \times (-3) \times (-2)$ **(l)** $(-8) \times (-3) \div (-6)$

5 Work out

(a) $(-1)^2$ **(b)** $(-1)^3$ **(c)** $(-1)^4$ **(d)** $(-1)^5$

(e) Is $(-1)^{51}$ positive or negative? Explain your answer.

6 Work out

(a) $(-2)^2$ **(b)** $(-3)^2$ **(c)** $3 \times (-2)^2$ **(d)** $(-3) \times (-2)^2$

Activity

Copy and complete the multiplication table.
Find two different ways to do it using whole numbers.

×		4	−8	
		−8	16	
−5		−20		
	21			−14
				12

Finishing off

Now that you have finished this chapter you should be able to:

- add and subtract negative numbers
- multiply and divide negative numbers
- work out powers of negative numbers.

Review exercise

1 Work out these additions.

(a) $(-8) + 8$
(b) $(-8) + 10$
(c) $(-8) + 6$
(d) $13 + (-8)$
(e) $(-8) + 13$
(f) $(-8) + (-13)$
(g) $20 + (-6)$
(h) $(-20) + (-6)$
(i) $(-20) + 6$

2 Work out these subtractions.

(a) $(-3) - 3$
(b) $(-3) - (-3)$
(c) $3 - (-3)$
(d) $0 - 3$
(e) $(-3) - 0$
(f) $0 - (-3)$
(g) $(-5) - 3$
(h) $5 - (-3)$
(i) $(-5) - (-3)$
(j) $10 - (-15)$
(k) $(-15) - (-10)$
(l) $(-10) - (-15)$
(m) $14 - (-6)$
(n) $6 - (-14)$
(o) $(-6) - (-14)$

3 Copy and complete this number wall.
The number in each brick is the sum of those in the two bricks beneath it.

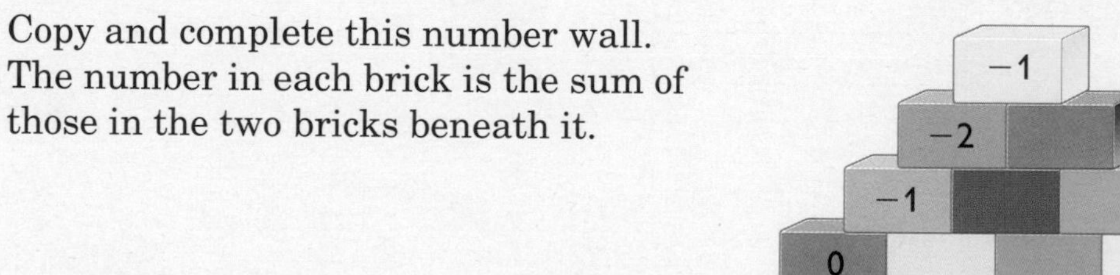

4 Copy these equations and fill in the missing numbers.

(a) $(-2) + \square = 7$
(b) $\square + (-3) = 13$
(c) $(-5) - \square = (-18)$
(d) $4 - \square = 7$
(e) $\square - (-3) = 4$
(f) $(-8) - \square = 10$

(g) Copy this equation and fill in the boxes. $\square - \square = (-5)$.
Give three different answers.

5 Look at the temperature graphs. Work out the mean temperature for each.

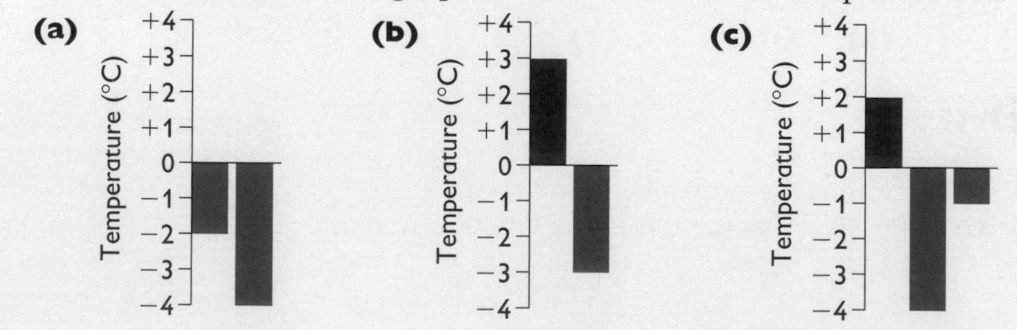

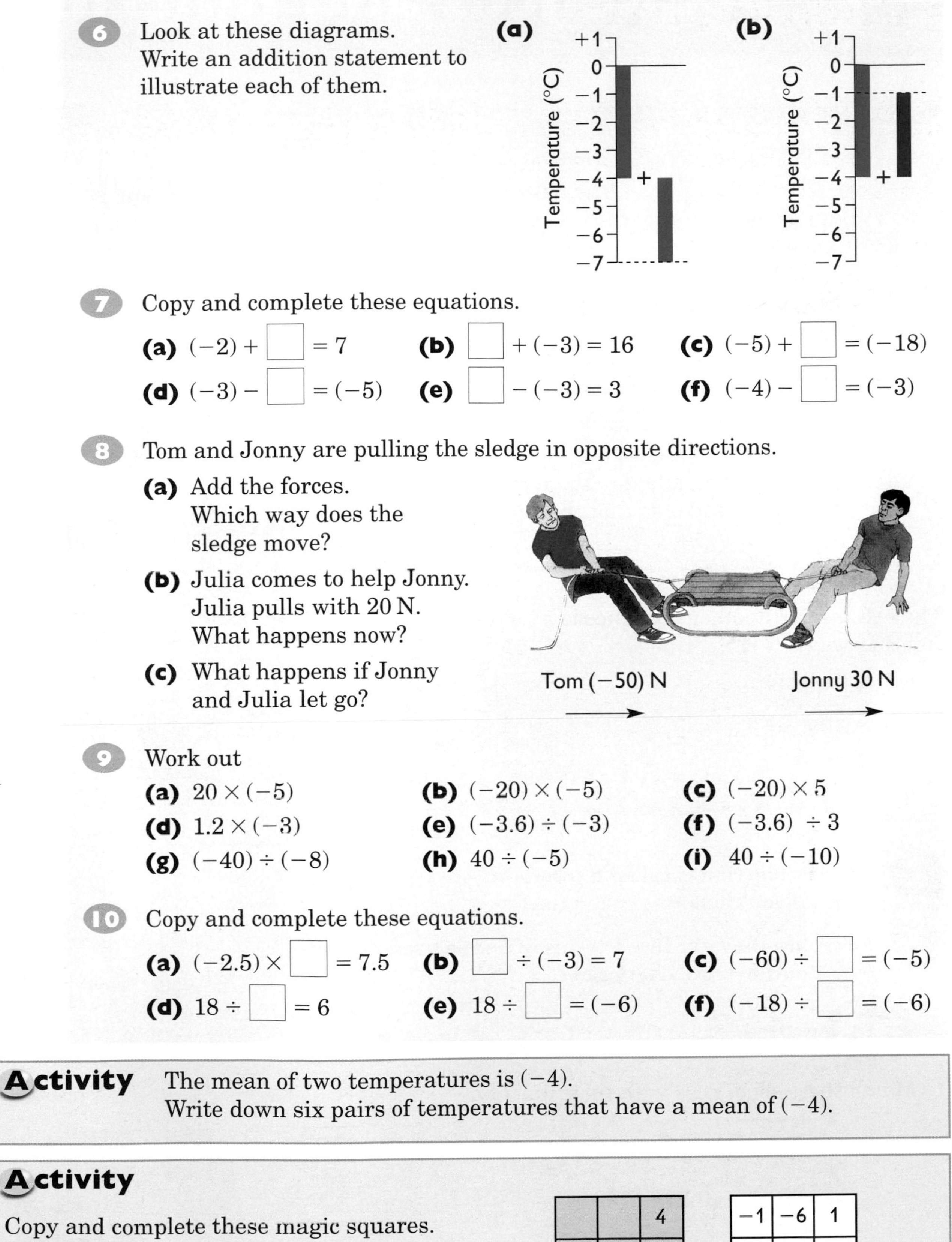

6 Look at these diagrams.
Write an addition statement to illustrate each of them.

7 Copy and complete these equations.

(a) $(-2) + \square = 7$ **(b)** $\square + (-3) = 16$ **(c)** $(-5) + \square = (-18)$

(d) $(-3) - \square = (-5)$ **(e)** $\square - (-3) = 3$ **(f)** $(-4) - \square = (-3)$

8 Tom and Jonny are pulling the sledge in opposite directions.

(a) Add the forces.
Which way does the sledge move?

(b) Julia comes to help Jonny.
Julia pulls with 20 N.
What happens now?

(c) What happens if Jonny and Julia let go?

Tom (−50) N → Jonny 30 N →

9 Work out

(a) $20 \times (-5)$ **(b)** $(-20) \times (-5)$ **(c)** $(-20) \times 5$

(d) $1.2 \times (-3)$ **(e)** $(-3.6) \div (-3)$ **(f)** $(-3.6) \div 3$

(g) $(-40) \div (-8)$ **(h)** $40 \div (-5)$ **(i)** $40 \div (-10)$

10 Copy and complete these equations.

(a) $(-2.5) \times \square = 7.5$ **(b)** $\square \div (-3) = 7$ **(c)** $(-60) \div \square = (-5)$

(d) $18 \div \square = 6$ **(e)** $18 \div \square = (-6)$ **(f)** $(-18) \div \square = (-6)$

Activity The mean of two temperatures is (−4).
Write down six pairs of temperatures that have a mean of (−4).

Activity

Copy and complete these magic squares.

All of the rows, columns and diagonals must add up to the same number.

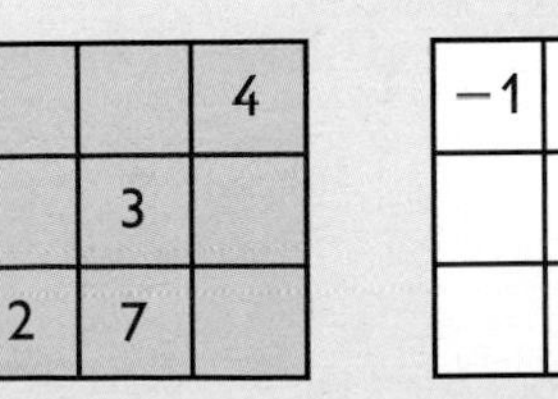

		4
	3	
2	7	

−1	−6	1
	2	

9 Sequences and functions

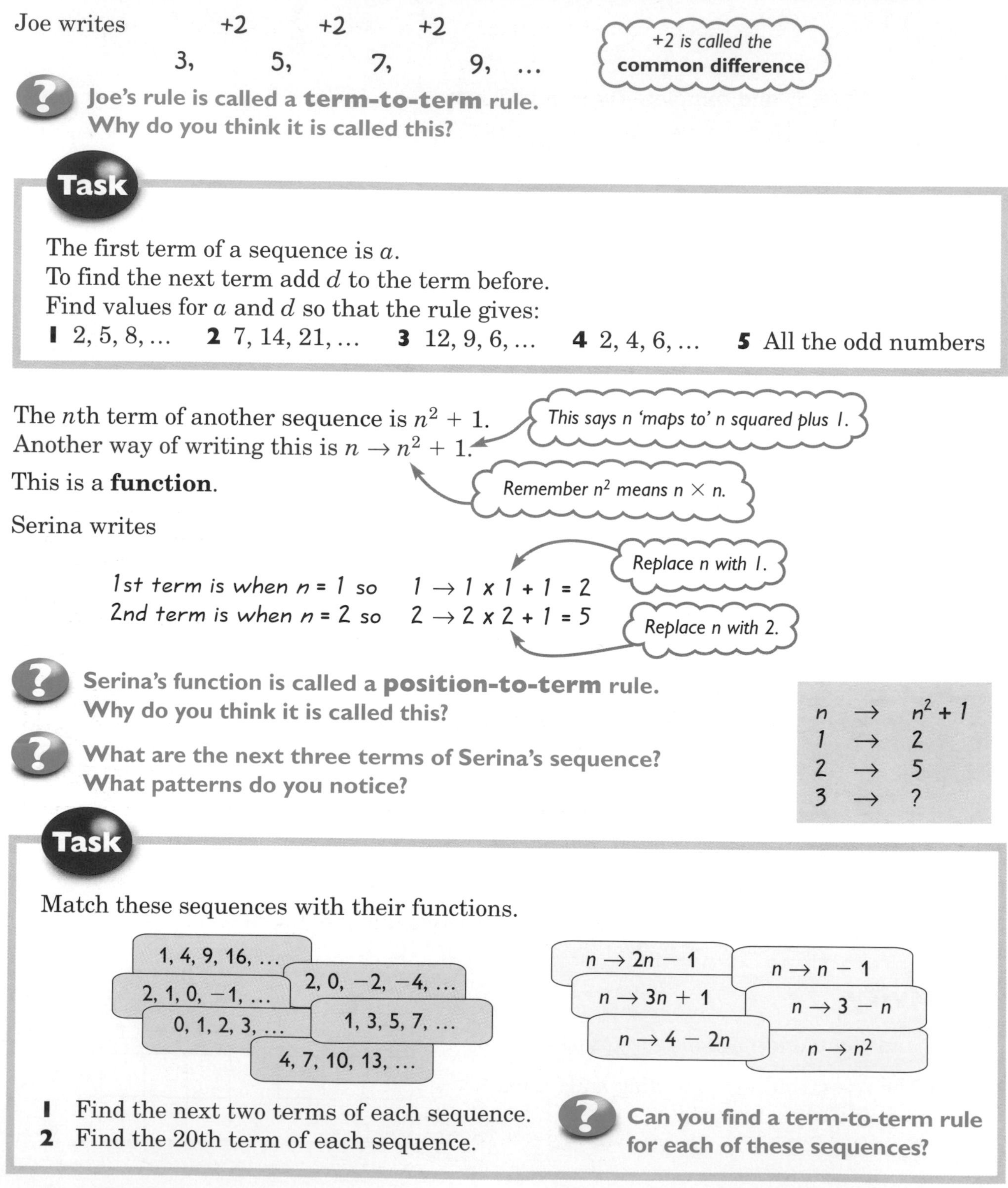

Generating sequences

Joe has a rule for a sequence. The first term is 3.

The next term is found by adding 2 to the term before.

Joe writes

+2 +2 +2

3, 5, 7, 9, …

+2 is called the **common difference**

Joe's rule is called a term-to-term rule. Why do you think it is called this?

Task

The first term of a sequence is a.
To find the next term add d to the term before.
Find values for a and d so that the rule gives:

1 2, 5, 8, … **2** 7, 14, 21, … **3** 12, 9, 6, … **4** 2, 4, 6, … **5** All the odd numbers

The nth term of another sequence is $n^2 + 1$.
Another way of writing this is $n \rightarrow n^2 + 1$.

This says n 'maps to' n squared plus 1.

Remember n^2 means $n \times n$.

This is a **function**.

Serina writes

1st term is when $n = 1$ so $1 \rightarrow 1 \times 1 + 1 = 2$

2nd term is when $n = 2$ so $2 \rightarrow 2 \times 2 + 1 = 5$

Replace n with 1.

Replace n with 2.

Serina's function is called a position-to-term rule. Why do you think it is called this?

What are the next three terms of Serina's sequence? What patterns do you notice?

n	$\rightarrow$	$n^2 + 1$
1	$\rightarrow$	2
2	$\rightarrow$	5
3	$\rightarrow$	?

Task

Match these sequences with their functions.

1, 4, 9, 16, …

2, 0, −2, −4, …

2, 1, 0, −1, …

0, 1, 2, 3, …

1, 3, 5, 7, …

4, 7, 10, 13, …

$n \rightarrow 2n - 1$

$n \rightarrow n - 1$

$n \rightarrow 3n + 1$

$n \rightarrow 3 - n$

$n \rightarrow 4 - 2n$

$n \rightarrow n^2$

1 Find the next two terms of each sequence.
2 Find the 20th term of each sequence.

Can you find a term-to-term rule for each of these sequences?

Exercise

1 Find the first five terms for each of the following sequences.

(a) The first term is 3.
To find the next term add 4 to the previous term.

(b) $n \rightarrow 2n + 4$

(c) The first term is 9.
To find the next term subtract 5 from the previous term.

(d) $n \rightarrow 3n - 3$ **(e)** $n \rightarrow n^2 + 2$

(f) Which of the rules in parts (a)–(e) are
(i) term-to-term rules **(ii)** position-to-term rules?

2 Find values for the missing terms in these sequences.

(a) 3, 5, ?, ?, ?, 13, … **(b)** 2, ?, 8, ?, 14, ?, … **(c)** 21, ?, 11, ?, 1, ?, …

3 The first five terms of a sequence are 6, ?, ?, 18, 22.

(a) Find values for the missing terms of the sequence.

(b) The 50th term is 202.
What is the 51st term?

4 Joanne is organising a tombola for her school quiz night.
To win a prize the number on the ticket must be a term in one of the following sequences.

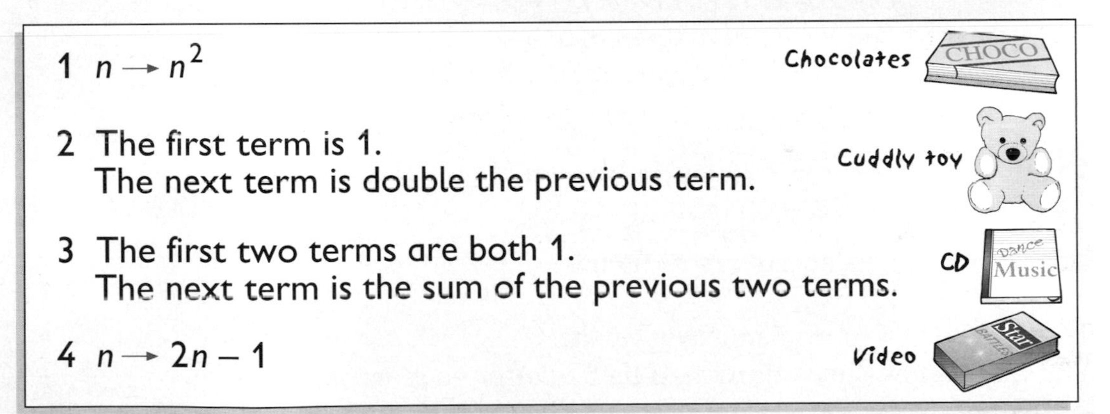

(a) Which of these tickets win a prize?

(b) Which tickets win more than one prize?

Linear sequences

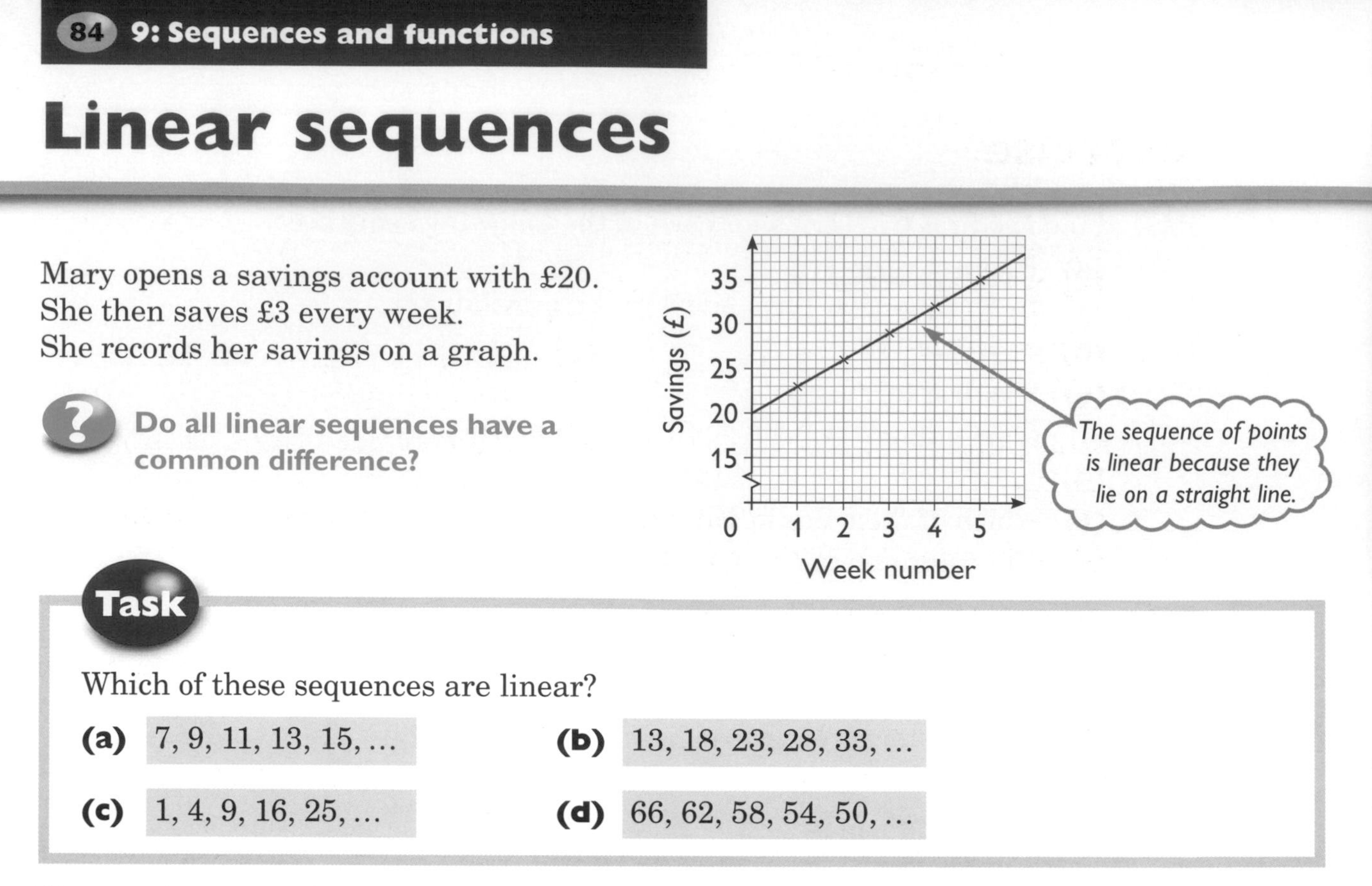

Mary opens a savings account with £20.
She then saves £3 every week.
She records her savings on a graph.

Do all linear sequences have a common difference?

Task

Which of these sequences are linear?

(a) 7, 9, 11, 13, 15, …

(b) 13, 18, 23, 28, 33, …

(c) 1, 4, 9, 16, 25, …

(d) 66, 62, 58, 54, 50, …

You can find a simple formula for the nth term of a linear sequence.

For Mary's sequence, the nth term $= 20 + 3n$. *This is the position-to-term rule*

She writes

For $n = 1$, 1st term $= 20 + 3 \times 1 = 23$
For $n = 2$, 2nd term $= 20 + 3 \times 2 = 26$
For $n = 3$, 3rd term $= 20 + 3 \times 3 = 29$

✓

Where does the 20 come from in Mary's formula?
What does the '$3n$' do?

David has a large aquarium at home.
He starts with 9 fish.
Then he buys 4 more fish every week.
The table shows how many fish he has after each week.

End of week number	1	2	3	4	5
Number of fish	13	17	21	25	29

The nth term of David's sequence has the formula $9 + 4n$.

For $n = 1$, $9 + 4 \times 1 = 9 + 4 = 13$
For $n = 2$, $9 + 4 \times 2 = 9 + 8 = 17$
For $n = 3$, $9 + 4 \times 3 = 9 + 12 = 21$

This is the number of fish he will have after 50 weeks.

What is the 50th term in David's sequence?
How long will it take for David to collect at least 100 fish? (assuming that none die).

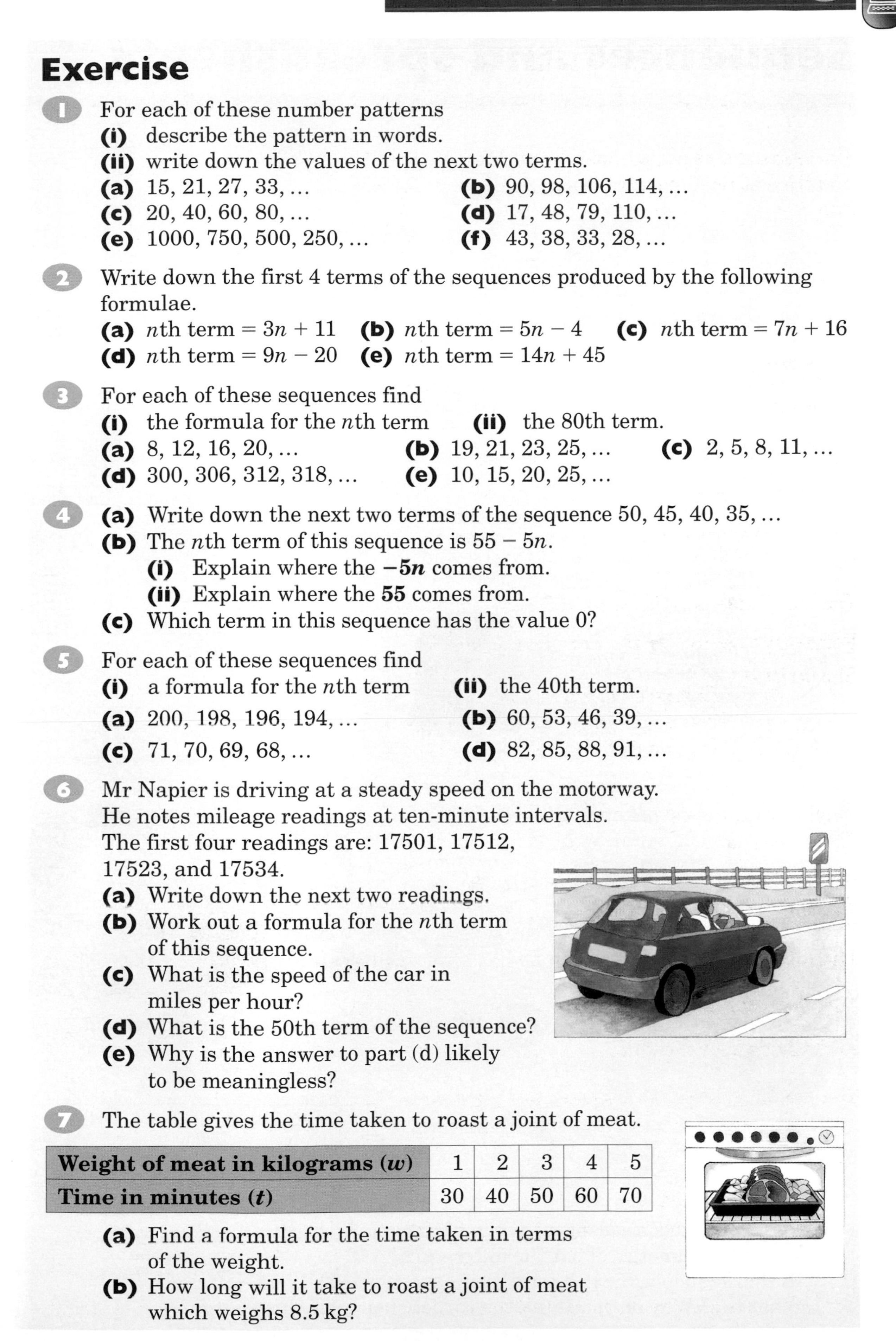

Exercise

1 For each of these number patterns

(i) describe the pattern in words.

(ii) write down the values of the next two terms.

(a) 15, 21, 27, 33, … **(b)** 90, 98, 106, 114, …

(c) 20, 40, 60, 80, … **(d)** 17, 48, 79, 110, …

(e) 1000, 750, 500, 250, … **(f)** 43, 38, 33, 28, …

2 Write down the first 4 terms of the sequences produced by the following formulae.

(a) nth term $= 3n + 11$ **(b)** nth term $= 5n - 4$ **(c)** nth term $= 7n + 16$

(d) nth term $= 9n - 20$ **(e)** nth term $= 14n + 45$

3 For each of these sequences find

(i) the formula for the nth term **(ii)** the 80th term.

(a) 8, 12, 16, 20, … **(b)** 19, 21, 23, 25, … **(c)** 2, 5, 8, 11, …

(d) 300, 306, 312, 318, … **(e)** 10, 15, 20, 25, …

4 **(a)** Write down the next two terms of the sequence 50, 45, 40, 35, …

(b) The nth term of this sequence is $55 - 5n$.

(i) Explain where the $\mathbf{-5n}$ comes from.

(ii) Explain where the **55** comes from.

(c) Which term in this sequence has the value 0?

5 For each of these sequences find

(i) a formula for the nth term **(ii)** the 40th term.

(a) 200, 198, 196, 194, … **(b)** 60, 53, 46, 39, …

(c) 71, 70, 69, 68, … **(d)** 82, 85, 88, 91, …

6 Mr Napier is driving at a steady speed on the motorway. He notes mileage readings at ten-minute intervals. The first four readings are: 17501, 17512, 17523, and 17534.

(a) Write down the next two readings.

(b) Work out a formula for the nth term of this sequence.

(c) What is the speed of the car in miles per hour?

(d) What is the 50th term of the sequence?

(e) Why is the answer to part (d) likely to be meaningless?

7 The table gives the time taken to roast a joint of meat.

Weight of meat in kilograms (w)	1	2	3	4	5
Time in minutes (t)	30	40	50	60	70

(a) Find a formula for the time taken in terms of the weight.

(b) How long will it take to roast a joint of meat which weighs 8.5 kg?

Sequences and spreadsheets

Christina uses her spreadsheet package to print the first 10 terms of the sequence whose nth term is $3n + 2$.

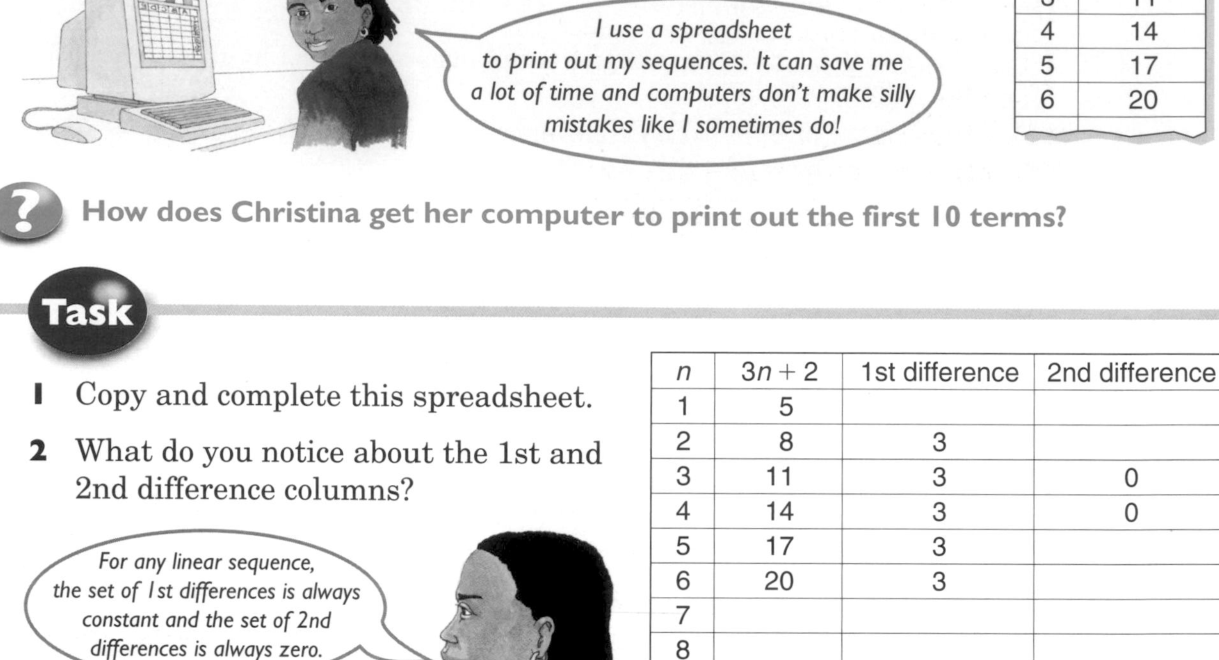

n	$3n + 2$
1	5
2	8
3	11
4	14
5	17
6	20

How does Christina get her computer to print out the first 10 terms?

Task

1 Copy and complete this spreadsheet.

2 What do you notice about the 1st and 2nd difference columns?

n	$3n + 2$	1st difference	2nd difference
1	5		
2	8	3	
3	11	3	0
4	14	3	0
5	17	3	
6	20	3	
7			
8			
9			
10			

For any linear sequence, the set of 1st differences is always constant and the set of 2nd differences is always zero.

Is Christina right?

David visits the carpet shop with his father.

What is the pattern that David spotted?

Task

Use a spreadsheet to find David's pattern.
Set up this spreadsheet.
Use columns for the 1st, 2nd and 3rd differences.

1 Complete the spreadsheet up to $n = 12$.

2 What do you notice about the differences?

3 In this sequence the nth term $= n^2 + c$.
What is c? Why do you think the carpet shop adds this charge?

n	Price (£s)	1st difference	2nd difference	3rd difference
1	11			
2	14			
3	19			
4	26			
5	35			

Exercise

1 **(a)** Use a spreadsheet to print the first 20 terms of each of these sequences.

(i) nth term $= 5n - 1$ **(ii)** nth term $= 7n + 98$

(iii) nth term $= 402 - 5n$ **(iv)** nth term $= \dfrac{n}{n+3}$

(b) Add a column on each of your spreadsheets to print the first differences of each sequence.

(c) Which of the sequences are linear?

2 **(a)** Use your spreadsheet to print the first 16 terms and 1st, 2nd and 3rd differences of each of these sequences. The first one is started for you.

(i) nth term $= n^2 + 7$

Term	Value	1st difference	2nd difference	3rd difference
1	8			
2	11	3		
3	16	5	2	
4	23	7	2	0
5	32	9	2	0
6				
⋮				
15				
16				

(ii) $5n^2$ **(iii)** $n^2 + 50n$

(iv) $200 - 3n^2$ **(v)** $n^2 + 3n - 11$

In a ***quadratic sequence*** *the nth term* $= an^2 + bn + c$.

(b) Mary says that a quadratic sequence always has a constant 2nd difference. Is she right?

3 Mr Jones is a gardener. He makes a table showing the lengths of square lawns and the cost of putting lawn food on them.

Length of lawn (metres)	1	2	3	4	5	6	7
Cost of lawn food (pence)	15	45	95	165	255	365	495

The formula for the nth term of this sequence is $10n^2 + k$.

(a) Find the value of k.

(b) What is the cost of lawn food for a square lawn of side 8 metres?

(c) Mrs Bell's lawn food costs £8.15. What size is her square lawn?

4 A flea sits in the middle of a circular table of radius 1 metre.
It takes a series of jumps towards the edge of the table.
On the first jump it jumps 0.5 metres.
On each following jump it jumps half of the remaining distance.

(a) Make a spreadsheet to show how far the flea travels with each jump and the total distance it has travelled at each stage.

(b) How far has the flea travelled after 20 jumps?

(c) How long will the flea take to reach the edge of the table?

Spatial patterns

Scott is investigating square jigsaw puzzles.

Pattern 1

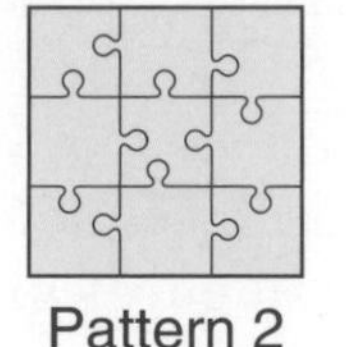

Pattern 2

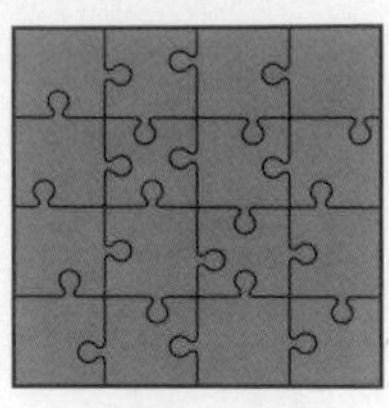

Pattern 3

Task

1 Draw the next two jigsaws.

2 How many **(a)** corner pieces **(b)** edge pieces **(c)** middle pieces does Scott need for each jigsaw?

Scott wants to know how many pieces, of each type, he needs for jigsaw 50. He makes a table.

Copy Scott's table and continue it for jigsaws 4 and 5.

Jigsaw number (n)	Corner pieces (c)	Edge pieces (e)	Middle pieces (m)
1	4	0	0
2	4	4	1
3	4	8	4

? Predict how many pieces of each kind are needed for jigsaws 6 and 7. How many corner pieces are needed in the *n*th pattern? Why?

? How many edge pieces, e, are there along *each side* for jigsaw
(a) 11 (b) 21 (c) 51 (d) 101?
How many edge pieces are there altogether for each jigsaw?

Scott writes *For the nth pattern, there are $n-1$ edge pieces along each side. Each jigsaw has 4 sides, so there are $4 \times (n-1)$ edge pieces.*

How else could you write $4 \times (n-1)$?

? What types of numbers are 0, 1, 4, 9, 16, 25, ...?
How would you work out how many middle pieces are needed for jigsaws
(a) 11 (b) 21 (c) 51 (d) 101?
Find an expression for the number of middle pieces for jigsaw *n*.

Task

Look at these jigsaws.
For each one investigate the number of
(a) corner pieces **(b)** edge pieces
(c) middle pieces needed.

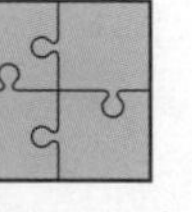

Pattern 1

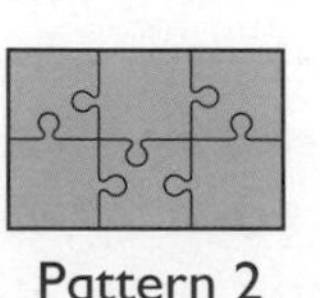

Pattern 2

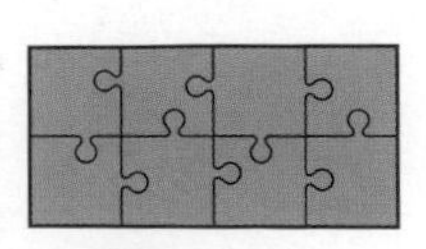

Pattern 3

Investigate other jigsaw patterns.

Exercise

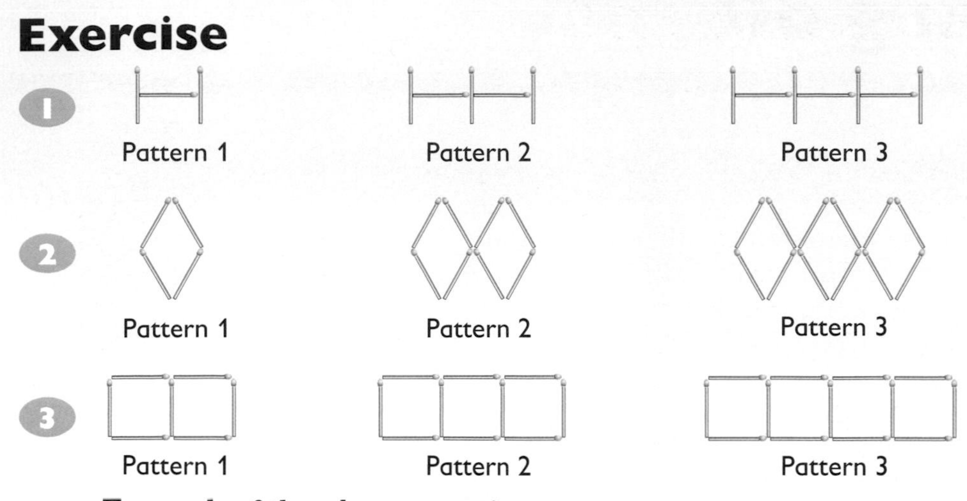

For each of the above questions

(a) Draw the next two patterns.

(b) Find how many matches

- **(i)** are added to form each new pattern
- **(ii)** there are in each pattern altogether
- **(iii)** there are in the 10th pattern.

(c) How many match sticks are added to the first pattern for the

(i) 2nd pattern **(ii)** 3rd pattern **(iii)** 4th pattern
(iv) 10th pattern **(v)** 20th pattern **(vi)** nth pattern?

(d) Find a rule for the number of match sticks in the nth pattern.

4 **(a)** Draw the next two patterns.

Pattern 1 Pattern 2 Pattern 3

(b) Copy and complete this table.

Pattern	2	3	4	5	6	7
Number of dots added to previous pattern	3	5	?	?	?	?

What type of number are these?
How many dots are added to the 9th pattern to make the 10th pattern?

(c) How many dots are in each pattern?
What type of number are these?

(d) How many dots are in the

(i) 5th pattern **(ii)** 6th pattern **(iii)** 7th pattern
(iv) 10th pattern **(v)** 20th pattern **(vi)** 100th pattern?

(e) What is the sum of the numbers $1 + 3 + 5 + 7 + 9 + 11 + 13 + 15$?
What pattern number has this number of dots?

(f) What is the sum of the first

(i) 3 **(ii)** 4 **(iii)** 5 **(iv)** 10 **(v)** 20 **(vi)** 100
odd numbers?

(g) Write an expression for the sum of the first n odd numbers?

Finishing off

Now that you have finished this chapter you should be able to:

- understand the terms sequence and function
- use term-to-term and position-to-term rules to generate sequences
- find the nth term of a linear sequence.

Review exercise

1 Write down the first 4 terms of the following sequences.

(a) (i) $n \rightarrow 2n + 1$ **(ii)** nth term $= 3n - 1$

(iii) $n \rightarrow n^2$ **(iv)** nth term $= n^2 + 1$

(b) Which of these sequences are linear?

2 The first term of a sequence is a.
The next term is found by adding d to the previous term.

Find values of a and d for the following sequences.

(a) 4, 8, 12, 16, … **(b)** multiples of 3

(c) 100, 95, 90, 85, … **(d)** −10, −8, −6, −4, …

(e) multiples of 5 **(f)** even numbers

3 Find the values of the missing terms of these linear sequences:

(a) 3, 5, ?, ?, 11, 13 **(b)** 6, ?, 14, ?, 22, 26

(c) 30, ?, 24, ?, 18, ? **(d)** 16, ?, ?, 22, ?, 26

(e) 50, ?, ?, 35, ?, 25, ? **(f)** −20, ?, −16, −14, ?, ?

4 Find a formula for the nth term of each of these sequences.

(a) 3, 6, 9, 12, 15, … **(b)** 4, 8, 12, 16, 20, …

(c) 3, 5, 7, 9, 11, 13, … **(d)** 4, 7, 10, 13, 16, …

(e) 1, 3, 5, 7, 9, 11, 13, … **(f)** 3, 8, 13, 18, 23, …

5 The first 5 terms of a linear sequence are 60, ?, ?, 51 and 48.

(a) Find the values of the missing terms.

(b) The 100th term is −237. What is the 101st term?

6 **(a)** Draw the next 3 patterns.

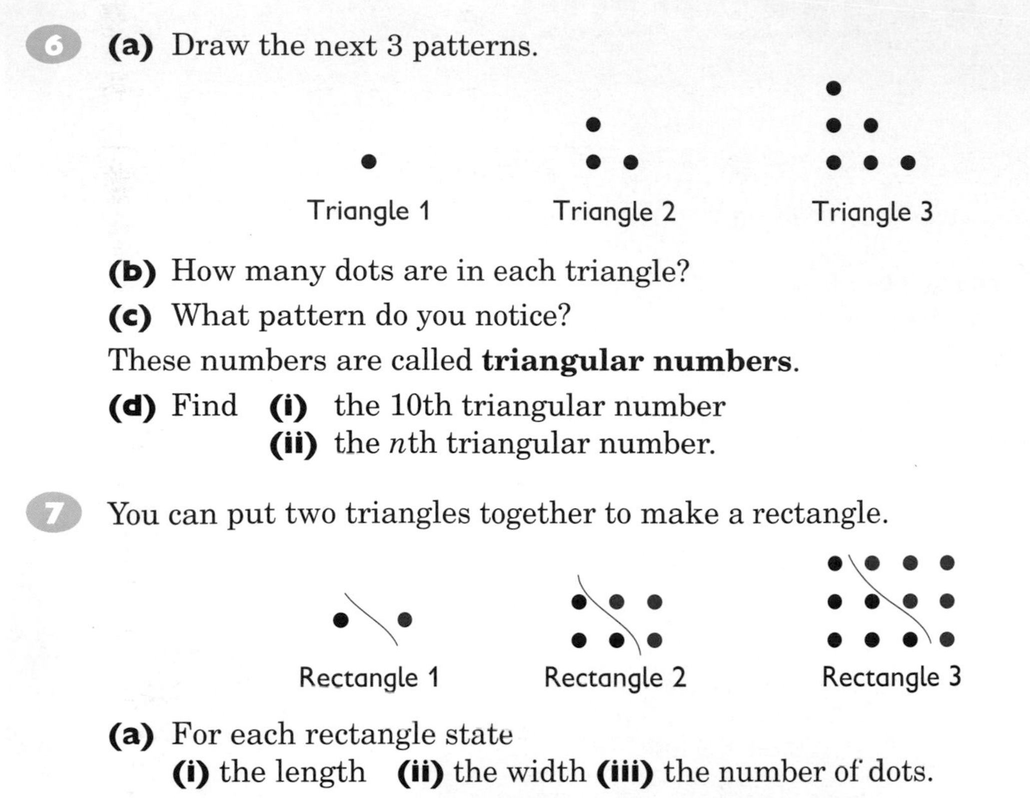

(b) How many dots are in each triangle?

(c) What pattern do you notice?

These numbers are called **triangular numbers**.

(d) Find **(i)** the 10th triangular number
(ii) the nth triangular number.

7 You can put two triangles together to make a rectangle.

(a) For each rectangle state
(i) the length **(ii)** the width **(iii)** the number of dots.

(b) Predict the length and width of the
(i) 10th **(ii)** 20th **(iii)** 100th rectangle.

(c) Predict the number of dots in the
(i) 10th **(ii)** 20th **(iii)** 100th rectangle.

(d) For the nth rectangle, give expressions for
(i) the length **(ii)** the width **(iii)** the number of dots.

(e) Use your answers to part **(d)** to show that the formula for the nth triangular number is $\frac{1}{2}n(n + 1)$.

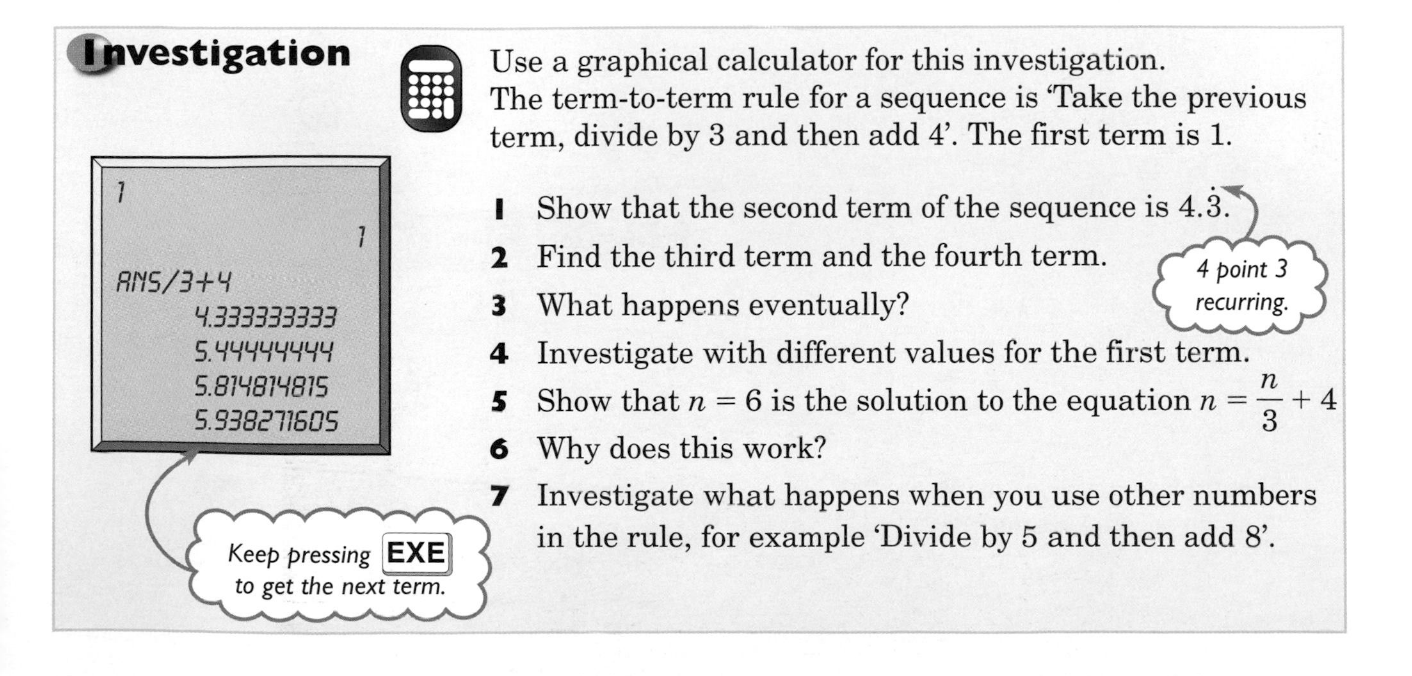

Investigation

Use a graphical calculator for this investigation.
The term-to-term rule for a sequence is 'Take the previous term, divide by 3 and then add 4'. The first term is 1.

1. Show that the second term of the sequence is $4.\dot{3}$.
2. Find the third term and the fourth term.
3. What happens eventually?
4. Investigate with different values for the first term.
5. Show that $n = 6$ is the solution to the equation $n = \frac{n}{3} + 4$
6. Why does this work?
7. Investigate what happens when you use other numbers in the rule, for example 'Divide by 5 and then add 8'.

10 Percentages

Proportion and percentages

A department store has ordered 250 copies of a CD at a cost of £3500.

They change the order to 350 copies. How much would these cost?

250 copies cost £3500

so 1 copy costs $\frac{£3500}{250} = £14$.

350 copies cost $£14 \times 350 = £4900$.

*Remember the **unitary** method. Find the cost of **one** first.*

Why is this called the unitary method?

Later in the month there is a sale in the store.

A television costs £240 normally. How much will it cost in the sale?

$$20\% \text{ of } £240 = \frac{20}{100} \times £240$$

Reduction $= £48$

Sale price $= £240 - £48 = £192$

100% is always used as the original amount.

The reduction is 20%. What percentage of the original is the sale price?

The sale price of a pair of designer sunglasses is £96. Find the original price.

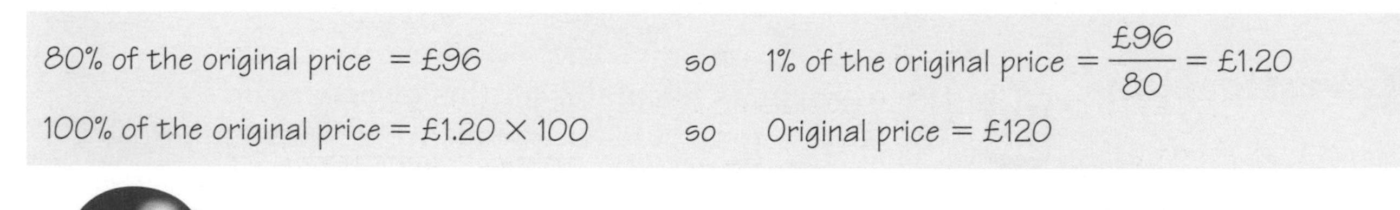

80% of the original price $= £96$ so 1% of the original price $= \frac{£96}{80} = £1.20$

100% of the original price $= £1.20 \times 100$ so Original price $= £120$

Task

20% off all items

1 Find the sale price of these items.

2 Find the original price of these sale items.

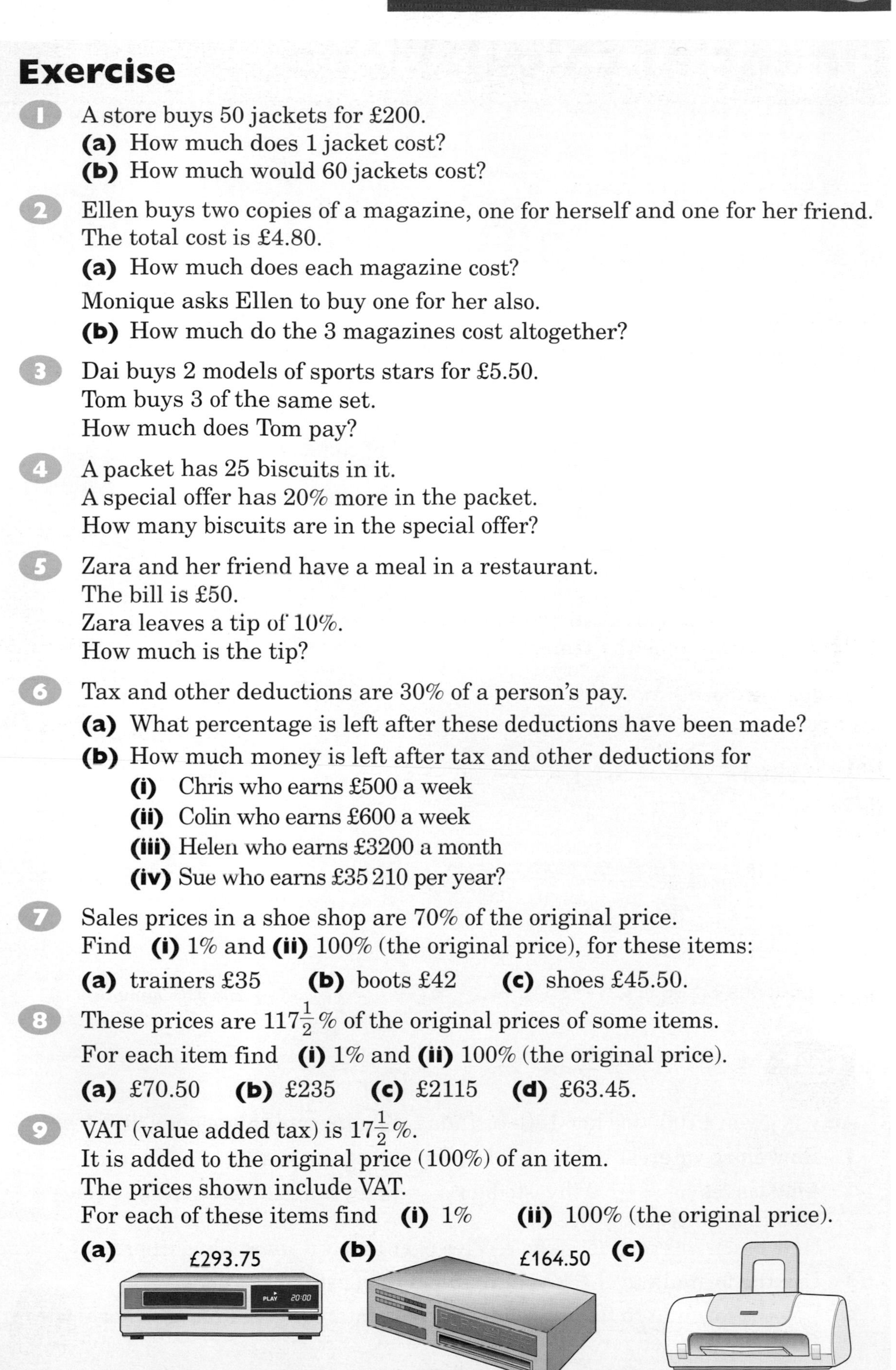

Exercise

1 A store buys 50 jackets for £200.

(a) How much does 1 jacket cost?

(b) How much would 60 jackets cost?

2 Ellen buys two copies of a magazine, one for herself and one for her friend. The total cost is £4.80.

(a) How much does each magazine cost?

Monique asks Ellen to buy one for her also.

(b) How much do the 3 magazines cost altogether?

3 Dai buys 2 models of sports stars for £5.50.
Tom buys 3 of the same set.
How much does Tom pay?

4 A packet has 25 biscuits in it.
A special offer has 20% more in the packet.
How many biscuits are in the special offer?

5 Zara and her friend have a meal in a restaurant.
The bill is £50.
Zara leaves a tip of 10%.
How much is the tip?

6 Tax and other deductions are 30% of a person's pay.

(a) What percentage is left after these deductions have been made?

(b) How much money is left after tax and other deductions for

(i) Chris who earns £500 a week

(ii) Colin who earns £600 a week

(iii) Helen who earns £3200 a month

(iv) Sue who earns £35 210 per year?

7 Sales prices in a shoe shop are 70% of the original price.
Find **(i)** 1% and **(ii)** 100% (the original price), for these items:

(a) trainers £35 **(b)** boots £42 **(c)** shoes £45.50.

8 These prices are $117\frac{1}{2}$% of the original prices of some items.
For each item find **(i)** 1% and **(ii)** 100% (the original price).

(a) £70.50 **(b)** £235 **(c)** £2115 **(d)** £63.45.

9 VAT (value added tax) is $17\frac{1}{2}$%.
It is added to the original price (100%) of an item.
The prices shown include VAT.
For each of these items find **(i)** 1% **(ii)** 100% (the original price).

(a) £293.75 **(b)** £164.50 **(c)** £199.75

Financial calculations

Simon saves £25 every month.

How much does he save in a year?

7.2% Tax-free Gross redemption yield, which may vary

14% Annual Growth Rate **

10%* p.a.* Maximum Income Bond Fund

He puts this total amount into an account which pays him 7% interest.

How much interest is he paid at the end of one year?

He receives the same interest again at the end of the next year.

How much interest has he received altogether over the two years?

This is known as **simple interest**.

Simon's £300 (the money he started with) is called the **principal**, £P.
He invests it at 7%. This is the **rate**, R% p.a.
The number of years is the **time**, T.

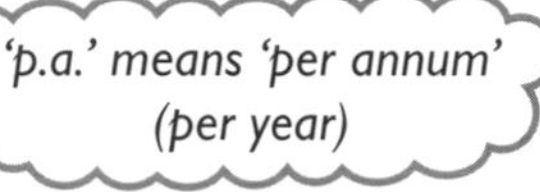

To calculate simple interest, you can use the formula

$$I = \frac{PRT}{100}$$

In this example

Principal, $P = £300$ Rate, $R = 7\%$ Time, $T = 2$ years

Simple interest, $I = \frac{£300 \times 7 \times 2}{100} = £42$

Simon now has £(300 + 42) = £342

*This is sometimes called the **amount**.*

Task

Amy is given £1000 for her 18th birthday. She invests this money at 8% per annum.

1 How much interest is Amy paid at the end of one year?

2 Amy leaves her £1000 invested for a second year and receives the same amount of interest.
How much interest does she receive over the two years altogether?

3 Use the formula to check your answers to questions **1** and **2**.

4 Use the formula to find how much simple interest is paid altogether if the £1000 is invested for

(a) 5 years **(b)** 7 years.

Exercise

1 A house costs £60 000.
It appreciates in value by 5% each year.
What is its value at the end of one year?

2 A motorbike costs £4000.
It depreciates in value by 10% in a year.
What is its value after one year?

3 A car costs £12 000.
It depreciates in value by 15% each year.
What is its value after one year?

4 Sean buys a stamp for £6 at a stamp fair.

(a) One year later it is worth £6.45.
By what percentage has it appreciated?

(b) The next year the stamp appreciates in value by 10% (from the start of the year).
What is its value 2 years after Sean buys it?

5 Find the simple interest paid on £650 invested at 6% p.a. for 5 years.

6 Find the simple interest paid on each of these investments.

(a) £780 at 11% p.a. for 3 years **(b)** £5000 at 8.5% p.a. for 4 years

(c) £220 at 5.5% p.a. for 2 years **(d)** £2500 at 8.75% for 20 years

7 **(a)** David invests £300 for 2 years.
He receives £36 simple interest.
At what rate does he invest?

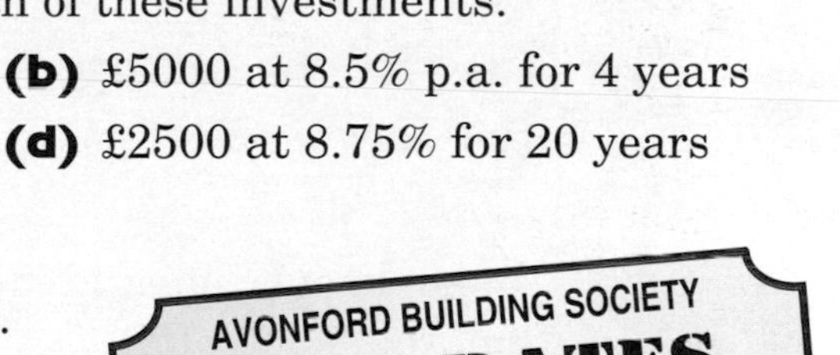

(b) Peter invests £400 for $2\frac{1}{2}$ years.
He receives £50 simple interest.
At what rate does he invest?

(c) Carolyn puts £300 into an account for 18 months.
She receives £36 simple interest.

(i) How many years is 18 months?

(ii) At what rate does she invest?

(d) Who receives the highest rate of interest, David, Peter or Carolyn?

8 Daniel invests £500 at 12% p.a. simple interest.
He receives £180 interest.
For how long does he invest his money?

9 Wendy receives £198 simple interest after investing her money at 11% p.a. for 6 years.
What is her principal?

The principal is the amount she invested at the beginning.

Finishing off

Now that you have finished this chapter you should be able to:

- use the unitary method in proportion
- find a percentage of an amount
- find the original amount once given a percentage of it
- find percentage increases and decreases
- calculate simple interest
- perform other financial calculations (e.g. appreciation and depreciation).

Review exercise

1 Matching pictures cost £25 for two.

(a) How much is each picture?

(b) How much would 3 pictures cost?

2 Daffodil bulbs are £6 for 30.

(a) How much is one bulb?

(b) How much are 50 bulbs?

3 A shopkeeper buys 12 identical umbrellas for £102.

(a) How much does one cost?

(b) How much should he pay for 20 of the same umbrellas?

4 Alex buys a trumpet. Its price is £200.
He pays a deposit of 30%.

(a) How much is the deposit?

Then he pays 12 monthly instalments of £15 each.

(b) How much are the instalments altogether?

(c) What is the total of the deposit and instalments?

(d) How much more than £200 does Alex pay by buying in this way?

5 Mr Stiles earns £1320 per month.
He pays 25% tax.
How much tax does he pay each month?

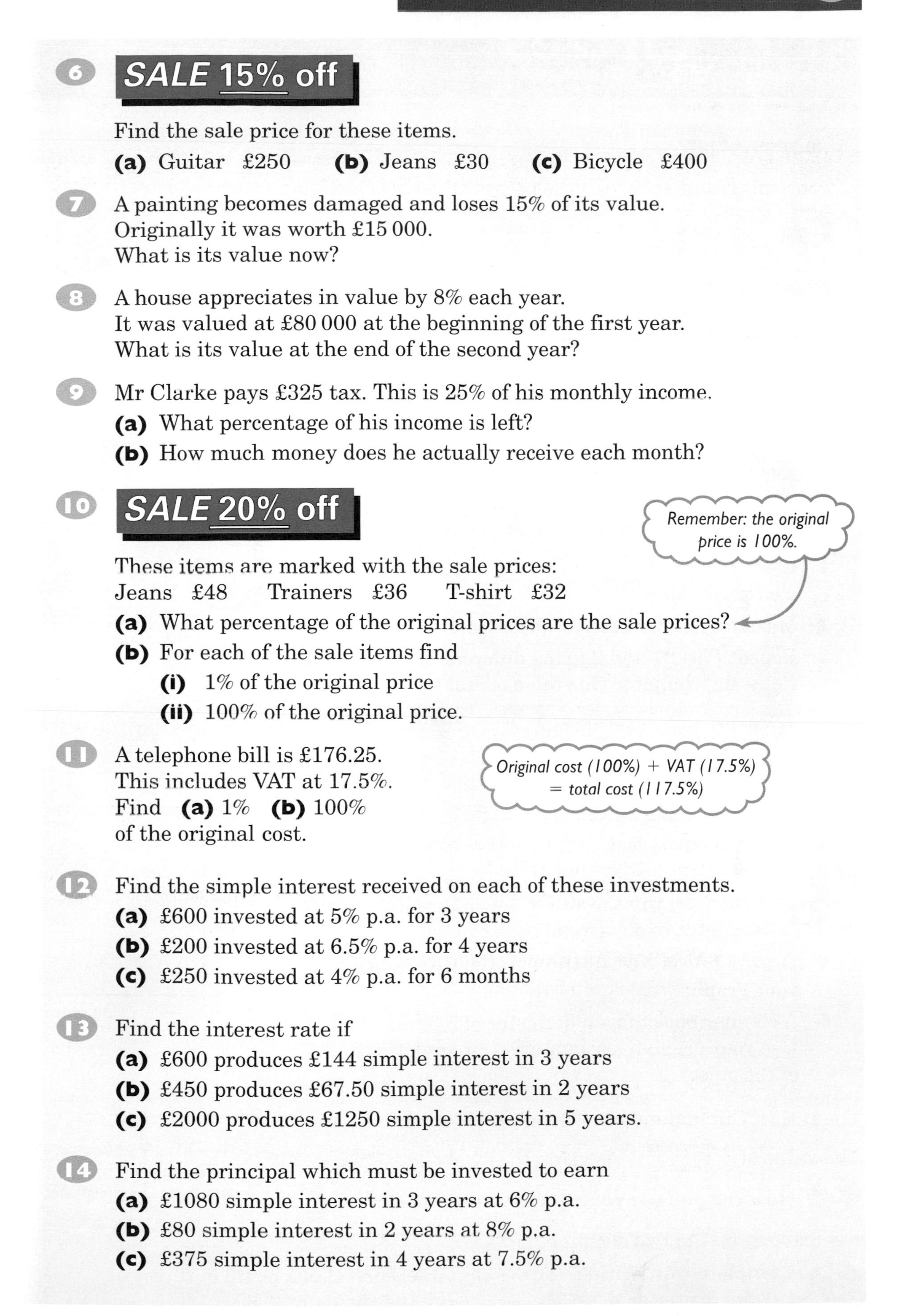

6 **SALE 15% off**

Find the sale price for these items.

(a) Guitar £250 **(b)** Jeans £30 **(c)** Bicycle £400

7 A painting becomes damaged and loses 15% of its value.
Originally it was worth £15 000.
What is its value now?

8 A house appreciates in value by 8% each year.
It was valued at £80 000 at the beginning of the first year.
What is its value at the end of the second year?

9 Mr Clarke pays £325 tax. This is 25% of his monthly income.

(a) What percentage of his income is left?

(b) How much money does he actually receive each month?

10 **SALE 20% off**

These items are marked with the sale prices:
Jeans £48 Trainers £36 T-shirt £32

(a) What percentage of the original prices are the sale prices?

(b) For each of the sale items find

(i) 1% of the original price

(ii) 100% of the original price.

11 A telephone bill is £176.25.
This includes VAT at 17.5%.
Find **(a)** 1% **(b)** 100%
of the original cost.

12 Find the simple interest received on each of these investments.

(a) £600 invested at 5% p.a. for 3 years

(b) £200 invested at 6.5% p.a. for 4 years

(c) £250 invested at 4% p.a. for 6 months

13 Find the interest rate if

(a) £600 produces £144 simple interest in 3 years

(b) £450 produces £67.50 simple interest in 2 years

(c) £2000 produces £1250 simple interest in 5 years.

14 Find the principal which must be invested to earn

(a) £1080 simple interest in 3 years at 6% p.a.

(b) £80 simple interest in 2 years at 8% p.a.

(c) £375 simple interest in 4 years at 7.5% p.a.

11 Circles

Do you remember?

These diagrams show you the names of parts of a circle.

The circumference is a locus of points. What is the *rule* for this locus?

***Part* of a circumference is called an arc. You often use arcs when you construct triangles. How do they help you?**

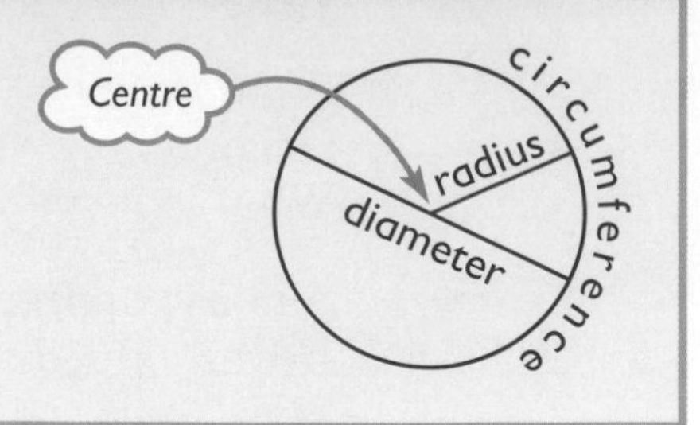

Circumference and diameter

For this task you will need: string, ruler, pencil, graph paper, a selection of circular objects.

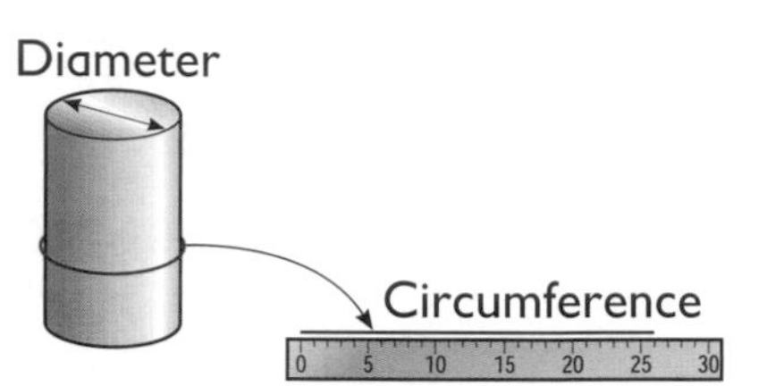

1 Use a piece of string to find the circumference of a circular object.

2 Measure the diameter of the circular face.

3 Repeat Parts 1 and 2 using different circular objects. Copy and complete this table of results.

Object	Circumference (cm)	Diameter (cm)	Circumference ÷ Diameter
Can			
10p piece			

4 For each object plot circumference against diameter. Arrange the axes as shown.

What can you say about the circumference and diameter of *any* circle?

5 Draw a line of best fit through the points on your graph.

6 A circular object has a diameter of 5.3 cm. Use your graph to estimate the circumference of the object.

The ratio Circumference : Diameter is the same for all circles.

The value of $\frac{\text{Circumference}}{\text{Diameter}}$ is given the symbol π.

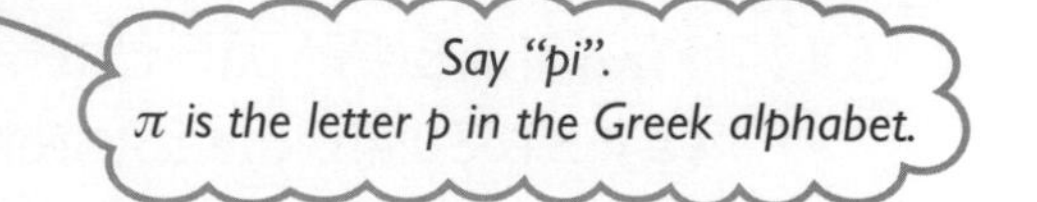

How can you use your graph to estimate π?

$\pi = 3.14$, correct to two decimal places.

Complete this formula for the circumference C of a circle in terms of
(a) the diameter, d $C =$ **(b) the radius, r** $C =$

Exercise

1 Use $\pi = 3.14$ to calculate the circumference of these circles.

(a) diameter = 10 cm **(b)** diameter = 8 cm **(c)** diameter = 2 m
(d) diameter = 23.6 cm **(e)** radius = 5 cm **(f)** radius = 3.8 m

2 The large wheel on Andy's wheelchair has a diameter of 60 cm.
Andy pushes the wheel round exactly once.

(a) How far has Andy moved?

Andy crosses a busy road, 20 m wide.

(b) How many times does he have to rotate the large wheel to do this?

3 A circle has a circumference of 6.28 m.
Calculate

(a) the diameter and **(b)** the radius of the circle.

4 π does not have an exact numerical value.
A rough approximation for π is 3.
Some calculators have a button for π.
A ten digit display calculator gives π as 3.141592654.

(a) Phil says 'π is 3.142'. Is Phil right?

Other approximations are $\frac{22}{7}$ and $\sqrt{10}$.

(b) (i) Use a calculator to change $\frac{22}{7}$ to a decimal.
(ii) What sort of decimal is your answer?
(iii) Compare $\frac{22}{7}$ and the value of π on your calculator.
How many decimal places are the same?

(c) (i) Use a calculator to change $\sqrt{10}$ to a decimal.
(ii) Compare your answer with the value given by the π button.

5 Use $\pi = \frac{22}{7}$ to calculate the circumference of circles with these dimensions:

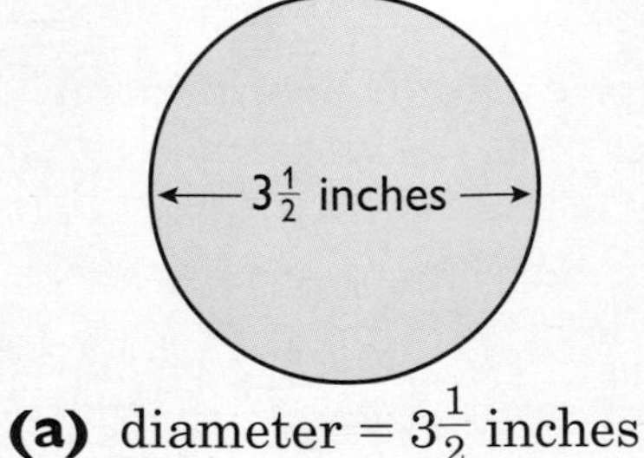

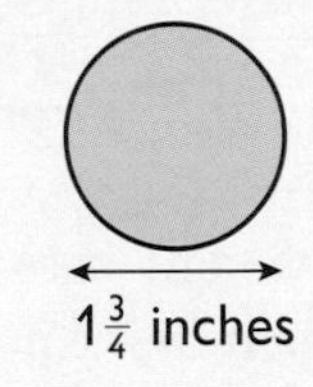

(a) diameter = $3\frac{1}{2}$ inches **(b)** diameter = $1\frac{3}{4}$ inches

6 A circular table has a diameter of 10 feet.
20 children are invited to a party tea.
Each child needs one foot six inches around the circumference of the table.
Can all the children sit around the table together?

Area of a circle

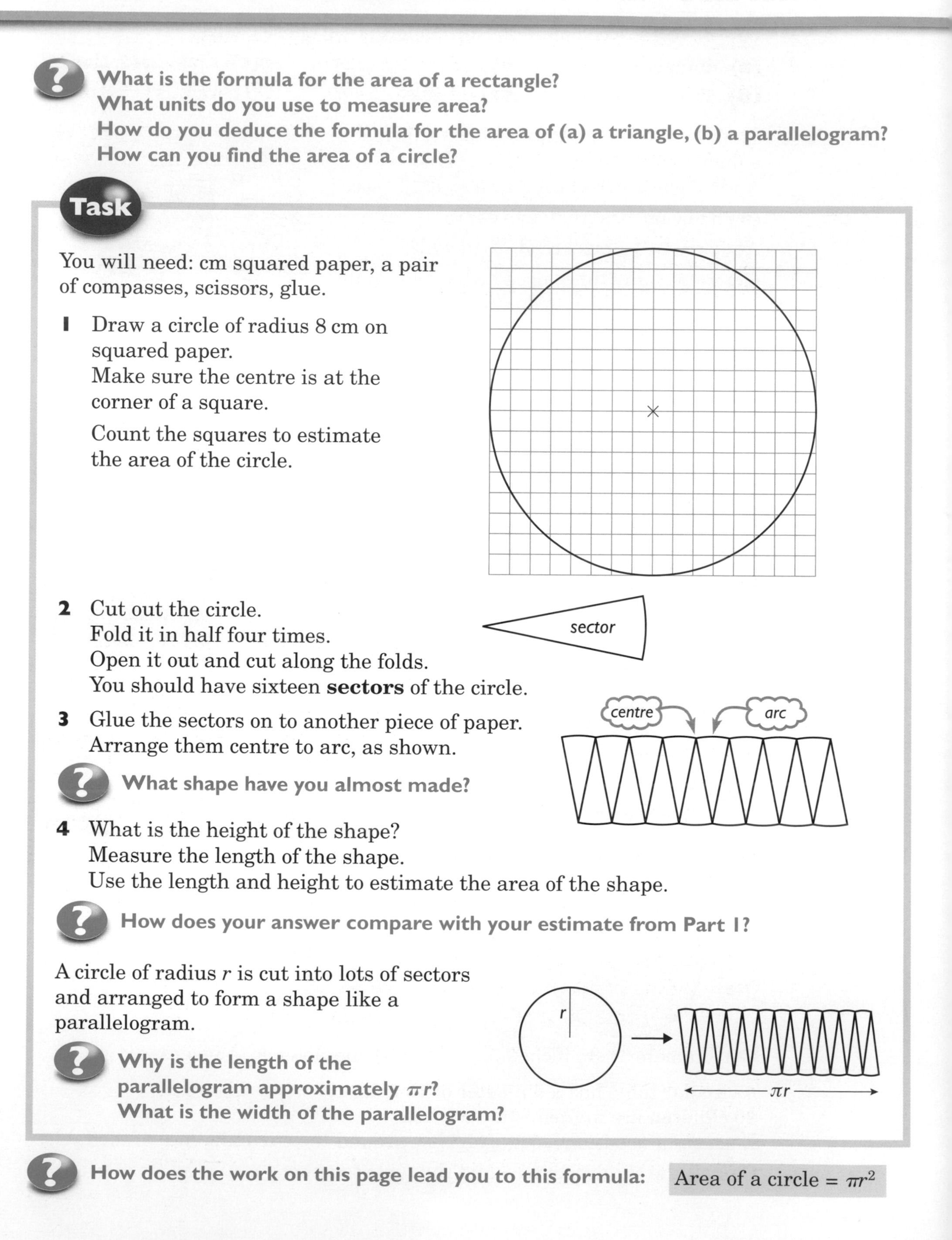

What is the formula for the area of a rectangle?
What units do you use to measure area?
How do you deduce the formula for the area of (a) a triangle, (b) a parallelogram?
How can you find the area of a circle?

Task

You will need: cm squared paper, a pair of compasses, scissors, glue.

1 Draw a circle of radius 8 cm on squared paper.
Make sure the centre is at the corner of a square.

Count the squares to estimate the area of the circle.

2 Cut out the circle.
Fold it in half four times.
Open it out and cut along the folds.
You should have sixteen **sectors** of the circle.

3 Glue the sectors on to another piece of paper.
Arrange them centre to arc, as shown.

What shape have you almost made?

4 What is the height of the shape?
Measure the length of the shape.
Use the length and height to estimate the area of the shape.

How does your answer compare with your estimate from Part 1?

A circle of radius r is cut into lots of sectors and arranged to form a shape like a parallelogram.

Why is the length of the parallelogram approximately πr?
What is the width of the parallelogram?

How does the work on this page lead you to this formula: Area of a circle = πr^2

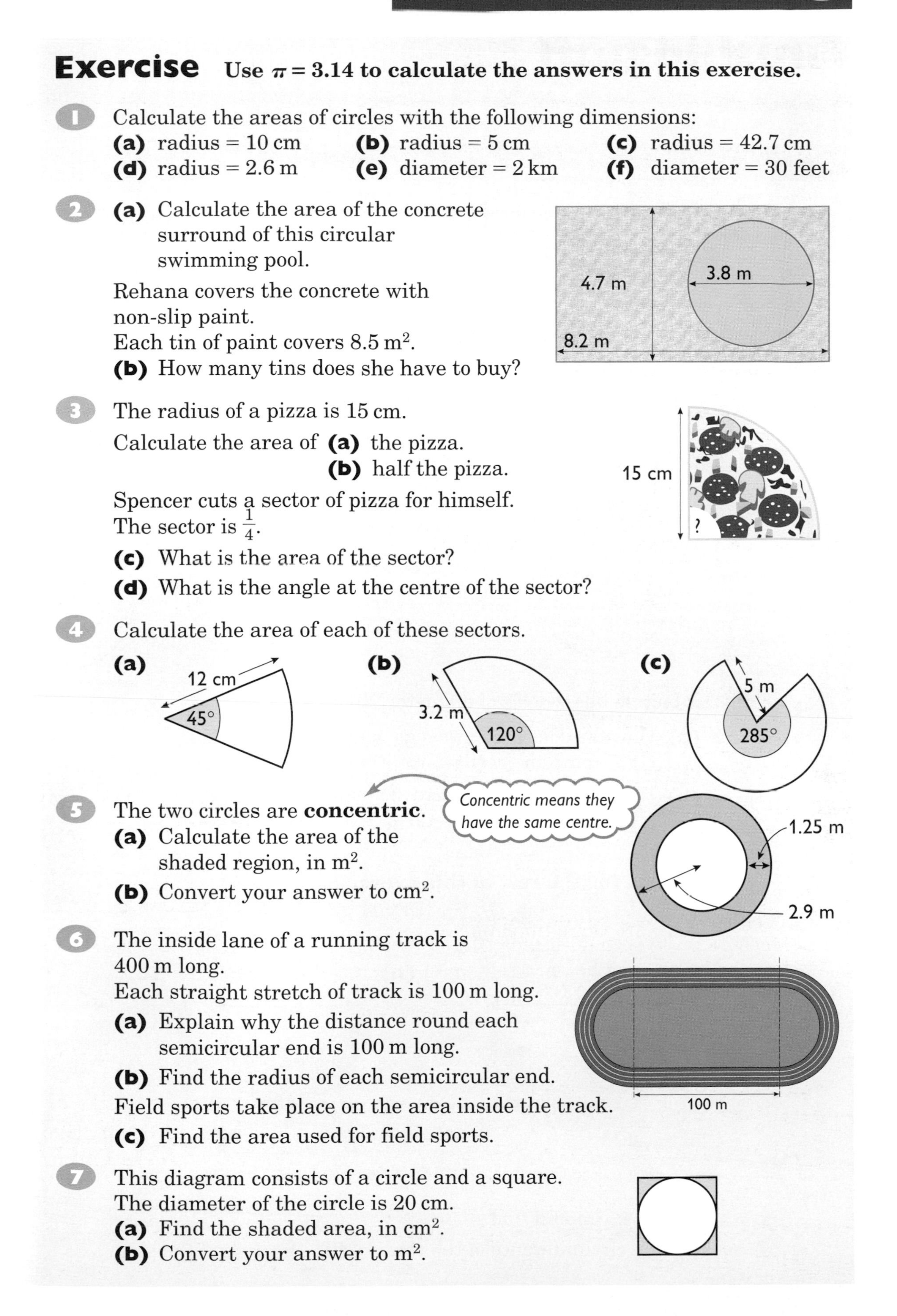

Exercise

Use $\pi = 3.14$ to calculate the answers in this exercise.

1 Calculate the areas of circles with the following dimensions:

(a) radius = 10 cm **(b)** radius = 5 cm **(c)** radius = 42.7 cm
(d) radius = 2.6 m **(e)** diameter = 2 km **(f)** diameter = 30 feet

2 **(a)** Calculate the area of the concrete surround of this circular swimming pool.

Rehana covers the concrete with non-slip paint.
Each tin of paint covers 8.5 m^2.

(b) How many tins does she have to buy?

3 The radius of a pizza is 15 cm.

Calculate the area of **(a)** the pizza.
(b) half the pizza.

Spencer cuts a sector of pizza for himself.
The sector is $\frac{1}{4}$.

(c) What is the area of the sector?
(d) What is the angle at the centre of the sector?

4 Calculate the area of each of these sectors.

(a) **(b)** **(c)**

5 The two circles are **concentric**.

(a) Calculate the area of the shaded region, in m^2.
(b) Convert your answer to cm^2.

6 The inside lane of a running track is 400 m long.
Each straight stretch of track is 100 m long.

(a) Explain why the distance round each semicircular end is 100 m long.
(b) Find the radius of each semicircular end.

Field sports take place on the area inside the track.

(c) Find the area used for field sports.

7 This diagram consists of a circle and a square.
The diameter of the circle is 20 cm.

(a) Find the shaded area, in cm^2.
(b) Convert your answer to m^2.

Finishing off

Now that you have finished this chapter you should know:

- that $\frac{\text{circumference}}{\text{diameter}}$ is the same for all circles, and is called π
- that two approximate values of π are 3.14 and $\frac{22}{7}$
- circumference $= \pi \times$ diameter $= 2\pi r$
- area of a circle $= \pi r^2$
- how to find the area of a sector of a circle.

Review exercise

Use $\pi = \frac{22}{7}$ for Questions 1 and 2.

1 Calculate the circumference of these circles.

(a) 14 cm **(b)** 1 inch **(c)** 5.6 m **(d)** $2\frac{6}{11}$ m

2 A bicycle has wheels of diameter 70 cm.

(a) How far does the bicycle go when a wheel makes:
(i) 1 rotation **(ii)** 5 rotations?

(b) How many rotations does the wheel make when the bicycle travels
(i) 10 m **(ii)** 100 m **(iii)** 10 km?

Use $\pi = 3.14$ for the rest of this exercise.

3 Here are three timpani.

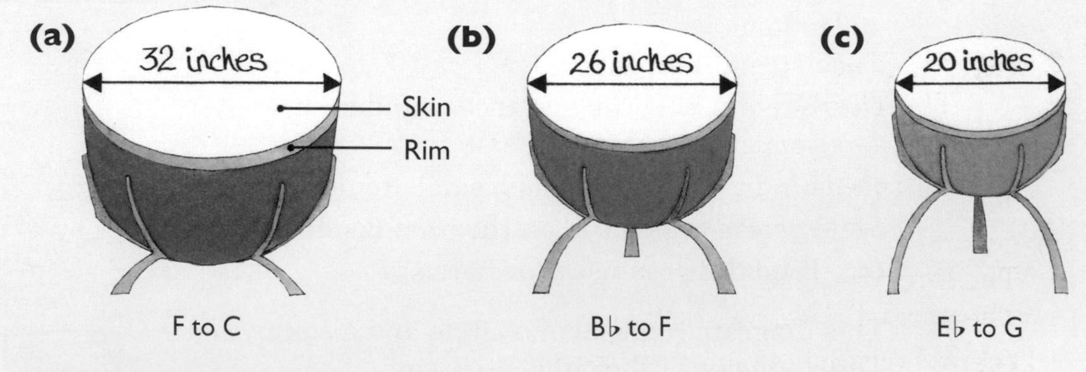

For each timpani find:
(i) the circumference of the rim **(ii)** the area of the skin.

4 Here is a church door.
The arch at the top is a semicircle.
Calculate the area of the door.

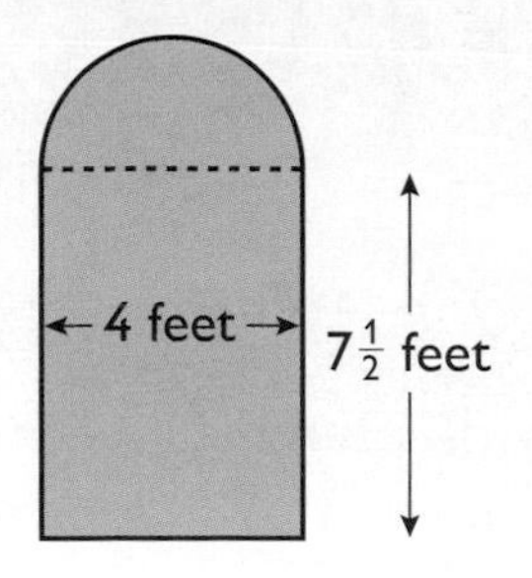

5 Here is a sketch of a wrought iron gate.
The curve at the top is a semicircle.
Calculate the total length of iron rod used to make the gate.

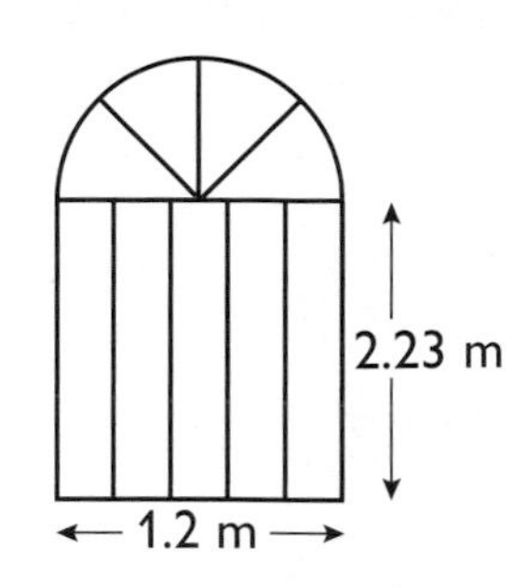

6 Here is a circular pond, surrounded by a path.
The path is 50 cm wide.
Calculate **(a)** the area of the pond
(b) the area of the path.
Give your answers in square metres, correct to 2 decimal places.

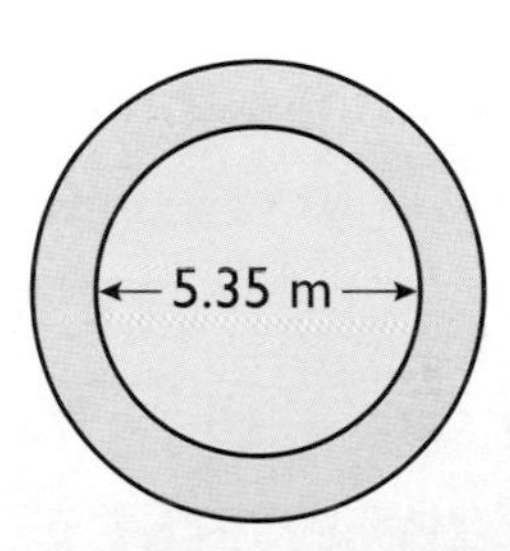

Investigation

1 The London Eye is a huge wheel.
The diameter of the wheel is 135 m.
One revolution of the wheel takes 30 minutes.

(a) Calculate the speed, in $\mathrm{m\,s^{-1}}$, at which the capsules travel.

(b) Compare your answer with the speed given at *www.british-airways.com/londoneye*

2 **(a)** Find out the markings of a football pitch. These are given at the Football Association's web site *www.the-fa.org/index.htm*

(b) Calculate the total length of lines used to mark out an international pitch of maximum size.

The lines are marked in chalk.
They are all 12 cm wide.

(c) What area of grass is covered by chalk?

12 Fractions

Adding and subtracting fractions

Billy and David walk along the footpath from the campsite to the village.

How far do they walk?

They return to the campsite along the road.

How much shorter is this route?

Task

Practise your fractions!

1 Equivalent fractions

(a) Fill in the missing numbers in the following to show fractions equivalent to $\frac{3}{5}$.

$$\frac{3}{5} = \frac{12}{\square} = \frac{\square}{25} = \frac{\square}{100} = \frac{3a}{\square}$$

(b) Cancel these fractions as far as possible.

(i) $\frac{4}{6}$ **(ii)** $\frac{12}{30}$ **(iii)** $\frac{5a}{6a}$

2 Adding fractions

Fill in the blank spaces to complete these additions.

(a) $\frac{3}{7} + \frac{2}{5} = \frac{\square + \square}{35} = \frac{29}{\square}$ **(b)** $\frac{3}{8} + \frac{5}{12} = \frac{\square + \square}{24} = \frac{\square}{\square}$

3 Subtracting fractions

Work out these subtractions.

(a) $\frac{3}{4} - \frac{2}{3}$ **(b)** $\frac{5}{6} - \frac{3}{4}$

(c) $\frac{3}{8} - \frac{3}{16}$

Explain how the common denominators have been chosen in part 2.

Mixed numbers

What is the total length of Billy and David's walk (from the campsite to the village and back)?

Change $\frac{1}{2}$, $\frac{1}{3}$ and $\frac{3}{4}$ into fractions with the same denominator. Add these together.

What do you notice?

Change the top-heavy fraction $\frac{19}{12}$ into a mixed number.

Task

1 Write these top-heavy fractions as mixed numbers.

$\frac{5}{4}$, $\frac{14}{5}$, $\frac{23}{8}$, $\frac{25}{12}$, $\frac{17}{3}$

2 Change these mixed numbers into top-heavy fractions.

$1\frac{1}{2}$, $2\frac{4}{5}$, $3\frac{1}{7}$, $4\frac{2}{9}$, $5\frac{3}{4}$

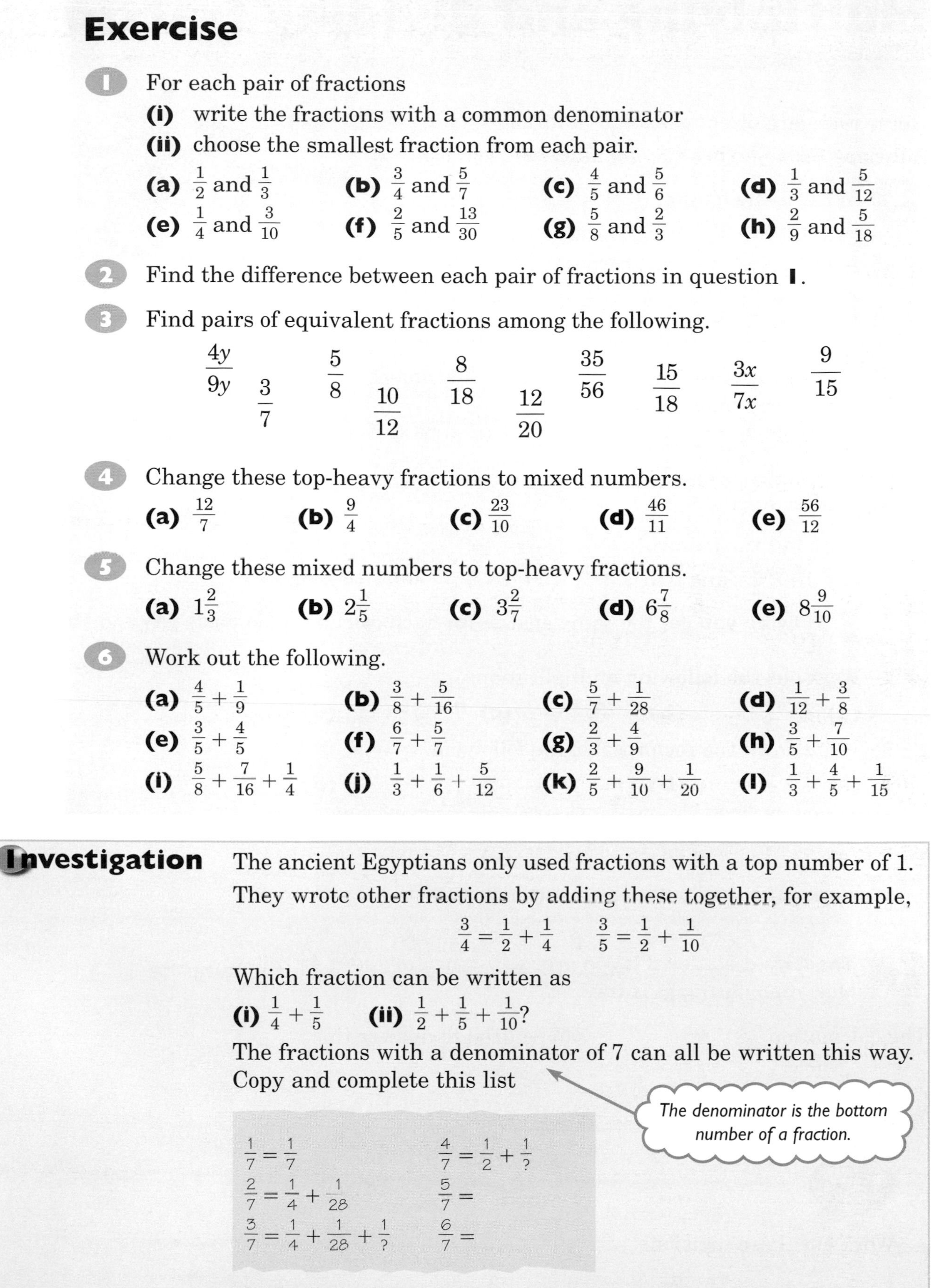

Exercise

1 For each pair of fractions

(i) write the fractions with a common denominator

(ii) choose the smallest fraction from each pair.

(a) $\frac{1}{2}$ and $\frac{1}{3}$ **(b)** $\frac{3}{4}$ and $\frac{5}{7}$ **(c)** $\frac{4}{5}$ and $\frac{5}{6}$ **(d)** $\frac{1}{3}$ and $\frac{5}{12}$

(e) $\frac{1}{4}$ and $\frac{3}{10}$ **(f)** $\frac{2}{5}$ and $\frac{13}{30}$ **(g)** $\frac{5}{8}$ and $\frac{2}{3}$ **(h)** $\frac{2}{9}$ and $\frac{5}{18}$

2 Find the difference between each pair of fractions in question **1**.

3 Find pairs of equivalent fractions among the following.

$$\frac{4y}{9y} \quad \frac{3}{7} \quad \frac{5}{8} \quad \frac{10}{12} \quad \frac{8}{18} \quad \frac{12}{20} \quad \frac{35}{56} \quad \frac{15}{18} \quad \frac{3x}{7x} \quad \frac{9}{15}$$

4 Change these top-heavy fractions to mixed numbers.

(a) $\frac{12}{7}$ **(b)** $\frac{9}{4}$ **(c)** $\frac{23}{10}$ **(d)** $\frac{46}{11}$ **(e)** $\frac{56}{12}$

5 Change these mixed numbers to top-heavy fractions.

(a) $1\frac{2}{3}$ **(b)** $2\frac{1}{5}$ **(c)** $3\frac{2}{7}$ **(d)** $6\frac{7}{8}$ **(e)** $8\frac{9}{10}$

6 Work out the following.

(a) $\frac{4}{5} + \frac{1}{9}$ **(b)** $\frac{3}{8} + \frac{5}{16}$ **(c)** $\frac{5}{7} + \frac{1}{28}$ **(d)** $\frac{1}{12} + \frac{3}{8}$

(e) $\frac{3}{5} + \frac{4}{5}$ **(f)** $\frac{6}{7} + \frac{5}{7}$ **(g)** $\frac{2}{3} + \frac{4}{9}$ **(h)** $\frac{3}{5} + \frac{7}{10}$

(i) $\frac{5}{8} + \frac{7}{16} + \frac{1}{4}$ **(j)** $\frac{1}{3} + \frac{1}{6} + \frac{5}{12}$ **(k)** $\frac{2}{5} + \frac{9}{10} + \frac{1}{20}$ **(l)** $\frac{1}{3} + \frac{4}{5} + \frac{1}{15}$

Investigation

The ancient Egyptians only used fractions with a top number of 1. They wrote other fractions by adding these together, for example,

$$\frac{3}{4} = \frac{1}{2} + \frac{1}{4} \qquad \frac{3}{5} = \frac{1}{2} + \frac{1}{10}$$

Which fraction can be written as

(i) $\frac{1}{4} + \frac{1}{5}$ **(ii)** $\frac{1}{2} + \frac{1}{5} + \frac{1}{10}$?

The fractions with a denominator of 7 can all be written this way. Copy and complete this list

The denominator is the bottom number of a fraction.

$\frac{1}{7} = \frac{1}{7}$ $\frac{4}{7} = \frac{1}{2} + \frac{1}{?}$

$\frac{2}{7} = \frac{1}{4} + \frac{1}{28}$ $\frac{5}{7} =$

$\frac{3}{7} = \frac{1}{4} + \frac{1}{28} + \frac{1}{?}$ $\frac{6}{7} =$

Write all the fractions with a bottom number of 5 in this way.

Multiplying and dividing fractions

Horse races are often measured in furlongs.
A furlong is one-eighth ($\frac{1}{8}$) of a mile.

? **What is the length, in furlongs, of a 3 mile race?**

? **Explain how the calculations $3 \div \frac{1}{8}$ or 3×8 help you answer this.**

What are the reciprocals of (a) 8 (b) $\frac{1}{8}$?

? **The Derby is run over a distance of 12 furlongs.**
How many miles is this?
Write down the multiplication you used to answer this question.

Task

1 Work out the following.

(a) $15 \times \frac{1}{3}$ and $15 \div 3$ **(b)** $18 \times \frac{1}{6}$ and $18 \div 6$

? **Why do you get the same answer for each calculation in parts (a) and (b)?**

2 Work out the following multiplications.

(a) $3 \times \frac{1}{2}$ **(b)** $6 \times \frac{1}{4}$ **(c)** $7 \times \frac{1}{3}$ **(d)** $8 \times \frac{2}{3}$

3 Write down the reciprocal of the following.

(a) 4 **(b)** $\frac{1}{5}$ **(c)** $\frac{3}{4}$ **(d)** $\frac{2}{7}$

4 Explain how to use reciprocals to calculate these divisions.

(a) $7 \div \frac{1}{5}$ **(b)** $6 \div \frac{3}{4}$ **(c)** $5 \div \frac{2}{7}$

5 Work out the calculations in question 4.

? **The Grand National is run over a distance of about $4\frac{1}{2}$ miles.**
How many furlongs is this?

The calculation $4\frac{1}{2} \div \frac{1}{8}$ can be used to answer this.

Write $4\frac{1}{2}$ as a top-heavy fraction.

$\frac{9}{2} \div \frac{1}{8} = \frac{9}{2} \times \frac{8}{1} = 36$

$4\frac{1}{2}$ miles is 36 furlongs

Task

Work out these divisions.

1 $4 \div \frac{1}{2}$ **2** $\frac{3}{4} \div \frac{1}{4}$ **3** $3\frac{1}{3} \div \frac{1}{3}$ **4** $1\frac{1}{2} \div \frac{3}{4}$

Exercise

1 Work out the following. Give your answer as mixed numbers.

(a) $5 \times \frac{1}{2}$ **(b)** $7 \times \frac{1}{4}$ **(c)** $8 \times \frac{2}{3}$ **(d)** $6 \times \frac{3}{4}$ **(e)** $9 \times \frac{2}{5}$

2 Write down the reciprocals of the following.

(a) 5 **(b)** $\frac{1}{7}$ **(c)** 15 **(d)** $\frac{5}{9}$ **(e)** $1\frac{1}{2}$

3 Work out these divisions.

(a) $\frac{3}{5} \div 3$ **(b)** $\frac{7}{8} \div 7$ **(c)** $\frac{10}{11} \div 5$ **(d)** $\frac{12}{13} \div 4$

(e) $5 \div \frac{1}{4}$ **(f)** $\frac{3}{4} \div \frac{1}{4}$ **(g)** $\frac{5}{7} \div \frac{1}{7}$ **(h)** $\frac{5}{9} \div \frac{2}{9}$

(i) $\frac{4}{5} \div \frac{1}{10}$ **(j)** $\frac{5}{12} \div \frac{1}{6}$ **(k)** $\frac{5}{6} \div \frac{2}{3}$ **(l)** $\frac{3}{4} \div \frac{5}{8}$

(m) $\frac{1}{3} \div \frac{1}{4}$ **(n)** $\frac{2}{5} \div \frac{2}{3}$ **(o)** $\frac{5}{8} \div \frac{2}{3}$ **(p)** $\frac{7}{9} \div \frac{5}{6}$

4 Ben eats $\frac{3}{4}$ of a tin of dog food each day.

(a) How long will

(i) 6 tins **(ii)** $1\frac{1}{2}$ tins last?

(b) How many tins will Ben eat in

(i) 12 days **(ii)** 5 days **(iii)** 7 days?

BEN

5 Change these distances to miles.

(i) 16 furlongs **(ii)** 20 furlongs **(iii)** 25 furlongs

6 Change these distances to furlongs.

(i) $2\frac{1}{2}$ miles **(ii)** $1\frac{3}{4}$ miles **(iii)** $2\frac{5}{8}$ miles

7 Erica is preparing a party for 30 people.
She estimates the amount of food each person will eat.

Pizza	$\frac{1}{4}$	*Tomato*	$\frac{3}{4}$
Garlic bread	$\frac{1}{3}$ *loaf*	*Salad cream*	$\frac{1}{12}$ *bottle*
Lettuce	$\frac{1}{6}$	*Coleslaw*	$\frac{1}{8}$ *tub*

How much of each type of food should Erica buy?

8 A candle will burn for a total of 10 hours.

(a) What fraction of the candle is burnt in

(i) 1 hour **(ii)** $2\frac{1}{2}$ hours **(iii)** $1\frac{3}{4}$ hours

(b) How long will it take to burn these fractions of the candle?

(i) $\frac{1}{2}$ **(ii)** $\frac{1}{3}$ **(iii)** $\frac{1}{6}$

(c) **(i)** Work out $\frac{1}{2} + \frac{1}{3} + \frac{1}{6}$.

(ii) What do you notice?

(iii) How can you use this to check your answers in part **(b)**?

Finishing off

Now that you have finished this chapter you should be able to:

- add and subtract fractions
- change top-heavy fractions into mixed numbers
- change mixed numbers into top-heavy fractions
- multiply and divide simple fractions.

Review exercise

1 Work out the following additions.

(a) $\frac{1}{2}+\frac{1}{5}$ **(b)** $\frac{1}{3}+\frac{1}{4}$ **(c)** $\frac{2}{5}+\frac{1}{10}$ **(d)** $\frac{1}{4}+\frac{3}{8}$

(e) $\frac{2}{3}+\frac{1}{6}$ **(f)** $\frac{3}{4}+\frac{1}{16}$ **(g)** $\frac{5}{8}+\frac{3}{16}$ **(h)** $\frac{2}{9}+\frac{2}{3}$

(i) $\frac{1}{2}+\frac{1}{4}+\frac{1}{8}$ **(j)** $\frac{1}{5}+\frac{1}{10}+\frac{1}{15}$ **(k)** $\frac{1}{3}+\frac{1}{6}+\frac{1}{9}$ **(l)** $\frac{2}{3}+\frac{3}{5}+\frac{1}{15}$

(m) $\frac{3}{4}+\frac{1}{8}+\frac{3}{16}$ **(n)** $\frac{1}{20}+\frac{3}{10}+\frac{4}{5}$

2 Work out the following subtractions.

(a) $\frac{1}{2}-\frac{1}{3}$ **(b)** $\frac{2}{5}-\frac{3}{10}$ **(c)** $\frac{3}{7}-\frac{4}{21}$ **(d)** $\frac{5}{6}-\frac{4}{9}$

(e) $\frac{5}{6}-\frac{3}{4}$ **(f)** $\frac{3}{10}-\frac{1}{7}$ **(g)** $\frac{4}{15}-\frac{1}{8}$ **(h)** $\frac{4}{9}-\frac{1}{4}$

3 Change these mixed numbers into top-heavy fractions.

(a) $1\frac{2}{3}$ **(b)** $1\frac{4}{5}$ **(c)** $2\frac{1}{3}$ **(d)** $2\frac{2}{5}$ **(e)** $3\frac{1}{7}$

4 Change these top-heavy fractions into mixed numbers.

(a) $\frac{4}{3}$ **(b)** $\frac{9}{7}$ **(c)** $\frac{7}{3}$ **(d)** $\frac{13}{4}$ **(e)** $\frac{24}{7}$

5 Write the following sets of fractions in order of size, starting with the smallest.

(a) $\frac{1}{5}, \frac{1}{2}, \frac{1}{7}, \frac{1}{9}, \frac{1}{3}$ **(b)** $\frac{2}{3}, \frac{5}{6}, \frac{7}{12}, \frac{3}{4}, \frac{5}{12}$ **(c)** $\frac{7}{5}, 1\frac{2}{3}, 1\frac{3}{10}, 1\frac{3}{5}, \frac{17}{15}$

6 Work out the following.

(a) $\frac{3}{5}\times\frac{4}{7}$ **(b)** $\frac{2}{9}\times\frac{3}{4}$ **(c)** $\frac{2}{5}\times\frac{5}{8}\times\frac{4}{7}$ **(d)** $\frac{5}{6}\times\frac{3}{7}\times\frac{14}{15}$

(e) $\frac{3}{4}$ of 24 **(f)** $\frac{2}{5}$ of 30 **(g)** $\frac{3}{7}$ of 14 **(h)** $\frac{7}{5}$ of 25

7 Write down the reciprocals of the following.

(a) $\frac{2}{3}$ **(b)** $\frac{4}{5}$ **(c)** $\frac{7}{3}$ **(d)** $1\frac{2}{3}$ **(e)** $3\frac{1}{2}$

8 Work out these divisions.

(a) $15\div\frac{1}{3}$ **(b)** $27\div\frac{3}{4}$ **(c)** $36\div\frac{4}{5}$ **(d)** $10\div\frac{2}{7}$

(e) $15\div1\frac{1}{2}$ **(f)** $20\div2\frac{1}{2}$ **(g)** $18\div4\frac{1}{2}$ **(h)** $8\div1\frac{1}{3}$

(i) $2\frac{1}{2}\div\frac{1}{2}$ **(j)** $3\frac{1}{4}\div\frac{1}{4}$ **(k)** $1\frac{2}{3}\div\frac{1}{3}$ **(l)** $3\frac{1}{2}\div\frac{1}{4}$

9 Jane eats $\frac{2}{5}$ of a bar of chocolate.
She divides the rest between her 3 friends.
What fraction of the bar does each friend have?

10 The map shows the distances in miles along a footpath.
Find the total length of the path.

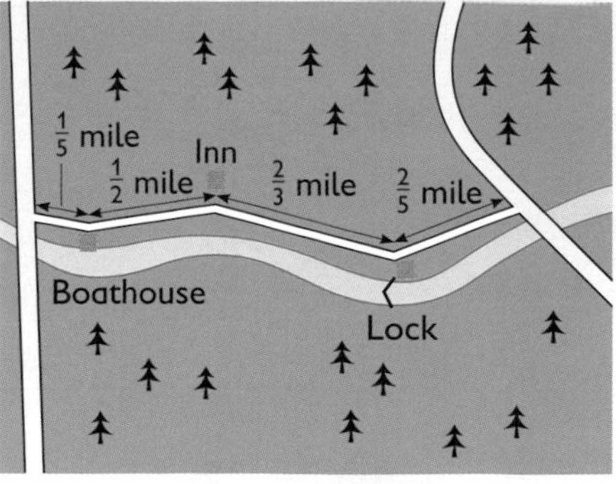

11 There are 3 candidates for team captain.
Darren receives $\frac{3}{7}$ of the votes and Brian $\frac{5}{14}$.
The rest of the votes are for John.
Who is elected?

12 Jenny records these programmes on a 3-hour tape.

News	$\frac{1}{2}$ *hr*
Quiz show	$\frac{3}{4}$ *hr*
Neighbours	$\frac{1}{2}$ *hr*
Snooker	$\frac{5}{6}$ *hr*

(a) What is the total time needed? **(b)** How much of the tape is unused?

Activity

When the number 5 bus leaves the terminus it is $\frac{1}{4}$ full.

1 At the first stop, $\frac{1}{8}$ of a bus load get on.
How full is the bus now?

2 At the second stop, $\frac{1}{16}$ of a bus load get off and $\frac{1}{2}$ get on.
What fraction of the bus is full now?

3 Nobody gets off at the next stop and the bus is full when it leaves. What fraction of a bus load got on?

4 The bus remains full until 8 people get off.
It is then $\frac{7}{8}$ full.
How many people does the bus hold?

5 At the next stop $\frac{5}{16}$ of a bus load get off.
How many people is this?

Write you own story of a bus journey.

13 Graphs

Mappings

Tony draws this table.

Number of tapes bought	0	1	2	3	4	5	6
Cost (£s)			6				

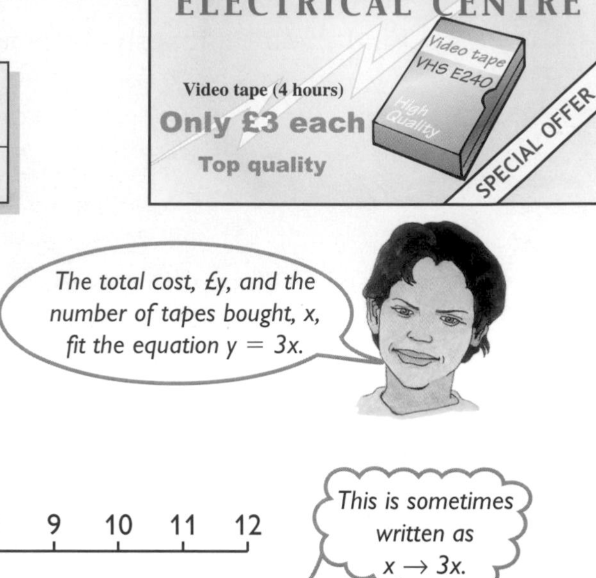

Copy and complete Tony's table.

Do you agree with Tony's comment?

Tony illustrates this equation with a **mapping diagram**.

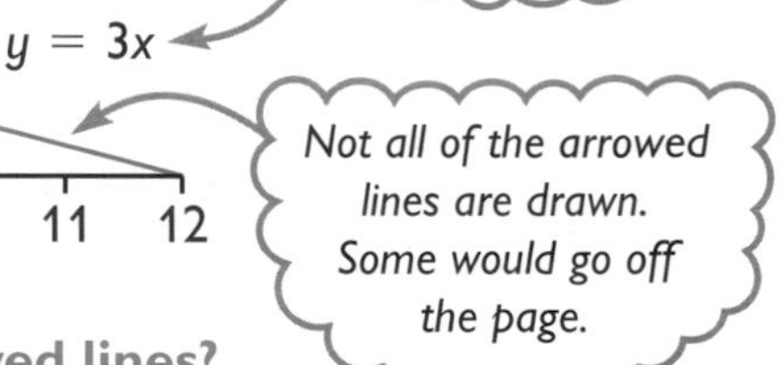

How could you change the y line to fit more arrowed lines?

Task

1 Copy and complete the table.

2 Draw a mapping diagram for the equation $y = 2x + 5$.

x	0	1	2	3	4	5
$2x$	0	2				
$+5$	5	5				
$y = 2x + 5$	5	7				

Look at the mapping diagram for the equation $y = x$.

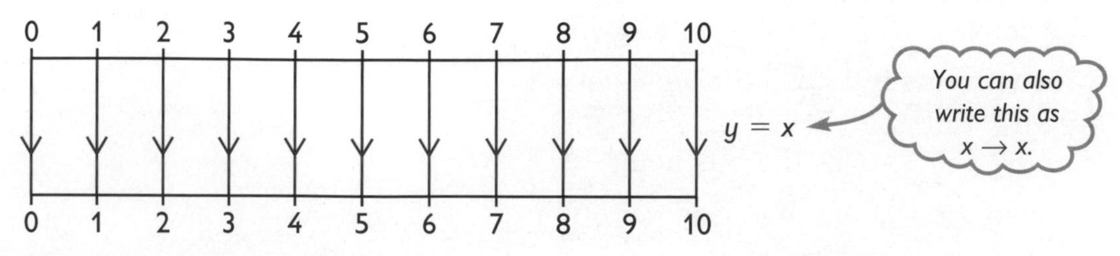

Every number maps onto itself.

$y = x$ is called the **identity function**.

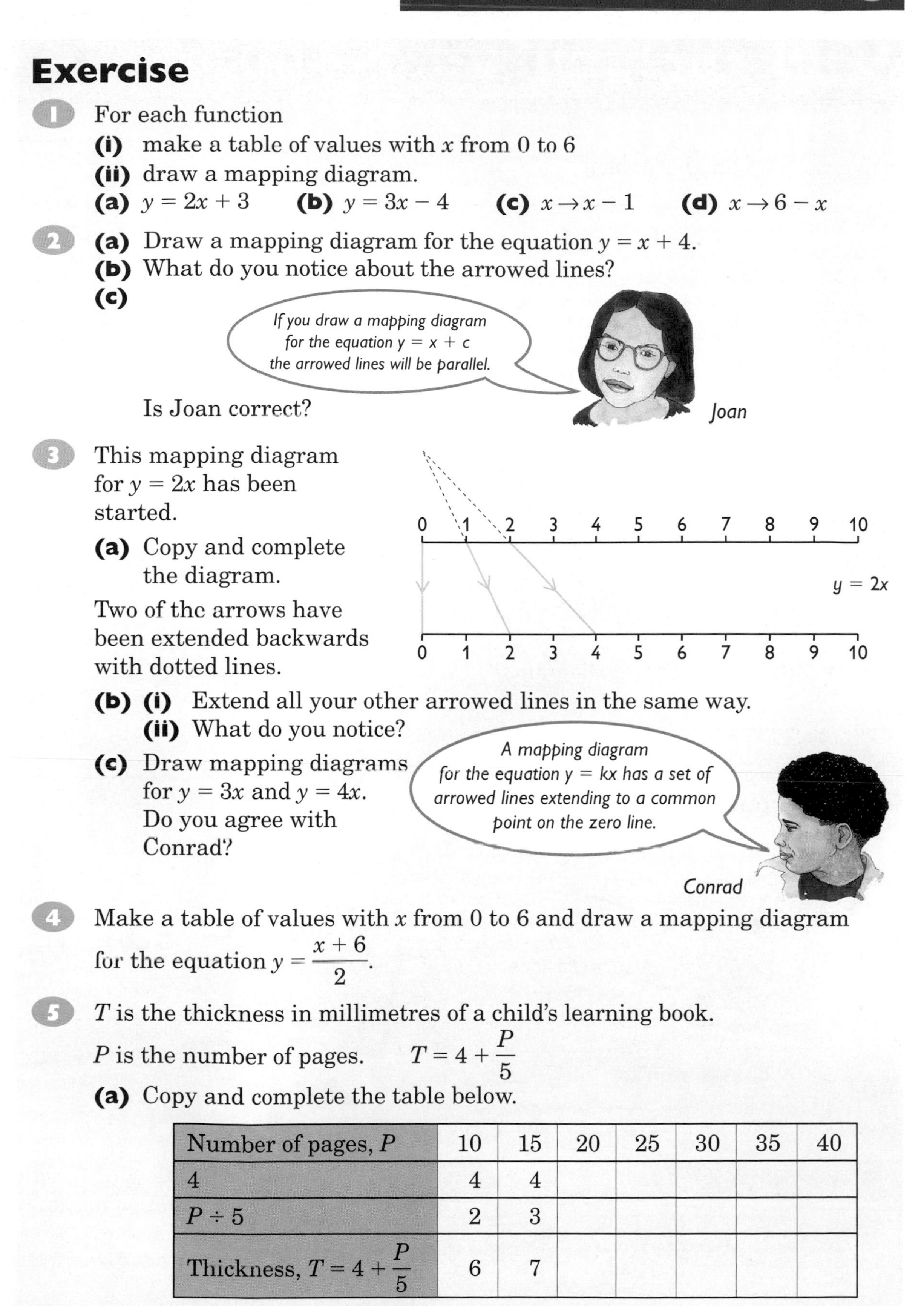

Exercise

1 For each function

(i) make a table of values with x from 0 to 6

(ii) draw a mapping diagram.

(a) $y = 2x + 3$ **(b)** $y = 3x - 4$ **(c)** $x \rightarrow x - 1$ **(d)** $x \rightarrow 6 - x$

2 **(a)** Draw a mapping diagram for the equation $y = x + 4$.

(b) What do you notice about the arrowed lines?

(c)

If you draw a mapping diagram for the equation $y = x + c$ the arrowed lines will be parallel.

Joan

Is Joan correct?

3 This mapping diagram for $y = 2x$ has been started.

0 1 2 3 4 5 6 7 8 9 10

$y = 2x$

0 1 2 3 4 5 6 7 8 9 10

(a) Copy and complete the diagram.

Two of the arrows have been extended backwards with dotted lines.

(b) **(i)** Extend all your other arrowed lines in the same way.

(ii) What do you notice?

(c) Draw mapping diagrams for $y = 3x$ and $y = 4x$. Do you agree with Conrad?

A mapping diagram for the equation $y = kx$ has a set of arrowed lines extending to a common point on the zero line.

Conrad

4 Make a table of values with x from 0 to 6 and draw a mapping diagram for the equation $y = \dfrac{x + 6}{2}$.

5 T is the thickness in millimetres of a child's learning book.

P is the number of pages. $T = 4 + \dfrac{P}{5}$

(a) Copy and complete the table below.

Number of pages, P	10	15	20	25	30	35	40
4	4	4					
$P \div 5$	2	3					
Thickness, $T = 4 + \dfrac{P}{5}$	6	7					

(b) Draw a mapping diagram to show this relationship.

(c) What is the thickness of one page?

(d) The front and back covers of the book are of equal thickness. How thick are the covers?

Using mapping diagrams

Tony draws up another table:

Amount paid (£s)	0	3	6	9	12
Number of video tapes bought	0	1	2	3	4

NIXON'S
ELECTRICAL CENTRE
Video tape (4 hours)
Only £3 each
Top quality
SPECIAL OFFER
Video tape VHS E240
High Quality

Explain Tony's comment.

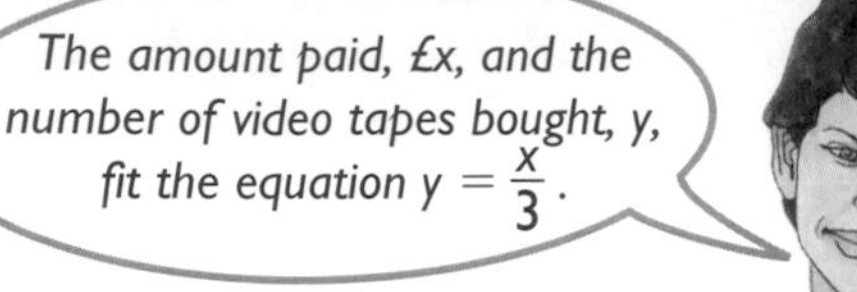

Look at these two mappings.

The top of the mapping diagram is the same as Tony used on page 110.

It is for $y = 3x$.

Underneath is the mapping diagram for $y = \frac{x}{3}$.

The mapping $y = \frac{x}{3}$ is called the **inverse mapping** of $y = 3x$.

0 1 2 3 4 5 6 7 8 9
$y = 3x$
0 1 2 3 4 5 6 7 8 9
$y = \frac{3}{x}$
0 1 2 3 4 5 6 7 8 9

What do you notice when the two mappings are combined?

An operation and its inverse are like a return ticket.

Simon

Eurostar London – Paris

Eurostar Paris – London

Why does Simon say this?

So the inverse of $y = x + 10$ is $y = x - 10$.

Do you remember?

Operation	Inverse operation
Add	Subtract
Subtract	Add
Multiply	Divide
Divide	Multiply

Task

For each of the following functions

(a) draw a mapping diagram with x ranging from 0 to 8

(b) find a formula for the inverse function.

(i) $y = x + 5$ **(ii)** $y = x - 7$

(iii) $y = 2x$ **(iv)** $y = \frac{x}{5}$

1 Write down the inverse function for each of these functions.

(a) $y = 7x$ **(b)** $y = x + 17$ **(c)** $x \to x - 9$ **(d)** $x \to \frac{x}{4}$

(e) $x \to 3x$ **(f)** $x \to x - 12$ **(g)** $x \to \frac{x}{5}$ **(h)** $x \to x + 11$

2 **(a)** Copy and complete this mapping diagram.

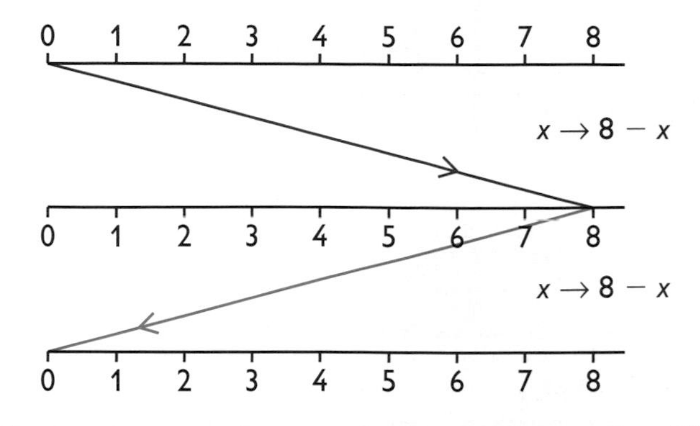

It shows that using $x \to 8 - x$ twice maps 0 onto itself: $0 \to 0$.

(b) What does each of 1, 2, … , 8 map onto?

(c) A mapping like $x \to 8 - x$ is called a **self-inverse** mapping. Why?

c is a constant

(d) Peter says that mapping of the form $y = c - x$, is a self-inverse mapping.
Make a similar mapping diagram for $x \to 5 - x$.
Is Peter's statement true?

3 Look at the table.

(a) Write down a formula for y in terms of x.

(b) What will be in the right-hand column opposite 7 in the left?

(c) Find the formula for x in terms of y.

(d) What will be in the left-column opposite 143 in the right?

x	y
1	13
2	26
3	39
4	52
⋮	⋮

Investigation For more complicated equations like $y = 3x + 5$ you can use the **flow diagram method** to find the inverse function.

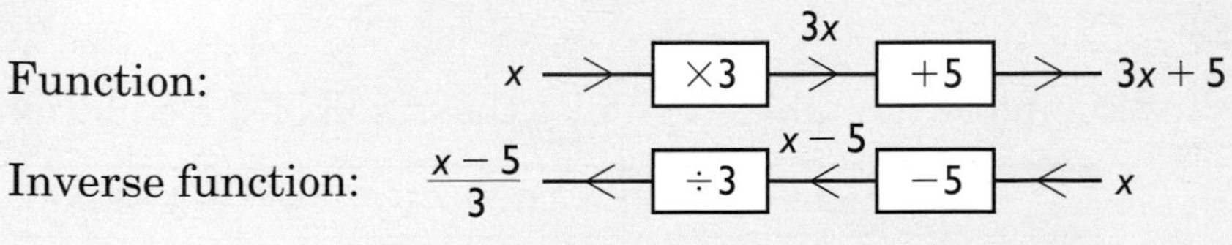

1 How does this method work?

2 Use the flow diagram method to find the inverse functions of

(a) $y = 2x + 9$ **(b)** $y = 8 + \frac{x}{3}$ **(c)** $y = 6x - 1$.

Plotting linear graphs

Annette is interested in this service for her party.

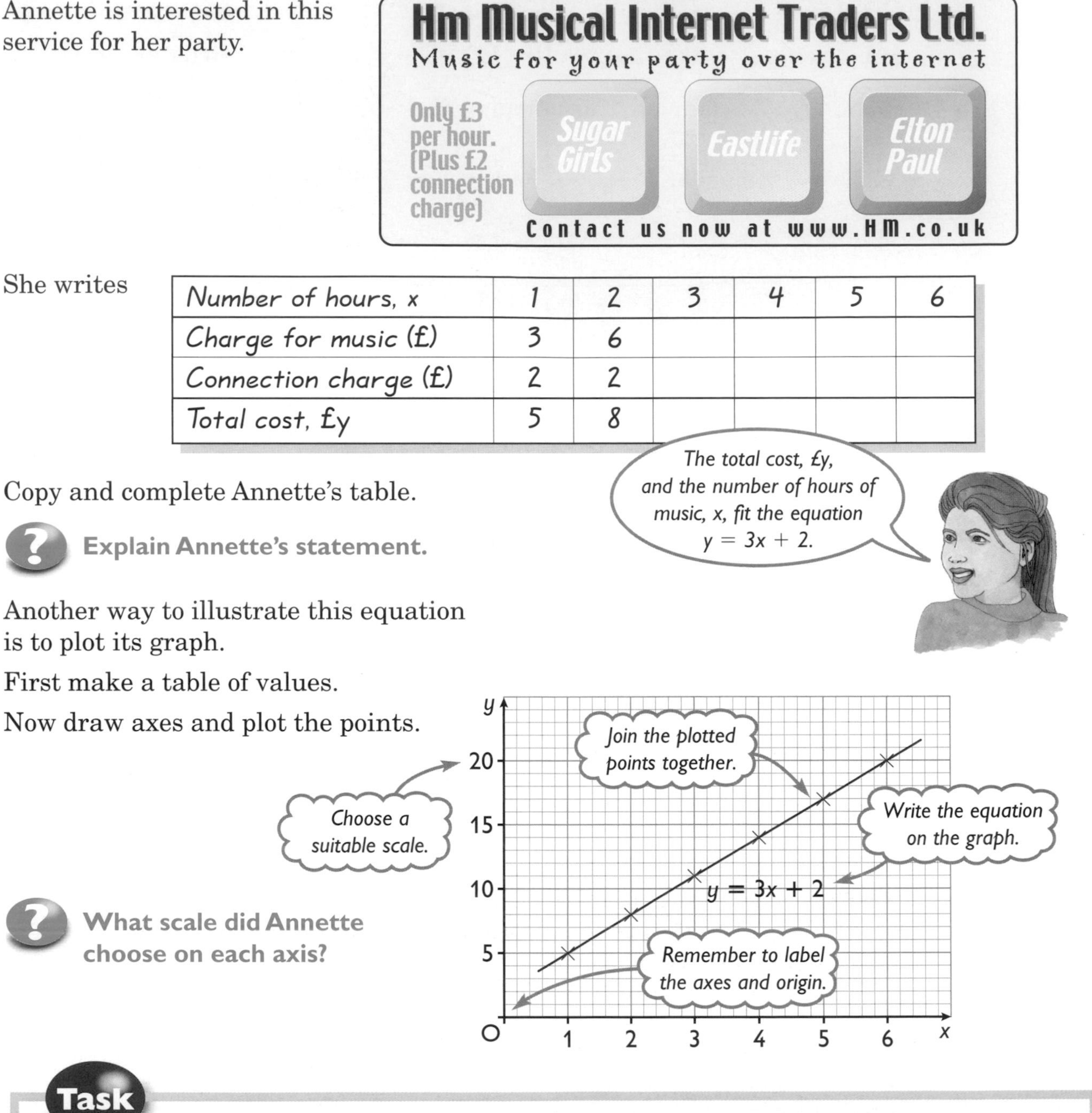

She writes

Number of hours, x	1	2	3	4	5	6
Charge for music (£)	3	6				
Connection charge (£)	2	2				
Total cost, £y	5	8				

Copy and complete Annette's table.

Explain Annette's statement.

Another way to illustrate this equation is to plot its graph.

First make a table of values.

Now draw axes and plot the points.

What scale did Annette choose on each axis?

Task

Copy and complete the table, then plot the graph of the equation $y = 2x + 7$.

x	-2	-1	0	1	2	3	4	5	6
$2x$	-4	-2	0						
$+7$	7	7	7						
$y = 2x + 7$	3	5	7						

Exercise

1 **(a)** Copy and complete the table.

(b) Use it to plot the graph of $y = 5x - 6$.

x	-1	0	1	2	3
$5x$			5		
-6			-6		
$y = 5x - 6$			-1		

(c) Find the value of y when $x = 2.2$.

2 **(a)** Copy and complete the table.

(b) Use it to plot the graph of $y = 2x + 5$.

(c) Find the value of x when $y = 8$.

x	-1	0	1	2	3	4	5	6
$2x$	-2	0						
$+5$	5	5						
$y = 2x + 5$	3	5						

3 **(a)** Copy and complete the table.

(b) Use it to draw the graph of $y = 12 - x$.

x	-1	0	1	2	3	4	5	6	7
12	12	12							
$-x$	$+1$								
$y = 12 - x$	13				9				5

(c) At which point on the graph are the values of x and y the same?

4 **(a)** Plot the line $y = 5x$. Take values of x from -1 to 6.

(b) Plot the line $y = -4x$. Take values of x from -3 to 5.

5 **(a)** Take values of x from -4 to 4 and y values from -8 to 8.
Plot the following lines on the same axes.
(i) $y = 2x + 8$ **(ii)** $y = 2x - 8$ **(iii)** $y = 8 - 2x$ **(iv)** $y = -8 - 2x$

(b) What shape have you drawn?

6 **(a)** Avonford Autos hire cars.
The basic charge is £20 plus a further 30p per mile.
Show that $C = 0.3x + 20$, where C is the cost of driving x miles.

(b) Plot a graph of this equation for values of x from 0 to 200.

(c) Use your graph to find the cost of hiring a car and driving 80 miles.

(d) Use your graph to find how many miles are driven when the hire cost is £56.

7 **(a)** Gas bills have a standing charge of £15 plus 10p per gas unit used.
The bill is £B.
Show this is related to the number of used gas units, x, by the equation $B = 15 + 0.1x$.

(b) Plot a graph of this equation with x from 0 to 800.

(c) Use your graph to find Mr Smith's bill when he uses 275 units of gas.

(d) The Butterworths' bill is £89.
Use your graph to find how many gas units they used.

Straight line graphs

The graph shows the straight line $y = 2x + 5$.

The **gradient** of a line tells you how steep it is.

The ***y*-intercept** of a line tells you where it crosses the y-axis.

$$\text{Gradient} = \frac{\text{change in } y}{\text{change in } x}$$

$$= \frac{13-9}{4-2} = \frac{4}{2} = 2$$

The line $y = 2x + 5$ has gradient 2 and y-intercept 5.

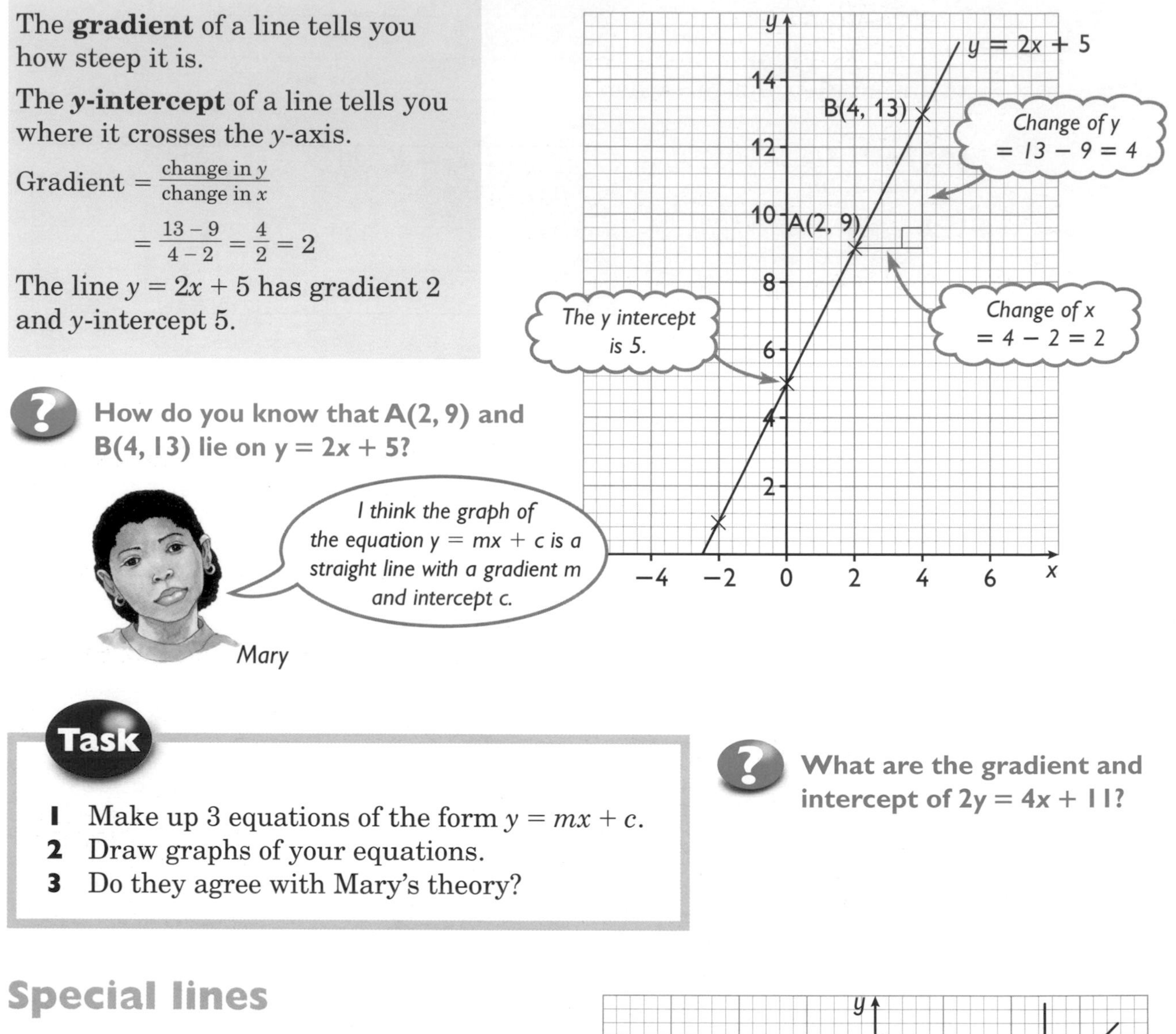

How do you know that A(2, 9) and B(4, 13) lie on $y = 2x + 5$?

Task

1. Make up 3 equations of the form $y = mx + c$.
2. Draw graphs of your equations.
3. Do they agree with Mary's theory?

What are the gradient and intercept of $2y = 4x + 11$?

Special lines

The lines $y = x$, $x = 5$ and $y = -2$ are drawn on the axes opposite.

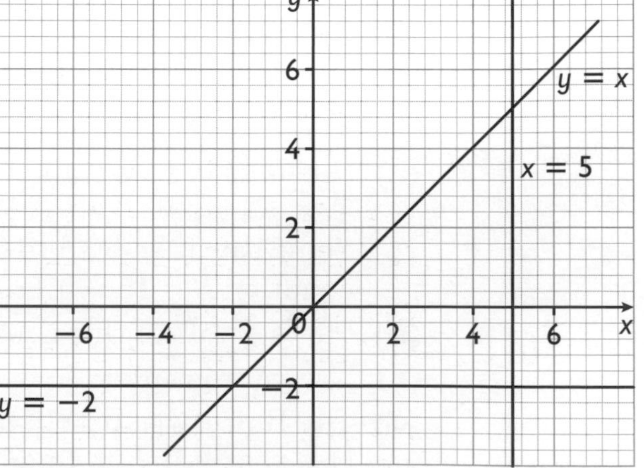

How can you recognise lines like these?

Complete these sentences.

(a) All lines with equation x = a number are …

(b) All lines with equation y = a number are …

Exercise

1 **(a)** Copy and complete the table of values for $y = 3x + 7$.

x	-2	-1	0	1	2	3	4	5	6
$3x$	-6	-3							
$+7$	7	7							
$y = 3x + 7$	1	4							

(b) Draw the graph of $y = 3x + 7$.

(c) Write down the co-ordinates of two points on the line.
Use them to calculate the gradient of the line.

(d) Write down the co-ordinates of the y-intercept.

2 State **(i)** the gradient **(ii)** the co-ordinates of the y-intercept for each of the following lines.

(a) $y = 8x + 3$ **(b)** $y = 5 + 6x$ **(c)** $y = 13x - 12$ **(d)** $y = 6x$
(e) $3y = 6x + 19$ **(f)** $y = 9$ **(g)** $y = 7 + 5x$ **(h)** $y = x + 10$

3 **(a)** Draw axes with values of both x and y from -5 to 5.

(b) Plot the lines **(i)** $y = x$ **(ii)** $y = -5$ **(iii)** $x = 5$ on the same axes.

(c) What is the area of the triangle formed by these lines?

4 **(a)** What is the angle between the lines $y = 2$ and $x = 3$?

(b) What are the co-ordinates of the point of intersection of these lines?

5 **(a)** Draw axes with values of both x and y from -2 to 6.
Draw the quadrilateral with vertices A$(-2, -2)$, B$(-1, 5)$, C$(3, 6)$ and D$(2, -1)$.

(b) Work out the gradient of the lines **(i)** AB **(ii)** BC **(iii)** DC **(iv)** AD.

(c) Describe the quadrilateral ABCD.

6 The graph opposite shows the line $y = 8 - 2x$.

(a) Complete the calculation of the gradient.

(b) This line has a negative $(-)$ gradient. What does this tell you about the line?

(c) **(i)** Plot the line $y = 10 - x$ for values of x from 0 to 5.

(ii) What is the gradient of this line?

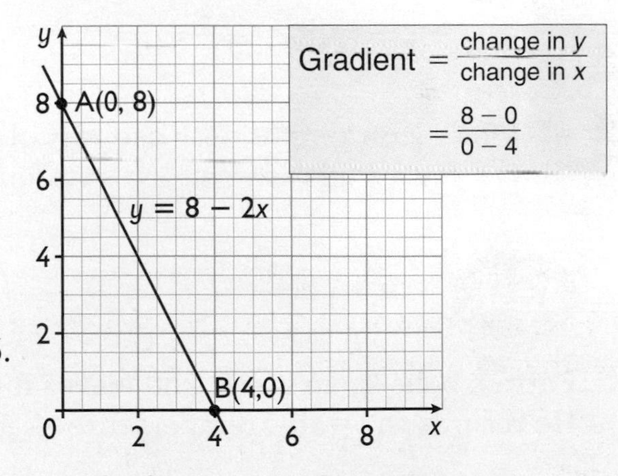

7 **(a)** Plot the line $y = 12 - 4x$ for values of x from -2 to 6.

(b) What is the gradient of this line?

8 **(a)** Make tables of values for the graphs **(i)** $y = 0.5x + 1$ and **(ii)** $y = 3 - 2x$.

(b) Draw axes with x from -1 to 4 and y from -5 to 5 using the same scale on each axis.

(c) Draw the two graphs on these axes.

(d) How are the lines related?

Curves

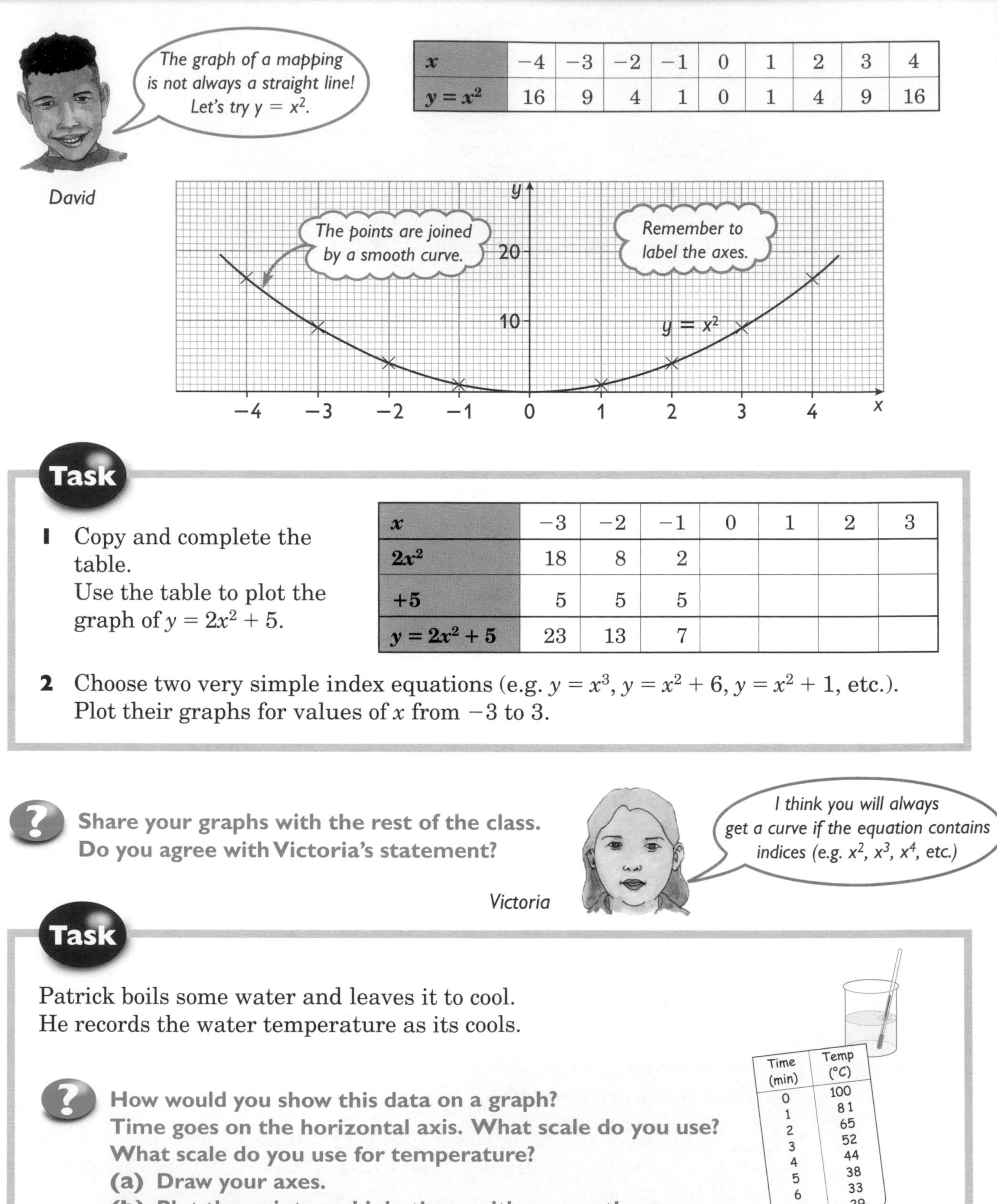

x	−4	−3	−2	−1	0	1	2	3	4
$y = x^2$	16	9	4	1	0	1	4	9	16

Task

1 Copy and complete the table.
Use the table to plot the graph of $y = 2x^2 + 5$.

x	−3	−2	−1	0	1	2	3
$2x^2$	18	8	2				
$+5$	5	5	5				
$y = 2x^2 + 5$	23	13	7				

2 Choose two very simple index equations (e.g. $y = x^3$, $y = x^2 + 6$, $y = x^2 + 1$, etc.).
Plot their graphs for values of x from −3 to 3.

? **Share your graphs with the rest of the class.**
Do you agree with Victoria's statement?

Task

Patrick boils some water and leaves it to cool.
He records the water temperature as its cools.

? **How would you show this data on a graph?**
Time goes on the horizontal axis. What scale do you use?
What scale do you use for temperature?

(a) Draw your axes.
(b) Plot the points and join them with a smooth curve.
(c) What is the temperature after 3.5 minutes?

Exercise

1 **(a)** Copy and complete the table.

x	−3	−2	−1	0	1	2	3
$y = x^3$	−27	−8					

(b) Plot the graph of $y = x^3$.

(c) Describe the symmetry of this graph.

2 **(a)** Copy and complete the table.

x	−4	−3	−2	−1	0	1	2	3	4
20	20	20	20						
$-x^2$	−16	−9	−4						
$y = 20 - x^2$	4	11	16						

(b) Plot the graph of $y = 20 - x^2$.

(c) This graph has a line of symmetry. What is the equation of this line?

3 **(a)** Copy and complete the table.

x	−4	−3	−2	−1	0	1	2	3	4
$3x^2$	48	27							
+2	2	2							
$y = 3x^2 + 2$	50	29							

(b) Plot the graph of $y = 3x^2 + 2$.

(c) Use your graph to estimate the value of y when $x = 2.5$.

4 **(a)** Copy and complete the table.

x	−1	0	1	2	3	4	5	6	7
$x - 3$	−4	−3	−2						
$y = (x - 3)^2$	16	9	4						

(b) Plot the graph of $y = (x - 3)^2$.

5 **(a)** Copy and complete the table.

x	−3	−2	−1	0	1	2	3
x^3	−27	−8					
+4	4	4					
$y = x^3 + 4$	−23	−4					

(b) Plot the graph of $y = x^3 + 4$.

6 **(a)** Draw the graphs of $y = x^2 + 2$ and $y = x^2 + 5$ for values of x from −3 to 3.

(b) Use your graphs to explain the effect of c in the equation $y = x^2 + c$.

7 **(a)** Draw the graphs of $y = 2x^2$ and $y = 3x^2$ for values of x from −3 to 3.

(b) Use your graphs to explain the effect of a in the equation $y = ax^2$.

8 A cargo ship transports an oil consignment at a speed of v kilometres per hour. At this speed the cost, £C, of a 300-kilometre trip is $C = 25v + \frac{8000}{v}$.

(a) Copy and complete this table.

(b) Draw a graph of C against v.

(c) Use your graph to estimate

(i) the cost of a journey at 27.5 km/hr

(ii) the speed at which the cost is least.

v	10	15	20	25	30	35	40
$\frac{8000}{v}$	800		400				200
$25v$	250		500				1000
$C = 25v + \frac{8000}{v}$	1050		900				1200

Finishing off

Now that you have finished this chapter you should be able to:

- draw mapping diagrams
- find a formula for a simple inverse function
- plot the graph of a straight line
- recognise the equation of a straight line
- plot graphs which result in a curve
- predict from the equation whether a graph is a straight line or a curve.

Review exercise

1 Copy and complete the table and draw a mapping diagram for the equation $y = 2x + 4$.

x	0	1	2	3	4	5	6
$2x$	0	2	4				
$+4$	4	4	4				
$y = 2x + 4$	4	6	8				

2 **(a)** Draw a mapping diagram for $y = x + 3$ for values of x from 0 to 7.
(b) What can you say about the arrowed lines of a mapping diagram with the form $y = x + c$?

3 Find the inverse function for **(i)** $y = x + 5$ **(ii)** $y = 9x$.

4 **(a)** Look at the mapping diagram for the equation $y = \frac{1}{2}x$.
Copy and complete the diagram.
Underneath that mapping diagram, join on and complete a mapping diagram for the equation $y = 2x$.

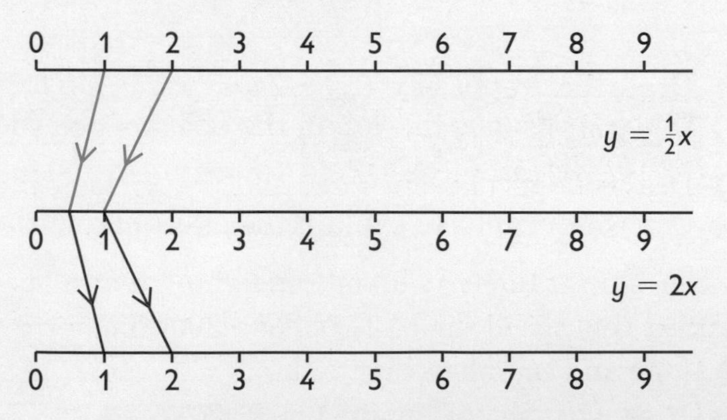

(b) How are these mappings related?

5 **(a)** Plot the graph of the equation $y = 3x - 11$ for values of x from -1 to 6.
(b) Solve the equation $3x - 11 = 0$.
How is the solution to this equation connected to part **(a)**?

6 Draw x and y axes, each from -6 to 6.
On these axes draw the lines
(a) $y = x$ **(b)** $y = -x$ **(c)** $x = -5$ **(d)** $y = 4$.

7 **(a)** Copy and complete the table.
(b) Draw axes, taking x from -3 to 3 and y from 0 to 50.
(c) Plot the graph of $y = 4x^2 + 9$.
(d) What is the minimum value that y takes?

x	-3	-2	-1	0	1	2	3
$4x^2$	36	16					
$+9$	9	9					
$y = 4x^2 + 9$	45	25					

8 **(a)** Copy and complete the table.

x	0	1	2	3	4	5	6	7	8
$x - 4$	-4	-3							
$y = (x - 4)^2$	16	9							

(b) Decide on suitable scales and draw x and y axes.
(c) Plot the graph of $y = (x - 4)^2$.
(d) What is the equation of the line of symmetry of this graph?

9 **(a)** Copy and complete the table.

x	-1	0	1	2	3	4	5	6	7	8
$7x$	-7	0								
$-x^2$	-1	0								
$y = 7x - x^2$	-8	0								

(b) Decide on suitable scales and draw x and y axes.
(c) Plot the graph of $y = 7x - x^2$.
(d) What is the equation of the line of symmetry of this graph?
What is the greatest value of y?

10 The table shows the value of £100 left in a savings account for 8 years.

Time (years)	0	1	2	3	4	5	6	7	8
Value of investment (£)	100	110	121	133	146	161	177	194	214

(a) Plot these points on a graph.
Join them with a smooth curve.
(b) How much is the investment worth after 2.5 years?
(c) After how long is the investment worth double its original value?

Investigation

Michael says 'reflecting the graph of a function in the line $y = x$ gives the graph of the inverse of the function'.
Plot the straight line graphs $y = x + 3$ and $y = 2x$.
On the same axes plot the inverse functions.

? **Do you agree or disagree with Michael's claim?**

14 Drawing and construction

Using scale drawings

Wes designs this wall bracket to support a hanging basket of flowers.

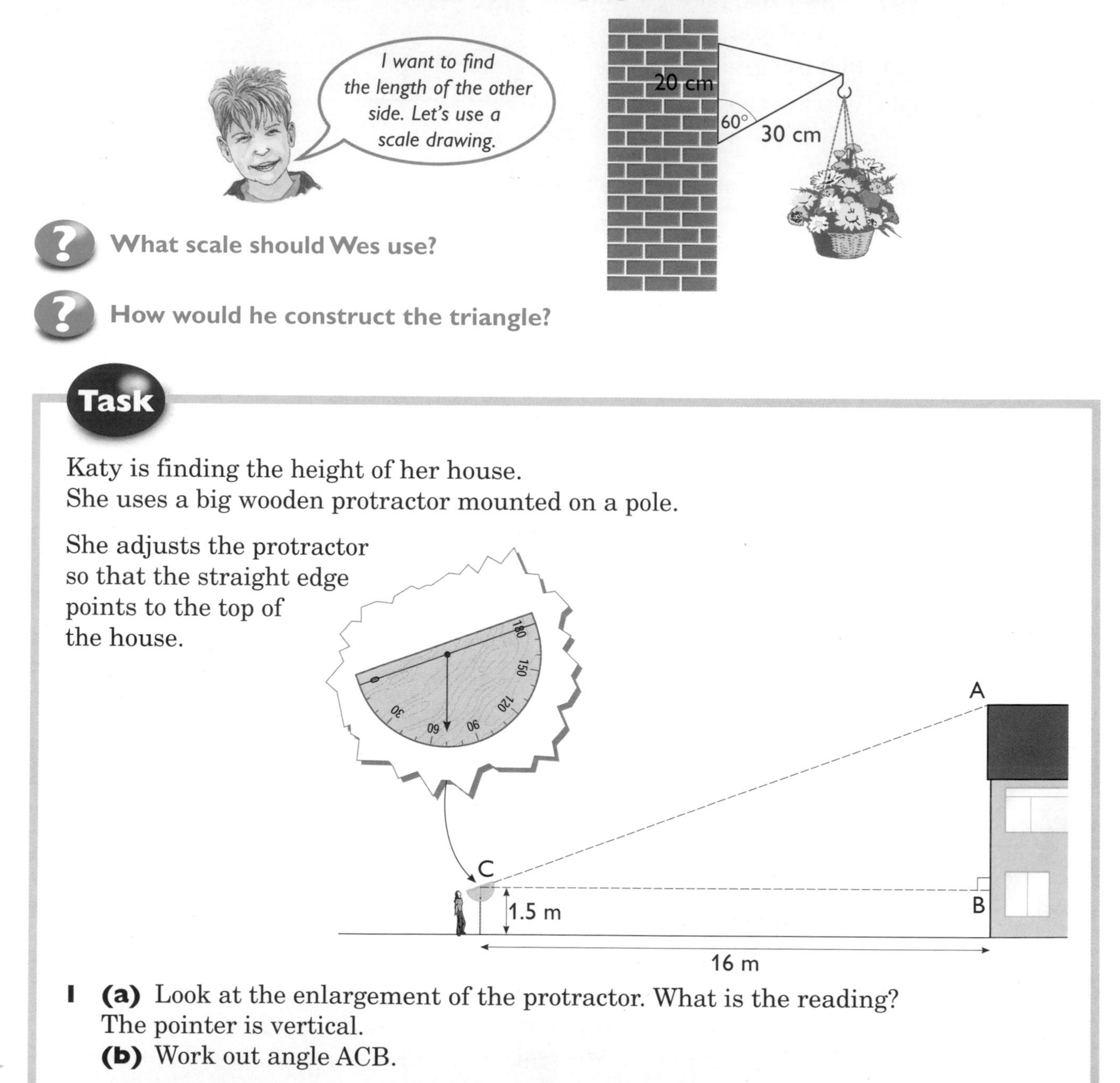

? **What scale should Wes use?**

? **How would he construct the triangle?**

Task

Katy is finding the height of her house.
She uses a big wooden protractor mounted on a pole.

She adjusts the protractor so that the straight edge points to the top of the house.

1 **(a)** Look at the enlargement of the protractor. What is the reading? The pointer is vertical.
(b) Work out angle ACB.

2 **(a)** Make an accurate scale drawing of triangle ACB. State the scale you are using.
(b) Measure the side AB.

3 Work out the height of Katy's house.

? **How accurate do you think Katy's method is?**

Exercise

Use a ruler, protractor and compasses as appropriate.

1 For each triangle

(i) make an accurate scale drawing

(ii) measure the unlabelled sides and angles.

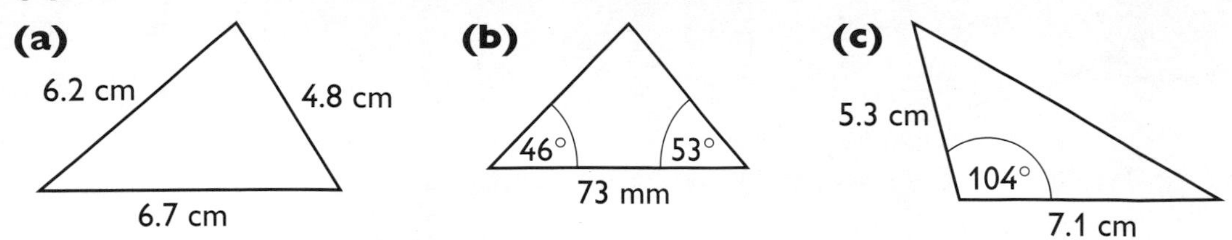

2 In triangle ABC, AB = 5.2 cm, BC = 6.8 cm and ∠B = 90°.

(a) Make an accurate drawing of triangle ABC.

(b) Measure the length AC.

3 Abigail is finding the height of a flagpole.
She stands 60 feet from the flagpole.
From this position the angle of elevation of the top is 26°.

Make a scale drawing to find the height of the flagpole.
Give your answer to the nearest foot.

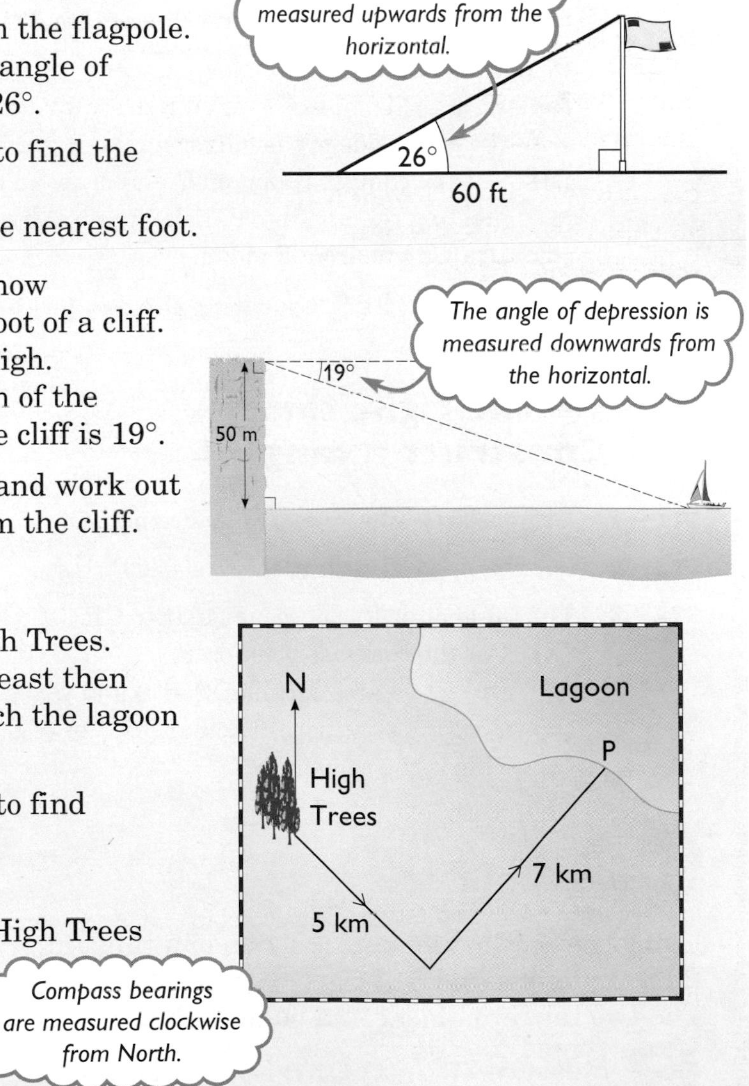

4 Robert is working out how far a boat is from the foot of a cliff.
The cliff is 50 metres high.
The angle of depression of the boat from the top of the cliff is 19°.

Make a scale drawing and work out how far the boat is from the cliff.

5 Lucy sets out from High Trees.
She walks 5 km south-east then 7 km north-east to reach the lagoon at point P.

Make a scale drawing to find

(a) how far point P is from High Trees

(b) the bearing from High Trees of point P.

Perpendiculars

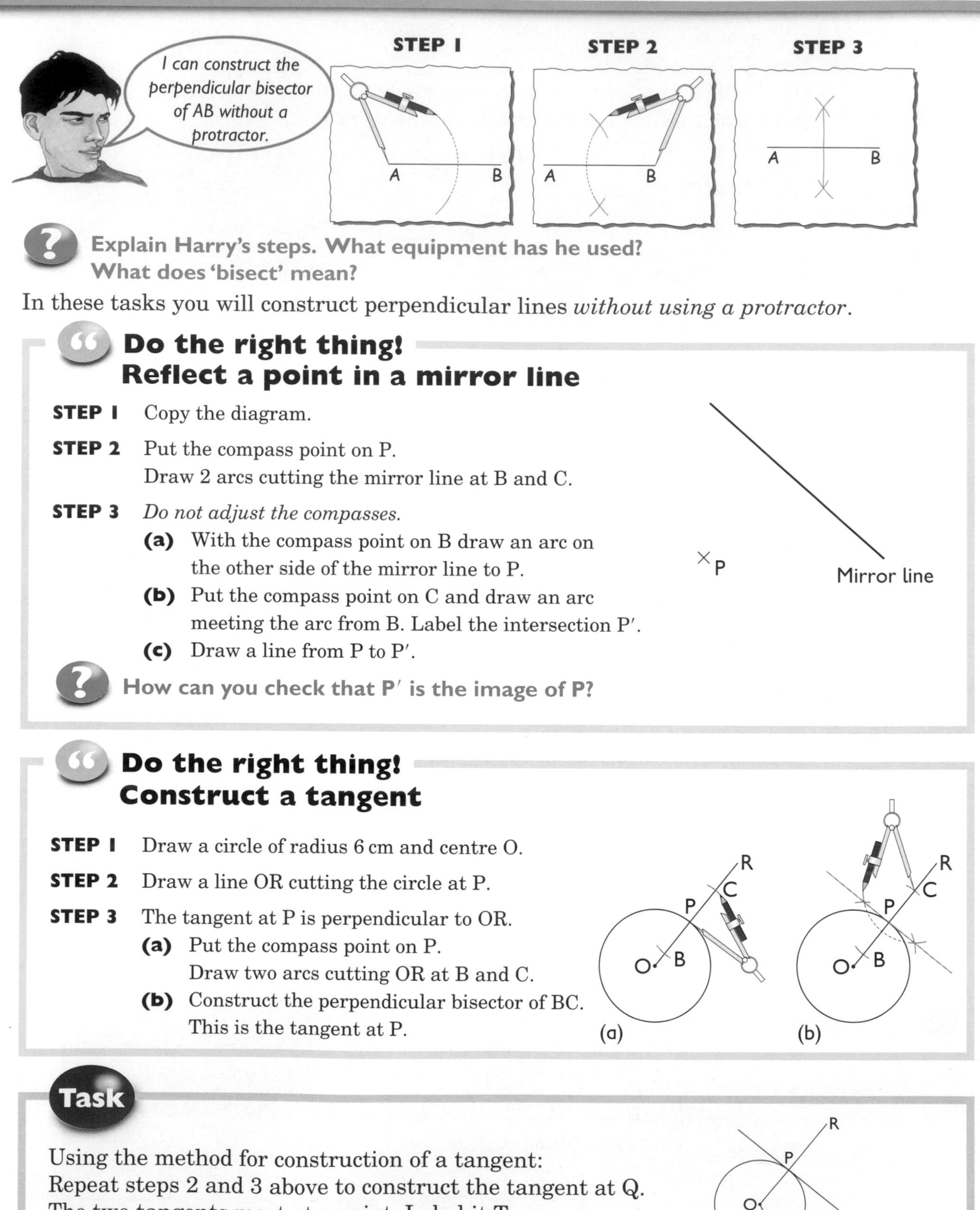

Explain Harry's steps. What equipment has he used?
What does 'bisect' mean?

In these tasks you will construct perpendicular lines *without using a protractor*.

Do the right thing! Reflect a point in a mirror line

STEP 1 Copy the diagram.

STEP 2 Put the compass point on P.
Draw 2 arcs cutting the mirror line at B and C.

STEP 3 *Do not adjust the compasses.*

(a) With the compass point on B draw an arc on the other side of the mirror line to P.

(b) Put the compass point on C and draw an arc meeting the arc from B. Label the intersection P′.

(c) Draw a line from P to P′.

How can you check that P′ is the image of P?

Do the right thing! Construct a tangent

STEP 1 Draw a circle of radius 6 cm and centre O.

STEP 2 Draw a line OR cutting the circle at P.

STEP 3 The tangent at P is perpendicular to OR.

(a) Put the compass point on P.
Draw two arcs cutting OR at B and C.

(b) Construct the perpendicular bisector of BC.
This is the tangent at P.

Task

Using the method for construction of a tangent:
Repeat steps 2 and 3 above to construct the tangent at Q.
The two tangents meet at a point. Label it T.

Measure TP and TQ. What do you notice?

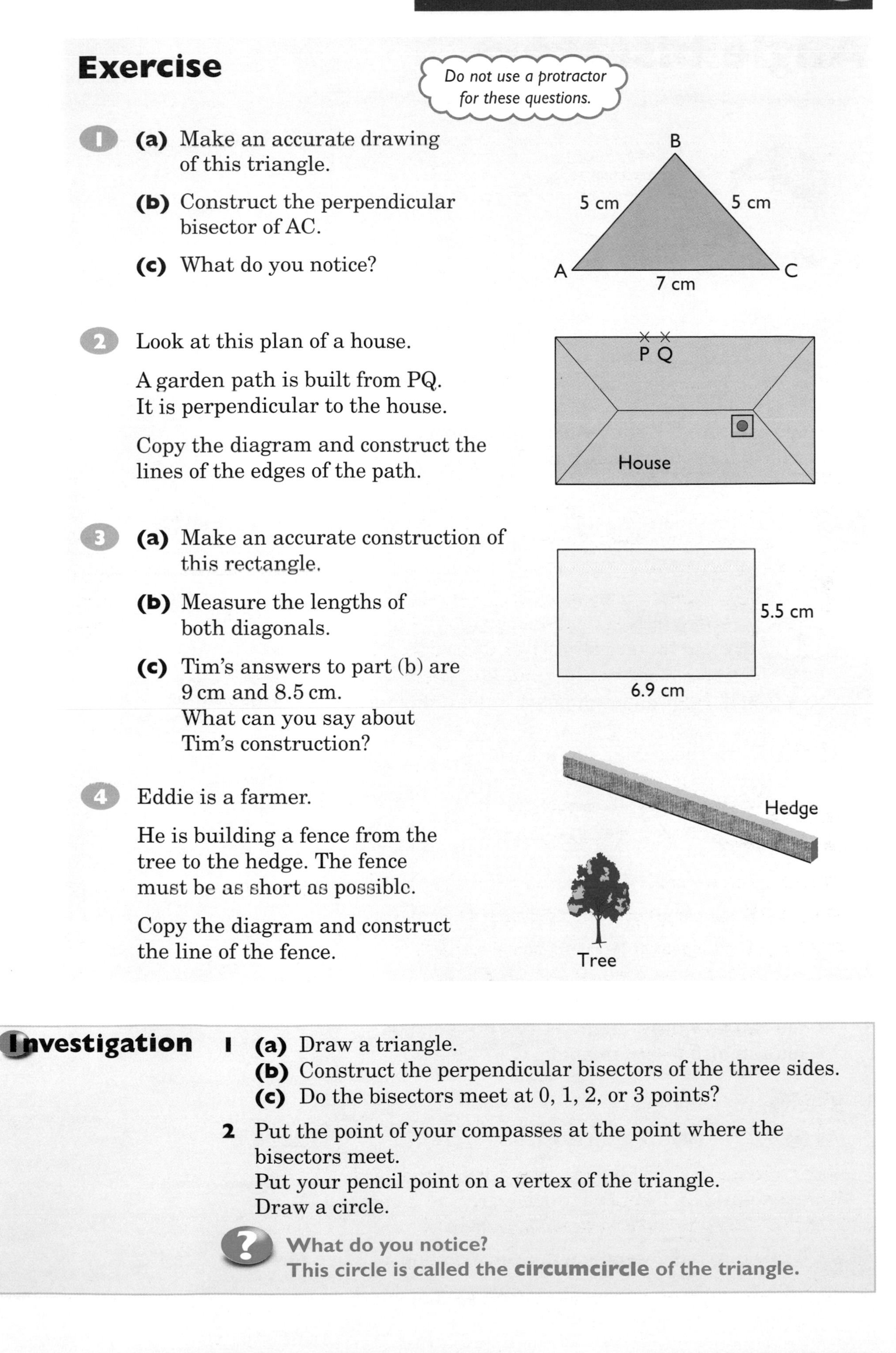

Exercise

Do not use a protractor for these questions.

1 **(a)** Make an accurate drawing of this triangle.

(b) Construct the perpendicular bisector of AC.

(c) What do you notice?

2 Look at this plan of a house.

A garden path is built from PQ.
It is perpendicular to the house.

Copy the diagram and construct the lines of the edges of the path.

3 **(a)** Make an accurate construction of this rectangle.

(b) Measure the lengths of both diagonals.

(c) Tim's answers to part (b) are 9 cm and 8.5 cm.
What can you say about Tim's construction?

4 Eddie is a farmer.

He is building a fence from the tree to the hedge. The fence must be as short as possible.

Copy the diagram and construct the line of the fence.

Investigation

1 **(a)** Draw a triangle.
(b) Construct the perpendicular bisectors of the three sides.
(c) Do the bisectors meet at 0, 1, 2, or 3 points?

2 Put the point of your compasses at the point where the bisectors meet.
Put your pencil point on a vertex of the triangle.
Draw a circle.

? What do you notice?
This circle is called the **circumcircle** of the triangle.

Angle bisectors

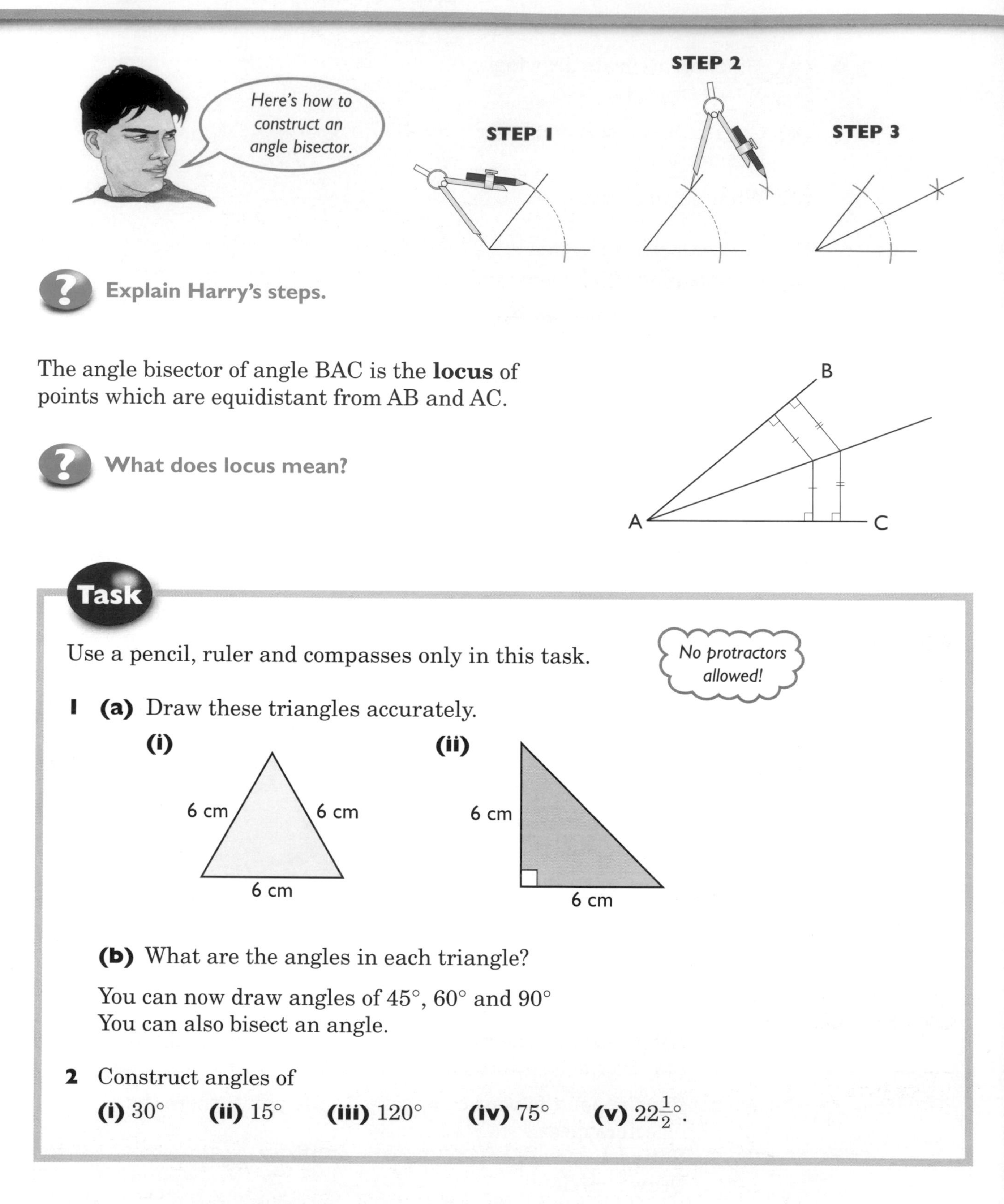

Explain Harry's steps.

The angle bisector of angle BAC is the **locus** of points which are equidistant from AB and AC.

What does locus mean?

Task

Use a pencil, ruler and compasses only in this task.

1 **(a)** Draw these triangles accurately.

(i) **(ii)**

(b) What are the angles in each triangle?

You can now draw angles of 45°, 60° and 90°
You can also bisect an angle.

2 Construct angles of

(i) 30° **(ii)** 15° **(iii)** 120° **(iv)** 75° **(v)** $22\frac{1}{2}$°.

Explain two different ways of constructing an angle of 45°.

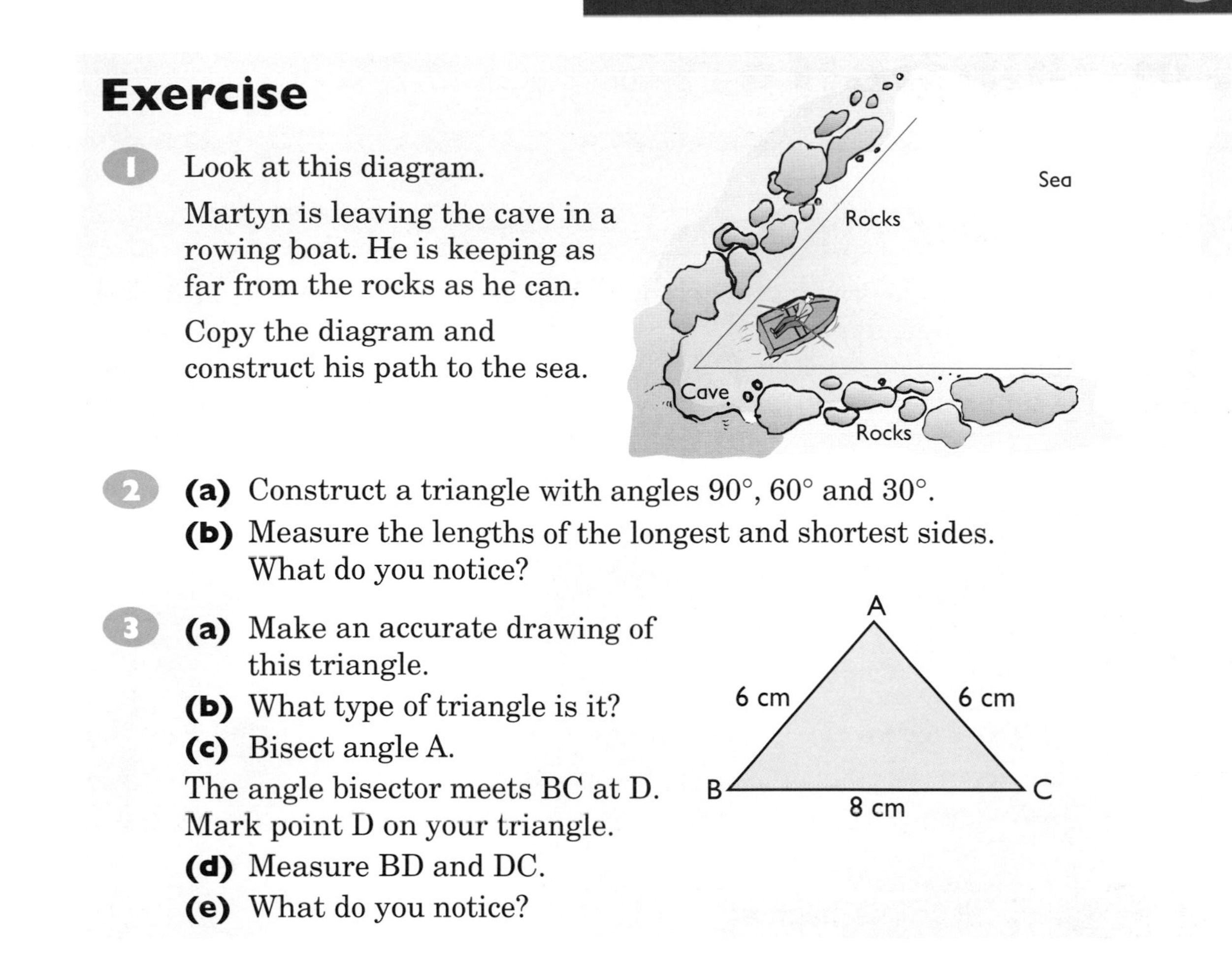

Exercise

1 Look at this diagram.

Martyn is leaving the cave in a rowing boat. He is keeping as far from the rocks as he can.

Copy the diagram and construct his path to the sea.

2 **(a)** Construct a triangle with angles 90°, 60° and 30°.

(b) Measure the lengths of the longest and shortest sides. What do you notice?

3 **(a)** Make an accurate drawing of this triangle.

(b) What type of triangle is it?

(c) Bisect angle A.

The angle bisector meets BC at D. Mark point D on your triangle.

(d) Measure BD and DC.

(e) What do you notice?

Investigation

1 **(a)** Draw a circle. Using the same radius make 6 marks round the circle.

(b) Join these points in order. What shape is formed?

2 **(a)** Repeat 1 (a).

(b) Bisect one side of the polygon (see diagram).

(c) Starting from where this bisector cuts the circle, make 6 more marks around the circle.

(d) Join the 12 points in order. What shape is formed?

Investigate what other regular polygons you can create in this way.

Activity

1 Draw a triangle and bisect each of the interior angles. What do you notice?

2 The point where the lines meet is O. Construct a perpendicular from O to one of the sides. The perpendicular cuts the line at P.

3 Put your compass point on O and pencil point on P and draw a circle.

? What do you notice? This circle is called the **incircle** of the triangle.

Finishing off

Now that you have finished this chapter you should be able to:

- make accurate scale drawings of triangles and use them to solve real-life problems
- construct perpendicular lines
- bisect angles and construct angles of 90°, 60°, 30° and 45°
- describe simple loci.

Review exercise

1 A 5-metre ladder rests against a wall.
The foot of the ladder is 1.5 metres from the wall.

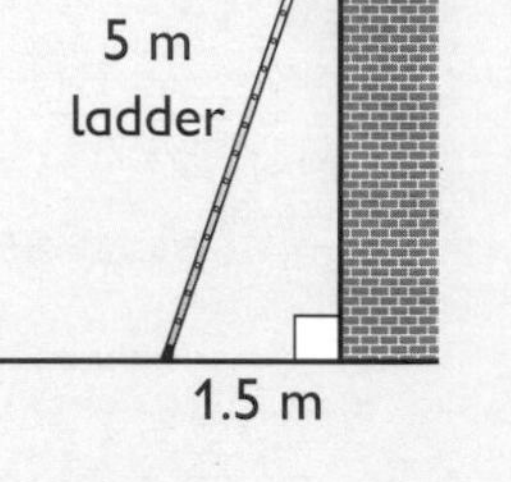

Make a scale drawing and use it to work out

(a) how far the ladder reaches up the wall

(b) the angle between the ladder and the ground.

2 Look at this diagram.

(a) Draw a sketch showing the locus of all points equidistant from AB and AC.

(b) Make an accurate drawing of this locus.

(c) What name is given to this locus?

3 **(a)** Make an accurate drawing of the quadrilateral ABCD.

(b) Measure **(i)** the angle ADC
(ii) the length AC.

(c) Work out the area of
(i) triangle BCD
(ii) the quadrilateral ABCD.

4 Look at this diagram.

(a) Draw a sketch showing the locus of all points equidistant from A and B.

(b) Make an accurate drawing of this locus.

(c) What name is given to this locus?

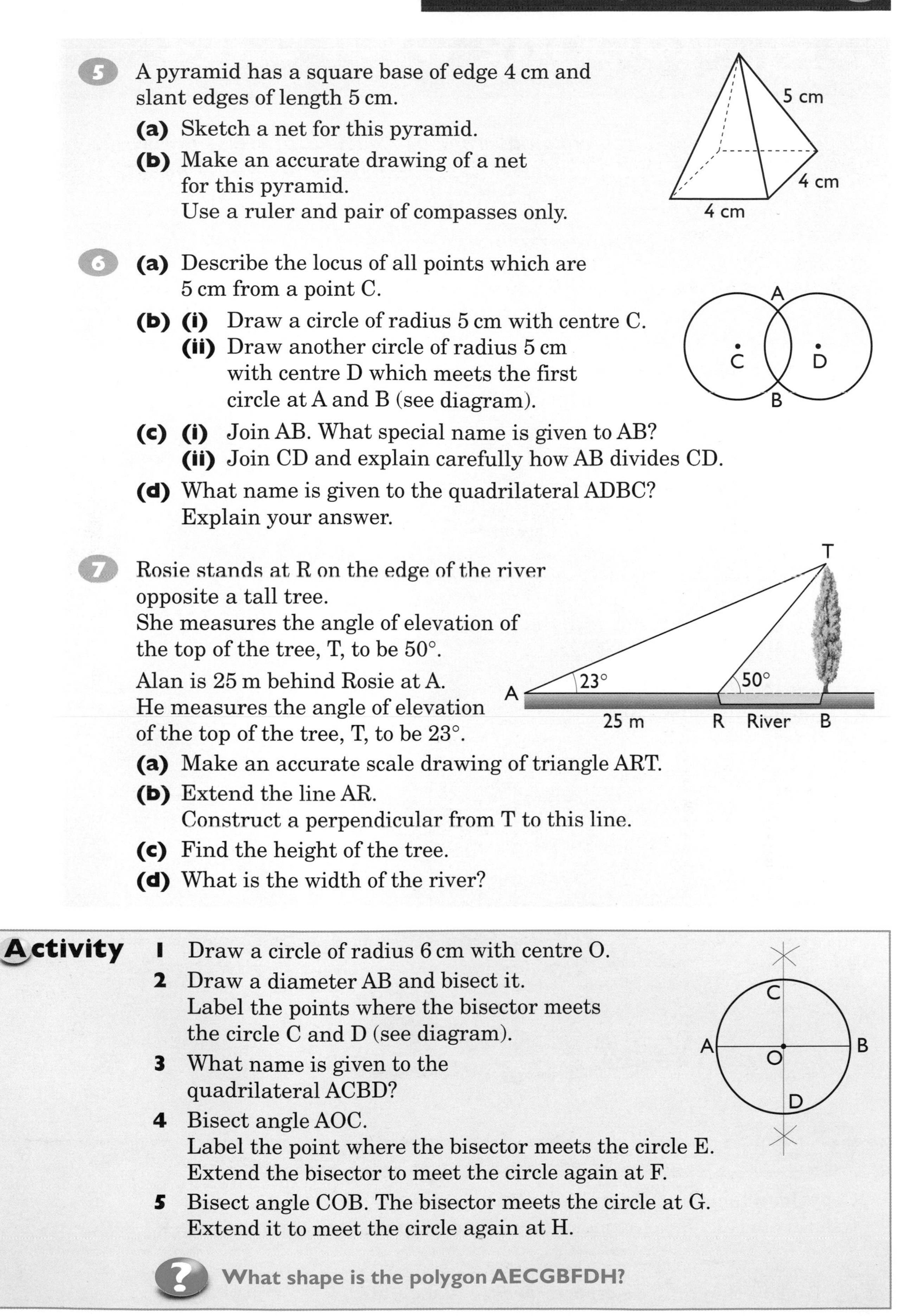

5 A pyramid has a square base of edge 4 cm and slant edges of length 5 cm.

(a) Sketch a net for this pyramid.

(b) Make an accurate drawing of a net for this pyramid.
Use a ruler and pair of compasses only.

6 **(a)** Describe the locus of all points which are 5 cm from a point C.

(b) **(i)** Draw a circle of radius 5 cm with centre C.

(ii) Draw another circle of radius 5 cm with centre D which meets the first circle at A and B (see diagram).

(c) **(i)** Join AB. What special name is given to AB?

(ii) Join CD and explain carefully how AB divides CD.

(d) What name is given to the quadrilateral ADBC? Explain your answer.

7 Rosie stands at R on the edge of the river opposite a tall tree.
She measures the angle of elevation of the top of the tree, T, to be 50°.
Alan is 25 m behind Rosie at A.
He measures the angle of elevation of the top of the tree, T, to be 23°.

(a) Make an accurate scale drawing of triangle ART.

(b) Extend the line AR.
Construct a perpendicular from T to this line.

(c) Find the height of the tree.

(d) What is the width of the river?

Activity

1 Draw a circle of radius 6 cm with centre O.

2 Draw a diameter AB and bisect it.
Label the points where the bisector meets the circle C and D (see diagram).

3 What name is given to the quadrilateral ACBD?

4 Bisect angle AOC.
Label the point where the bisector meets the circle E.
Extend the bisector to meet the circle again at F.

5 Bisect angle COB. The bisector meets the circle at G.
Extend it to meet the circle again at H.

? **What shape is the polygon AECGBFDH?**

15 Working with data

Schoolchildren eat an average of 7 packets of crisps per week.

The average family goes to Spain on holiday.

Children get an average of £4 pocket money per week.

What does the word *average* mean in each of the statements above?

The mean

Joe does not believe the statement about crisps.
He asks some friends how many packets
of crisps they ate last week.
Here are his results.

6	2	8	7	5
0	5	1	7	7
2	4	5	6	4

Joe finds the mean.

Total number of packets of crisps = 69
Number of people = 15
Mean number of bags of crisps = $\frac{69}{15}$ = 4·6

Compare Joe's result with the newspaper report.
Is Joe's result sensible?

Joe decides to a do a bigger survey.
This time he asks 80 people.

No. of packets	Frequency	Total no. of packets
0	8	0 × 8 = 0
1	12	1 × 12 = 12
2	9	2 × 9 = 18
3	4	
4	11	
5	10	
6	8	
7	15	
8	2	
9	1	
Total	80	

Task

Copy Joe's table and finish his working out.
Use this to calculate the mean number of crisp packets eaten per week.

Do you think the newspaper article is accurate?

Exercise

1 Find the mean of each of the following sets of numbers.

(a) 6, 8, 2, 4, 10, 6, 1, 3
(b) 0, 4, 6, 2, 3, 2, 0, 1, 2, 2, 3, 5
(c) 11, 29, 82, 23, 44, 25, 61, 72, 55, 42

2 James and Joanne are ten-pin bowling.

The table shows the number of pins they each knock down in each turn.

James	0	2	1	5	6	8
	4	10	4	6	5	9
Joanne	4	6	8	7	4	5
	5	2	3	6	5	

(a) Find James's mean score.
(b) Find Joanne's mean score.
(c) Who do you think is the better bowler?

3 James thinks he can improve his bowling scores.
After practicing for several days, he records his scores over 100 turns.

Here are his results.

Score	0	1	2	3	4	5	6	7	8	9	10
Frequency	2	4	2	6	8	7	13	23	21	9	5

Calculate James's new mean score.

4 A packet of toffees is labelled 'Average contents 12'.

Stuart is a Quality Control Inspector.
He checks the number of toffees in 50 packets.

Here are his results.

Number of toffees	9	10	11	12	13	14	15
Frequency	1	4	9	17	12	5	2

(a) Find the mean number of toffees in a packet.
(b) Is the label on the packet accurate?

5 Every week Ellie's class has a mental maths test.
There are always 20 questions.
After half a term (7 weeks), Ellie works out that her mean score is 14.

These are Ellie's scores in the second half of the term (6 weeks).

16 15 12 14 18 15

(a) What is Ellie's mean score for the second half of term?
(b) Did Ellie do better in the first or second half of term?
(c) What is Ellie's total score after the first half of term?
(d) What is her total score at the end of term?
(e) What is Ellie's mean score for the whole term?

The median

What does Sam mean when he says he is average height?

Amy gets £3.50 pocket money per week.
She asks six of her friends how much they get.

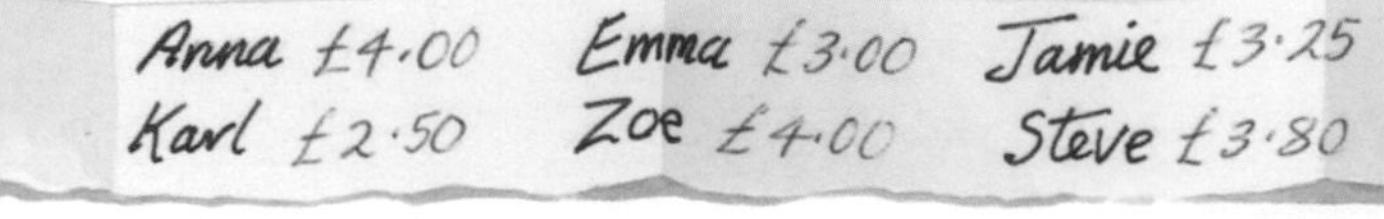
Anna £4.00 Emma £3.00 Jamie £3.25
Karl £2.50 Zoe £4.00 Steve £3.80

Amy writes the results in order.

Karl	Emma	Jamie	Amy	Steve	Anna	Zoe
£2.50	£3.00	£3.25	£3.50	£3.80	£4.00	£4.00

Amy's amount is in the middle: three people get less than Amy and three people get more.
The middle value of a set of data is called the **median**.

In this example, there are 7 data values. The median is the 4th data value.

Which is the median out of 19 data values?
What is the rule for finding the median of an odd number of data values?

Here are the data again with Kevin included:

Karl	Emma	Jamie	Amy
£2.50	£3.00	£3.25	£3.50
Steve	Anna	Zoe	Kevin
£3.80	£4.00	£4.00	£4.00

Now there are 8 data items.
8 is an even number so there is no single middle value.
The median is halfway between Amy's and Steve's values.

The median is £3.65

£2.50 £3.00 £3.25 £3.50 £3.80 £4.00 £4.00 £4.00

In this example, the median is the mean of the 4th and 5th pieces of data.
What is the rule for finding the median for an even number of data values?

Task

Find out the shoe size of everyone in the room, including your teacher.
Find the mean and the median shoe sizes. Are they close together?

Do you think the mean and the median will usually be close together?

What are the advantages and disadvantages of the mean and the median?

Exercise

1 Find the median of each of these sets of data.

(a) 4, 6, 0, 5, 2, 9, 3, 2, 4

(b) 12, 11, 10, 12, 17, 15, 16, 17, 11, 13

(c) 15, 24, 11, 62, 54, 36, 25, 18, 72, 38, 49, 50

Remember to write the data in order first.

2 Paul measures the heights of the boys and girls in his class.
Here are his results, in centimetres.

Boys	144	132	165	128	145	139	152
	137	161	150	148	152	141	
Girls	152	158	143	147	161	156	159
	164	143	156	138	148	157	152

(a) Find the median for **(i)** the boys **(ii)** the girls.

(b) On the whole, who are taller the boys or the girls?

3 The number of times each child in Class 9C was absent during one term is as follows.

0	0	2	5	0	0	1	0	3	4	0
28	1	2	0	8	5	2	1	1	0	0
7	0	10	3	8	1	0				

(a) Find the median and the mean of these data.

(b) Which average do you think describes the data best?
Explain your answer.

4 Claire and Lucy are both football fans.
They each keep a goal count for their favourite team in each match of the season.

Claire's team

2	3	3	2	4	3	1	4	3	2	3
3	2	1	0	2	3	2	4	1	2	

Lucy's team

1	0	0	3	5	2	6	0	0	2	4
3	0	5	4	0	1	1	5	4		

(a) For each team make a frequency table showing the number of goals scored in each match.

(b) Draw a bar chart of the same data.

(c) Find the mean and the median goal score for each team.

(d) Which team do you think is better at scoring goals?
Explain your answer.

Activity Choose two newspapers, one broadsheet and one tabloid.

For each newspaper, count the number of letters in each word for 50 words. Make a frequency table. Calculate the mean and median word length for each newspaper.

Comment on your results.

The median for large sets of data

Mrs Patel teaches French in evening classes.
All her students take the same test.

Here are their results, as percentages.

57	82	73	64	67	41	75	33	88	79	48	66
58	71	60	64	73	52	66	35	31	81	93	52
67	70	45	56	94	51	69	49	55	71	81	56
62	50	73	85	91	47	56	64	67			

Mrs Patel wants to find the median mark.

There are 45 marks altogether. When the data have been put in order, which is the median out of 45 data values?

One way to make this easier is put the data in a **stem-and-leaf** diagram.
This is a bit like a grouped tally chart, but you write the actual data in the table.

Here is the start of Mrs. Patel's stem-and-leaf diagram:

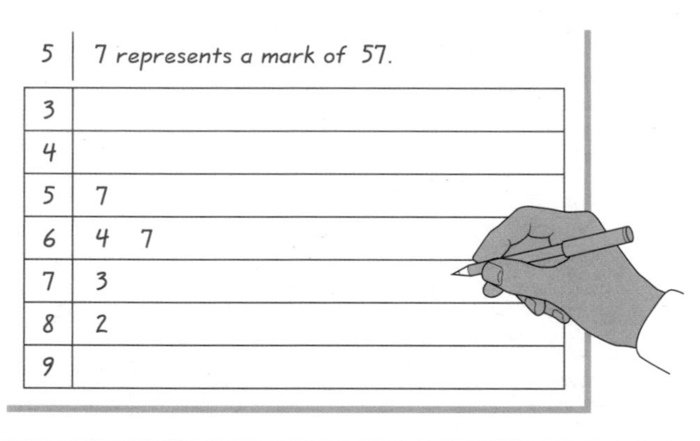

5 | 7 represents a mark of 57.

3	
4	
5	7
6	4 7
7	3
8	2
9	

Task

Copy Mrs Patel's stem-and-leaf diagram above and finish filling it in.
Now rewrite the table putting the data in each row in order.
Check your table is the same as the one below.

Here is Mrs Patel's finished diagram, with the data put in order.

Now Mrs Patel can count along each row to find the median mark.

5 | 7 represents a mark of 57.

3	1 3 5
4	1 5 7 8 9
5	0 1 2 2 5 6 6 6 7 8
6	0 2 4 4 4 6 6 7 7 7 9
7	0 1 1 3 3 3 5 9
8	1 1 2 5 8
9	1 3 4

What is the median mark?

Catherine says 'I am in the top half of the class.' What can you say about her mark?

Exercise

1 The ages of the people on a package holiday are as follows.

45	38	27	32	30	7	4	28	48	42
13	10	56	58	37	24	25	69	34	26
32	35	64	51	60	49	46	17	38	62
52	53	29	24	27	23	57	45	43	26
31	46	52	31	15	44	57	37		

(a) Copy and complete this stem-and leaf diagram.

(b) Rewrite the diagram with each row sorted into order.

(c) Use your table to find the median age of the people on the holiday.

4 | 5 represents age 45.

0	
1	
2	7
3	8 2 0
4	5
⋮	

2 Here are the times a group of students took to run 100 metres.

15.5	18.3	21.2	19.4	19.5	16.2	18.1	17.4
16.2	14.1	15.8	18.5	21.7	17.6	16.2	18.3
15.6	17.4	16.8	14.7	20.0	15.1	14.9	16.3
16.1	19.8	14.6	16.1	18.4	17.0	15.5	19.2
18.7	16.4	15.2	18.5	17.1	15.8	18.6	16.4
14.9	16.3	15.7	15.9	17.8			

(a) Copy and complete this stem-and-leaf diagram.

(b) Rewrite the diagram with each row sorted into order.

(c) Use your diagram to find the students' median time taken to run 100 metres.

15 | 5 represents 15.5 seconds.

14	
15	5
16	
17	
18	3
⋮	

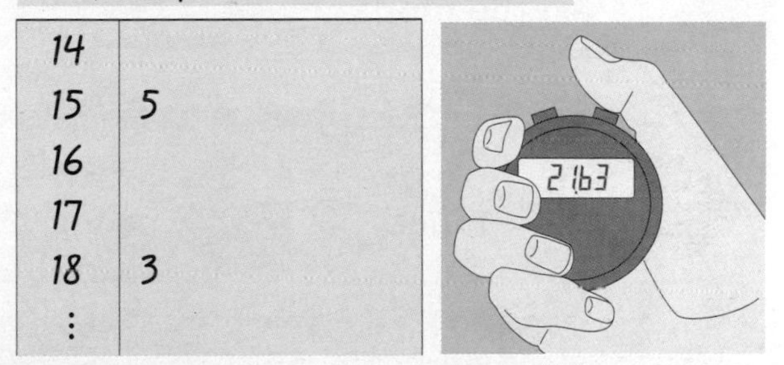

3 40 students took a practice examination.
These are their marks (out of 60) in each of the two papers.

Paper 1

44	52	31	49	35	27	43	56	22	51	44	32	40	17
58	53	26	34	46	31	28	58	14	24	39	50	43	47
55	41	49	33	15	26	37	34	49	46	30	53		

Paper 2

28	35	17	22	59	36	24	6	42	39	44	23	16	29
13	28	37	28	24	48	42	51	31	52	40	43	37	32
19	27	31	29	35	44	38	52	41	38	22	43		

(a) Make a sorted stem-and-leaf diagram for each set of results.

(b) Find the median mark for each paper.

(c) On the whole which paper do you think was harder?

The mode

Tessa wants to find out where the average family spends their holiday.

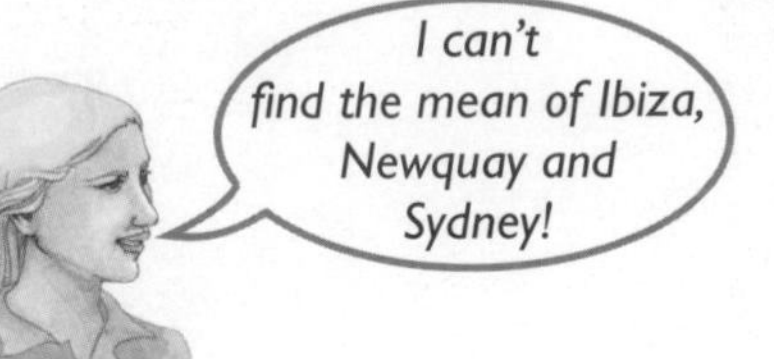

You cannot add up holiday resorts!
Instead, you find out the most popular destination.
The most popular, or most common, item is called the **mode**.

Look back to Joe's data about the crisps on page 130. What is the mode of the data from Joe's survey of 80 people?

The mode for grouped data

Look back to Mrs Patel's data on page 134. What is the mode of this set of data?

Is the mode useful in this case?

Mrs Patel's data can be written in a grouped frequency table:

Mark	20-29	30-39	40-49	50-59	60-69	70-79	80-89	90-99
Frequency	1	2	5	10	11	8	5	3

The most common group is 60–69.
This is called the **modal class**.

Task

Find out the number of pets owned by everyone in your class.
Work out the mean, the median and the mode for your data.

Which do you think is the most useful average for the number of pets: the mean, the median or the mode? Why?

The range

The **range** of a set of data is the difference between the biggest and the smallest data values.
The range is not an average.
It tells you how spread out the data are.

In Amy's pocket money survey on page 132, the biggest data value is £4.00 and the smallest is £2.50.
So the range is £4.00 − £2.50 = £1.50.

What is the range of Mrs Patel's data on page 134?

Exercise

1 A safari park keeps a record of the number of cars visiting each day during the month of August.

Here are the results.

354	296	398	462	257	331	346	103
471	229	337	358	406	511	379	253
317	352	460	488	425	350	399	176
286	216	339	318	376	269	414	

(a) Make a grouped tally chart to show these data.
Use the groups 100–149, 150–199, 200–249, etc.

(b) Draw a bar chart to illustrate these data.

(c) What is the modal class?

(d) What is the range of the data?

2 The maximum temperatures, in °C, are recorded in London and Edinburgh each day during the month of June one year.

June

London

21	22	21	24	23	21
20	18	18	19	23	23
24	26	27	25	24	24
23	24	22	22	21	22
24	23	23	22	21	20

Edinburgh

17	19	22	21	22	20
17	16	16	15	18	21
23	26	25	25	23	19
18	18	20	21	22	20
19	21	20	18	19	21

(a) Find the mode, the mean and the range for each city.

(b) Which city was warmer on the whole?

(c) Which city had more varied temperatures?

3 Alice, Ben and Charlotte are all different ages.

The range of their ages is 8 years.
The median of their ages is 7.
The mean of their ages is 9.

What are their ages?

Activity

Work with a friend.

Ask your friend to estimate 30 seconds while you keep the time with a stopwatch.

Record how many seconds actually passed.
Do this 5 times, then swap and have your friend time you while you estimate 30 seconds.

Use a suitable average to find out who was better at estimating 30 seconds.
Use the range to find out whose estimates were more consistent.

Finishing off

Now that you have finished this chapter you should be able to:

- find the mean, median and mode of a set of data
- use a frequency table to find the mean of a large set of data
- use a stem-and-leaf diagram to find the median of a large set of data
- find the modal class of a set of grouped data
- find the range of a set of data.

Review exercise

1 Jim is a milkman.
He delivers milk to 32 houses in Chestnut Avenue.
Here are the numbers of pints of milk he delivers to each house one morning.

2	4	3	3	5	1	1	4
3	5	4	4	1	2	2	3
2	2	3	1	2	1	4	3
4	1	2	2	3	2	4	2

(a) Make a tally chart to illustrate these data.
(b) Draw a bar chart.
(c) Find the mode.
(d) Find the mean.
(e) Find the median.

2 These are the times, in seconds, for a group of students to swim one length of the swimming pool.

Boys	52	41	72	56	43	37	75	63	52	47		
Girls	46	51	32	64	66	57	84	71	56	49	68	63

(a) Find the mean time for **(i)** the boys **(ii)** the girls.
(b) Who swam faster on average, the boys or the girls?
(c) Find the range for **(i)** the boys **(ii)** the girls?
(d) Which times were more spread out, the boys or the girls?

3 Lisa carries out a survey about reading.
She asks 50 Year 9 students how many books they read last month.
Here are her results:

Number of books	0	1	2	3	4	5
Frequency	12	16	13	5	3	1

Find the mean number of books read.

4 55 children take a spelling test consisting of 20 words.
Here are the scores for the boys and the girls.

Boys	15	12	16	19	18	11	14	13
	12	16	14	18	17	15	9	12
	11	16	15	13	17	12	10	8
Girls	14	18	16	11	20	18	5	12
	10	18	19	14	17	13	14	6
	8	20	18	13	19	16	17	16
	14				15			

(a) Find the median mark for **(i)** the boys **(ii)** the girls.
(b) Find the range of marks for **(i)** the boys **(ii)** the girls.
(c) Who did better, the boys or the girls?
(d) Whose results were more consistent?

5 Hassan is collecting data on the heights of adult men for a clothing company.
He measures the heights, in centimetres, of 50 adult men.

168	179	164	182	175	178	185	163	165	181
178	175	172	179	173	162	177	176	184	191
169	170	172	188	180	177	167	156	194	168
174	178	186	174	166	165	159	173	185	162
169	170	176	161	182	173	184	180	174	163

(a) Copy and complete Hassan's stem-and leaf diagram.

16 | 8 represents 168 cm.

15*	
15	
16*	4
16	8
17*	
17	9
18*	
18	
19*	
19	

(b) Rewrite the table with each row sorted into order.
(c) Use your table to find the median height of the men Hassan measures.
(d) Use your table to make a grouped frequency table using groups 155–159, 160–164, etc.
(e) What is the modal class?

16 Index notation

Powers of ten

How do you write 10 million and 1 billion as powers of 10?

RECORD BREAKERS

£10 million paid for racehorse

Largest TV audience
Each week over 1 billion viewers world-wide watch 'Bay View'

Task

The table shows the powers of 10 from 10 to 10 000.

Look at the figures in the middle column.

What operation will take you from one number to the number below it?

Number	In figures	As a power of 10
Ten thousand	10 000	10^4
Thousand	1000	10^3
Hundred	100	10^2
Ten	10	10^1

The table is continued.
Copy the table and continue the patterns to complete it.

One	1	10^0
One tenth	0.1	10^{-1}
One hundredth	0.01	10^{-2}

Explain why one millionth can be written as 10^{-6}?

Name all the powers of ten from 10^{-9} to 10^9.

Multiplying and dividing by powers of 10

Work out 7×10^1, 4×10^3 and 2.1×10^2.

Explain how to multiply by a positive power of 10.
What do the words 'positive' and 'negative' mean in mathematics?

Write the answers to the following as decimals.

$6 \div 100 \qquad 6 \times \frac{1}{100} \qquad 6 \times 0.01 \qquad 6 \times 10^{-2}$

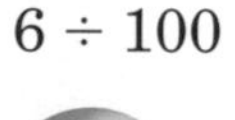

They all have the same answer. Why?

Do you agree with Mark and Jane?

What is meant by 10^{-3}?
Write 4×10^{-3} as a decimal.

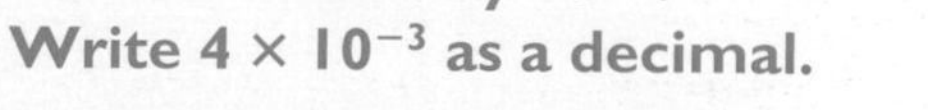

Explain how to multiply by a negative power of 10.

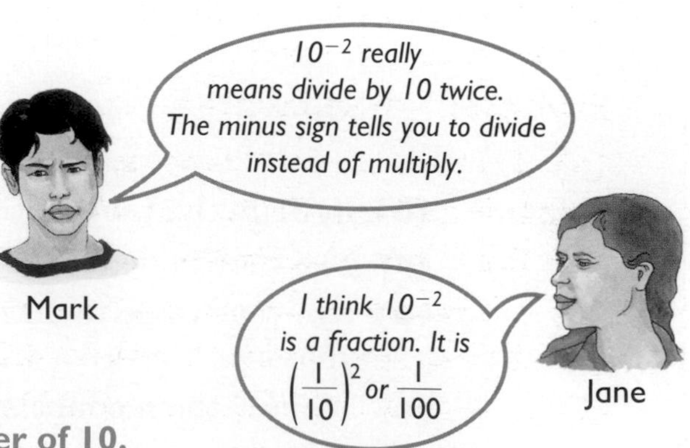

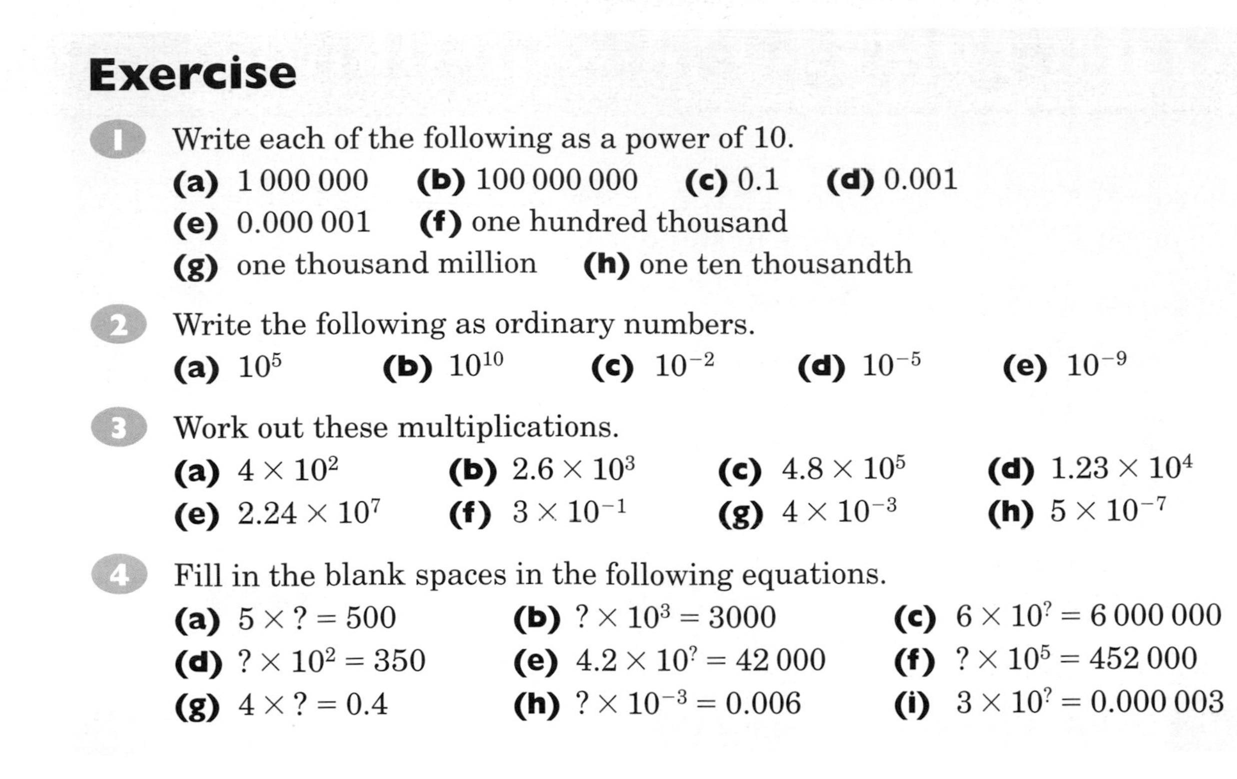

Exercise

1 Write each of the following as a power of 10.

(a) 1 000 000 **(b)** 100 000 000 **(c)** 0.1 **(d)** 0.001
(e) 0.000 001 **(f)** one hundred thousand
(g) one thousand million **(h)** one ten thousandth

2 Write the following as ordinary numbers.

(a) 10^5 **(b)** 10^{10} **(c)** 10^{-2} **(d)** 10^{-5} **(e)** 10^{-9}

3 Work out these multiplications.

(a) 4×10^2 **(b)** 2.6×10^3 **(c)** 4.8×10^5 **(d)** 1.23×10^4
(e) 2.24×10^7 **(f)** 3×10^{-1} **(g)** 4×10^{-3} **(h)** 5×10^{-7}

4 Fill in the blank spaces in the following equations.

(a) $5 \times ? = 500$ **(b)** $? \times 10^3 = 3000$ **(c)** $6 \times 10^? = 6\,000\,000$
(d) $? \times 10^2 = 350$ **(e)** $4.2 \times 10^? = 42\,000$ **(f)** $? \times 10^5 = 452\,000$
(g) $4 \times ? = 0.4$ **(h)** $? \times 10^{-3} = 0.006$ **(i)** $3 \times 10^? = 0.000\,003$

Activity

In the metric system there are special names for some powers of ten.

*A **giga**byte is 10^9 bytes (bits of computer data)*

1 Copy and complete the table to remind you of these.

Some new names have been included for you to learn.

*A **centi**metre is a hundredth of a metre.*

Name (or prefix)	Power of 10
tera	10^{12}
giga	10^9
mega	
	10^3
hecta	
	10^1
	10^{-1}
centi	
	10^{-3}
	10^{-6}
nano	
pico	10^{-12}

2 Use powers of ten to write the following measures in the units indicated in brackets.

Example: 4 kilometres is 4×10^3 metres.

(a) The annual world production of sugar cane is 1 megatonne. (tonnes)
(b) Infrared light has a wavelength of 100 micrometres. (metres)
(c) Blue light has a wavelength of 400 nanometres. (metres)
(d) 1 kilogram of nuclear fuel will release 70 terajoules of energy. (joules)
(e) A grain of sand has a mass of 0.1 milligrams. (grams)
(f) A stroboscope flashes every 12 microseconds. (seconds)

Writing large and small numbers

Mercury is 5.79×10^7 kilometres away from the sun.
The number 5.79×10^7 is written in **standard form**.

? Explain why 5.79×10^7 is the same as 57 900 000.

The probability of winning the lottery jackpot is $\frac{1}{14 \text{ million}}$ or about 7.1×10^{-8}.

7.1×10^{-8} is written in **standard form**.

*This is also known as **standard index form**.*

? Write 7.1×10^{-8} as an ordinary number.

> A number in standard form is written as
> (a number between 1 and 10) × (a power of 10)

**? 32 000 000 people watched the 1966 World Cup final.
In standard form this is 3.2 × *a power of 10*. What power of 10 is needed?**

**? The radius of a human hair is 0.053 mm.
Write this as 5.3 × *a power of 10*. Why do you need a negative power?**

Ordering numbers in standard form

**? Which is smaller, 3.46×10^7 or 9.8×10^6?
Explain why you only have to look at the powers of 10 to answer this.**

**? Which is smaller, 2.56×10^8 or 3.1×10^8?
Explain how you made your choice.**

? Explain why 4.2×10^{-5} is smaller than 3×10^{-4}.

Task

1 Write these numbers in order starting with the smallest.
3.4×10^4, 563 000, 7.4×10^6, 820 000, 4.56×10^5

2 These numbers are all smaller than 1. Write them in order starting with the smallest.
0.000 004 1, 6.7×10^{-6}, 0.000 002 47, 3.78×10^{-5}, 4×10^{-6}

3 Write down a number that lies between each of these pairs of numbers.
(a) 2.3×10^4 and 2.3×10^6 **(b)** 1.2×10^3 and 2.0×10^3 **(c)** 4×10^{-3} and 3×10^{-4}

? Do you agree with Beverly?

? What does the number part tell you?

When a number is written in standard form the power of 10 tells you how big it is.

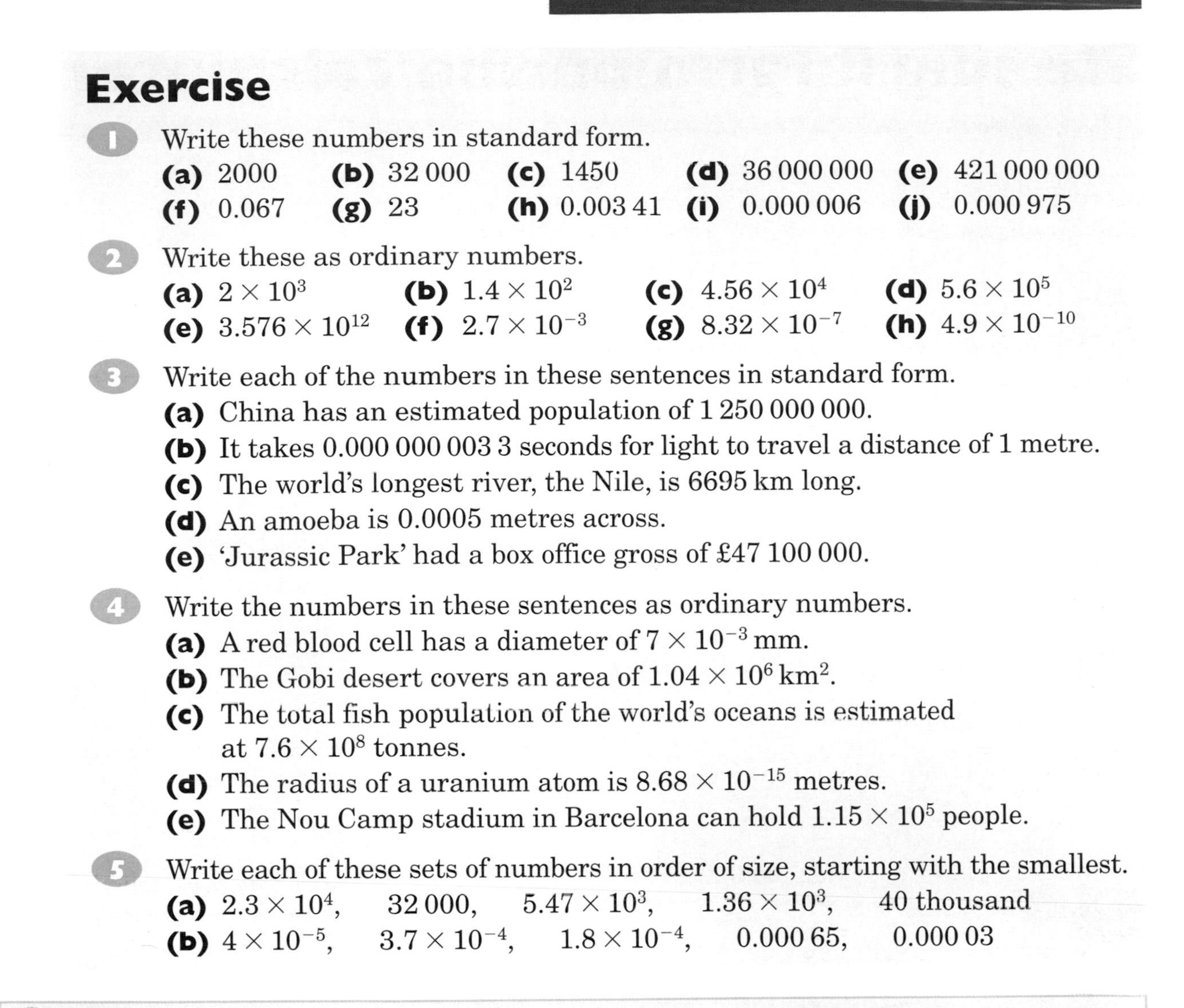

Exercise

1 Write these numbers in standard form.

(a) 2000 **(b)** 32 000 **(c)** 1450 **(d)** 36 000 000 **(e)** 421 000 000
(f) 0.067 **(g)** 23 **(h)** 0.003 41 **(i)** 0.000 006 **(j)** 0.000 975

2 Write these as ordinary numbers.

(a) 2×10^3 **(b)** 1.4×10^2 **(c)** 4.56×10^4 **(d)** 5.6×10^5
(e) 3.576×10^{12} **(f)** 2.7×10^{-3} **(g)** 8.32×10^{-7} **(h)** 4.9×10^{-10}

3 Write each of the numbers in these sentences in standard form.

(a) China has an estimated population of 1 250 000 000.
(b) It takes 0.000 000 003 3 seconds for light to travel a distance of 1 metre.
(c) The world's longest river, the Nile, is 6695 km long.
(d) An amoeba is 0.0005 metres across.
(e) 'Jurassic Park' had a box office gross of £47 100 000.

4 Write the numbers in these sentences as ordinary numbers.

(a) A red blood cell has a diameter of 7×10^{-3} mm.
(b) The Gobi desert covers an area of 1.04×10^6 km^2.
(c) The total fish population of the world's oceans is estimated at 7.6×10^8 tonnes.
(d) The radius of a uranium atom is 8.68×10^{-15} metres.
(e) The Nou Camp stadium in Barcelona can hold 1.15×10^5 people.

5 Write each of these sets of numbers in order of size, starting with the smallest.

(a) 2.3×10^4, 32 000, 5.47×10^3, 1.36×10^3, 40 thousand
(b) 4×10^{-5}, 3.7×10^{-4}, 1.8×10^{-4}, 0.000 65, 0.000 03

Investigation

How do you know that the number 3.2×10^3 is about 3 thousand?

What power of 10 do you use to write $\frac{4}{1000}$ in standard form?

What power of 10 do you use to write 1.2 billion in standard form?

In the number 4.12×10^6, the first digit is worth 4 million or 4 000 000.
The numbers in the table are written in standard form.
Copy the table and fill in the value of the first digit for each number.

Number	Value of first digit	Number	Value of first digit
6.1×10^4		2×10^{-2}	
3.62×10^4		1.46×10^{-2}	
2.9×10^7		3×10^{-4}	
1.352×10^7		6.2×10^{-4}	
4.5×10^9		5×10^{-6}	
1.236×10^9		3.21×10^{-6}	

? **A number is written in standard form. Part of it is 10^8.
What can you say about the size of that number?**

Standard form on the calculator

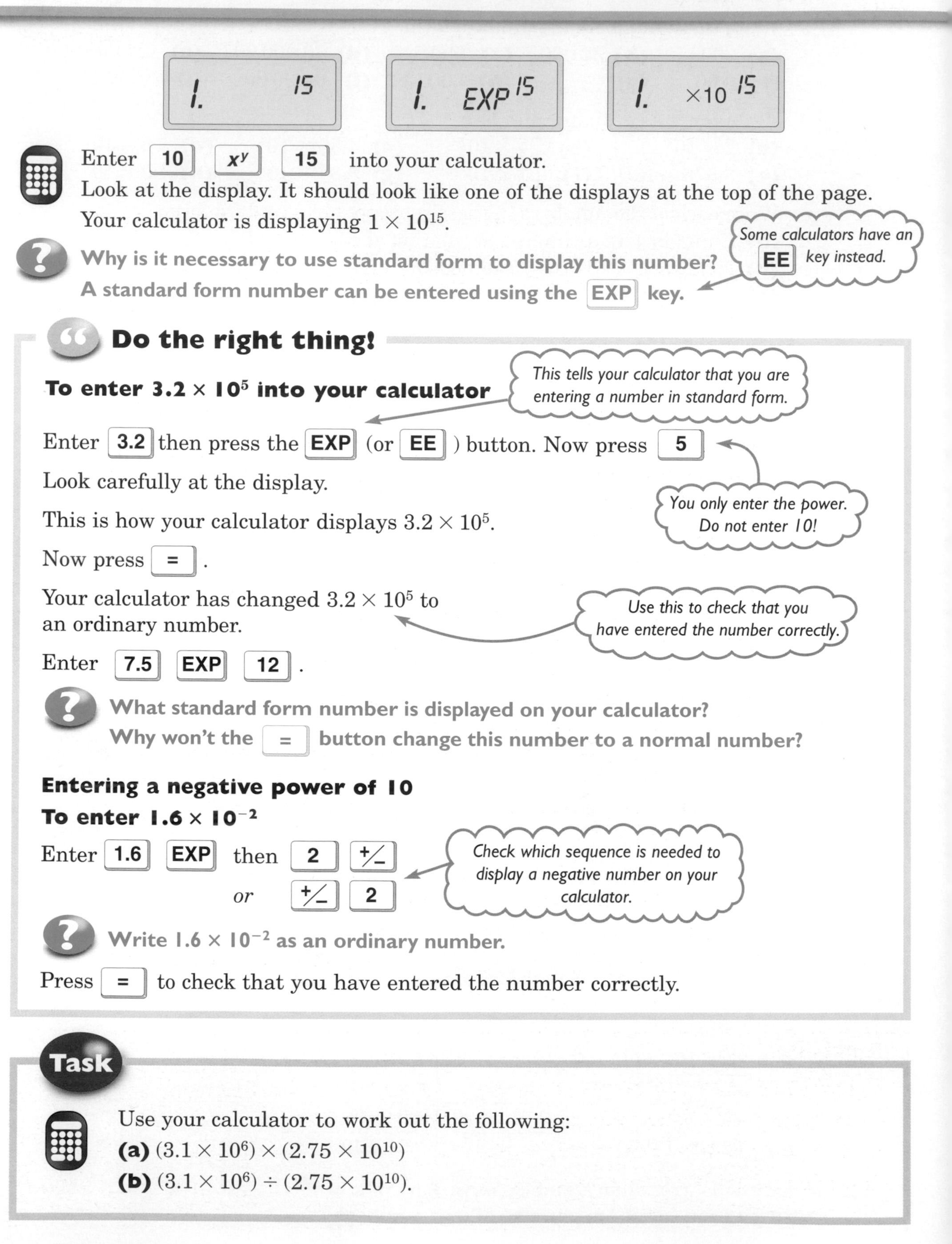

Enter 10 x^y 15 into your calculator.
Look at the display. It should look like one of the displays at the top of the page.
Your calculator is displaying 1×10^{15}.

Why is it necessary to use standard form to display this number?
A standard form number can be entered using the EXP key.

Some calculators have an EE key instead.

Do the right thing!

To enter 3.2×10^5 into your calculator

This tells your calculator that you are entering a number in standard form.

Enter 3.2 then press the EXP (or EE) button. Now press 5

You only enter the power. Do not enter 10!

Look carefully at the display.

This is how your calculator displays 3.2×10^5.

Now press = .

Your calculator has changed 3.2×10^5 to an ordinary number.

Use this to check that you have entered the number correctly.

Enter 7.5 EXP 12 .

What standard form number is displayed on your calculator?
Why won't the = button change this number to a normal number?

Entering a negative power of 10

To enter 1.6×10^{-2}

Enter 1.6 EXP then 2 +/–
or +/– 2

Check which sequence is needed to display a negative number on your calculator.

Write 1.6×10^{-2} as an ordinary number.

Press = to check that you have entered the number correctly.

Task

Use your calculator to work out the following:

(a) $(3.1 \times 10^6) \times (2.75 \times 10^{10})$

(b) $(3.1 \times 10^6) \div (2.75 \times 10^{10})$.

Exercise

1 **(a)** Write down the standard form numbers shown on these calculator displays.

4. 5 | 6.2 7 | 1.4 -6

(b) Enter each number into your calculator.

2 For each of the following standard form numbers
(i) write it as an ordinary number
(ii) display it in standard form on your calculator, then
(iii) use the = button to convert the display to an ordinary number.
Check your answers by comparing **(i)** and **(ii)**.

(a) 2×10^3 **(b)** 1.4×10^6 **(c)** 7.56×10^4 **(d)** 1.234×10^5
(e) 5×10^{-2} **(f)** 3.7×10^{-3} **(g)** 4.76×10^{-2} **(h)** 1.34×10^{-5}

3 Use your calculator to work out the following.

(a) $(7 \times 10^9) \times (6.4 \times 10^8)$ **(b)** $(7 \times 10^9) \div (6.4 \times 10^8)$
(c) $(7 \times 10^9) + (6.4 \times 10^8)$ **(d)** $(7 \times 10^9) - (6.4 \times 10^8)$

4 These problems can be solved either by multiplying or by dividing. Choose the correct operation for each one and then answer the question, using your calculator.

(a) A mouse weighs 1.5×10^{-2} kg.
An owl eats 3 mice in one night. What weight is this?

(b) The speed of sound is 3.3×10^2 metres per second.
How far does sound travel in a minute?

(c) A grain of salt weighs 2×10^{-5} grams.
How many grains of salt are there in a 750 gram packet?

(d) A packet of 500 sheets of paper is 55 mm thick.
How thick is each sheet of paper?

(e) The average number of clover leaves in a square metre of lawn is 1.5×10^3.
Estimate the number of clover leaves in a park with 5×10^4 m^2 of lawns.

5 Find the difference between these pairs of numbers.
Give your answers in standard form.

(a) 1.6×10^4 and 3.1×10^5 **(b)** 4×10^{-2} and 3×10^{-1}
(c) 6.7×10^5 and 8.2×10^5 **(d)** 4.65×10^{19} and 3.7×10^{20}

Activity

Light travels at 3×10^8 metres per second.
A light year is the distance that light will travel in one year.
Find the length, in metres, of a light year.
Our galaxy is 10^5 light years in diameter.
How many metres is this?

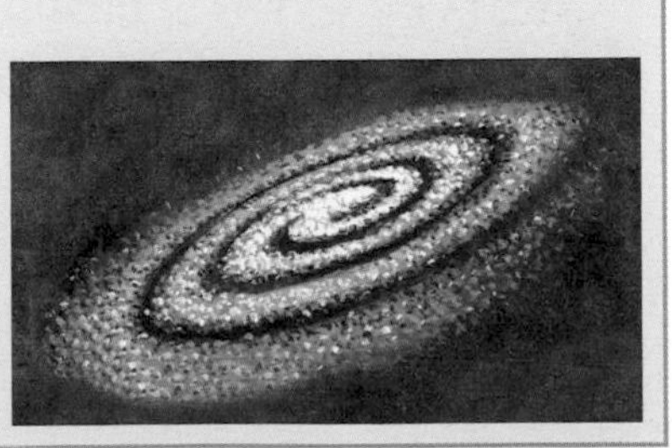

Finishing off

Now that you have finished this chapter you should be able to:

- understand numbers when they are written as powers of 10
- understand the meaning of the prefixes used in the metric system
- write and interpret large and small numbers in standard form
- use your calculator to deal with numbers in standard form.

Review exercise

1 Write the following numbers in standard form.

(a) 367 **(b)** 25 000 **(c)** 34 000 000 **(d)** 1 230 000
(e) 0.06 **(f)** 0.003 76 **(g)** 0.000 014 **(h)** 0.000 321 4

2 Write the following as ordinary numbers.

(a) 4×10^2 **(b)** 6.21×10^5 **(c)** 1×10^6 **(d)** 6.5×10^{10}
(e) 3×10^{-1} **(f)** 4.2×10^{-3} **(g)** 1×10^{-4} **(h)** 5.32×10^{-9}

3 Write the numbers below in (correct) standard form.

(a) 36×10^2 **(b)** 0.27×10^3 **(c)** 123×10^4 **(d)** 0.8×10^4
(e) 45 million **(f)** 120 thousand **(g)** 2.4 billion **(h)** 0.6 million
(i) $\frac{2}{1000}$ **(j)** 25×10^{-5} **(k)** 0.4×10^{-2} **(l)** 0.035×10^{-4}

4 Write the following numbers as powers of 10.

(a) 1 million **(b)** one hundred thousand **(c)** 1 billion
(d) one thousand million **(e)** 1 thousandth **(f)** $\frac{1}{1\,000\,000}$

5 Choose the larger number from each of these pairs of numbers.

(a) 2.3×10^7 and 1.9×10^8 **(b)** 5.7×10^6 and 4.31×10^6
(c) 3×10^{-4} and 2×10^{-5} **(d)** 1.23×10^{-6} and 4.6×10^{-7}
(e) 2 570 000 and 3.6×10^5 **(f)** 7.8 billion and 1.9×10^{10}
(g) 0.004 89 and 2.3×10^{-4} **(h)** 4.7×10^{-6} and 0.000 006 7

6 Write a number that lies between each of the following pairs of numbers. (Write your answer in standard form.)

(a) 4×10^4 and 5×10^6 **(b)** 1.23×10^6 and 4.6×10^6
(c) 6×10^{-3} and 5.8×10^{-5} **(d)** 6.3×10^{-4} and 2.9×10^{-4}

7 Write the standard form numbers in these sentences as ordinary numbers.

(a) The length of a human chromosome is 5×10^{-6} metres.
(b) An oil film is 5×10^{-7} cm thick.
(c) The CN Tower in Toronto, Canada, is 5.53×10^2 m tall.
(d) The mass of an electron is 9.11×10^{-31} kg.
(e) The earth has a radius of 6.37×10^6 metres.

8 Write these very large or very small numbers in standard form.

(a) It is estimated that by the year 2005 there will be 37 million mobile phone users in the UK.

(b) The distance between the earth and the moon is 239 000 miles.

(c) A £5 note is 0.000 22 metres thick.

(d) Quartz fibre has a diameter of 0.000 001 metres.

(e) The universe is estimated to be 13 600 000 000 years old.

9 Change these measurements into the units indicated.

(a) 2 kilometres (metres) **(b)** 5 megatons (tonnes)

(c) 8 nanoseconds (seconds) **(d)** 4 picometres (metres)

(d) 8 micrograms (grams) **(f)** 3 terajoules (joules)

10 The speed of light is 3×10^8 metres per second.
The distance from the earth to the sun is 1.53×10^8 km.
How long does it take light from the sun to reach the earth?

11 The speed of sound is 3.3×10^2 metres per second.
How long does it take for sound to travel a distance of 5000 metres?

Investigation

1 Explain how to work out 2000×300 in your head.
Write your answer in standard form.

2 Write the answer to $(2 \times 10^3) \times (3 \times 10^2)$ in standard form.

Look carefully at your answer.

? **How is the number worked out? How is the power of 10 worked out?**

3 Work out

(a) 400×2000 and $(4 \times 10^2) \times (2 \times 10^3)$

(b) $12\,000 \times 300$ and $(1.2 \times 10^4) \times (3 \times 10^2)$

? **Explain an easy way to multiply numbers when they are written in standard form.**

4 Use your method to work out $(3 \times 10^2) \times (4 \times 10^3)$.

Write your answer in standard form.

Activity

The diagram is known as an *ouraborus*.

It shows the size of everything from the smallest particle to the known universe.

Draw a number line showing powers of 10 from 10^{-20} to 10^{20}.

Use your line to show the size of very small and very large objects.

10^{25} cm
10^{-20} cm
10^{20} cm
10^{-15} cm
10^{15} cm
10^{-10} cm
10^{10} cm
10^{-5} cm
10^{5} cm

17 Formulae

Barry asks some of his friends round for a meal.
He decides to cook a turkey.
His cookery book gives these instructions.

 **How long will it take to cook a 6 kg turkey?**

Barry writes a formula for the time.

$T = 40 + 30 \times W + 30$

$T = 30W + 70$

 **What do T and W stand for?**

Barry writes

Substitute $W = 6$ into the formula:

$T = 30 \times 6 + 70$

$T = 180 + 70$

$T = 250$

 **Is this the same as your answer?**

Tina cooks a turkey for 400 minutes.

 **How heavy is it?**

You need to solve the **equation** $400 = 30W + 70$ to find W.

 **What is the difference between an equation and a formula?**

Task

The total cost for hiring a car is £50 per day plus 20p for every mile travelled.

(a) Write this as a formula in the simplest way.

(b) Vronnie hires a car for 3 days and travels 150 miles.
How much does it cost?

(c) Sam hires a car for 2 days. His bill is £180.
Write an equation to find how far he drives.

(d) Does the advertisement tell you the whole truth?
Design a more honest poster.

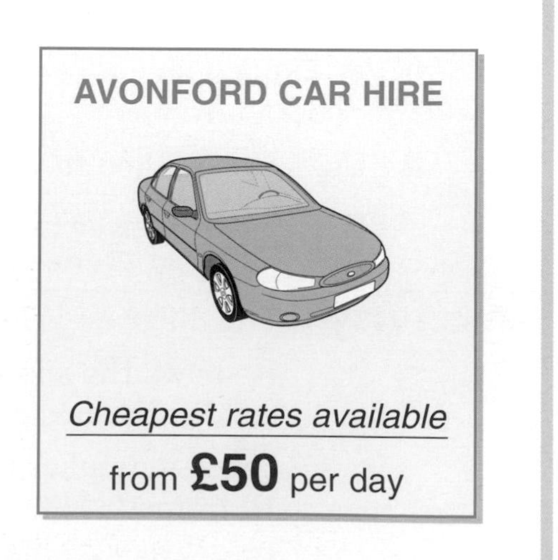

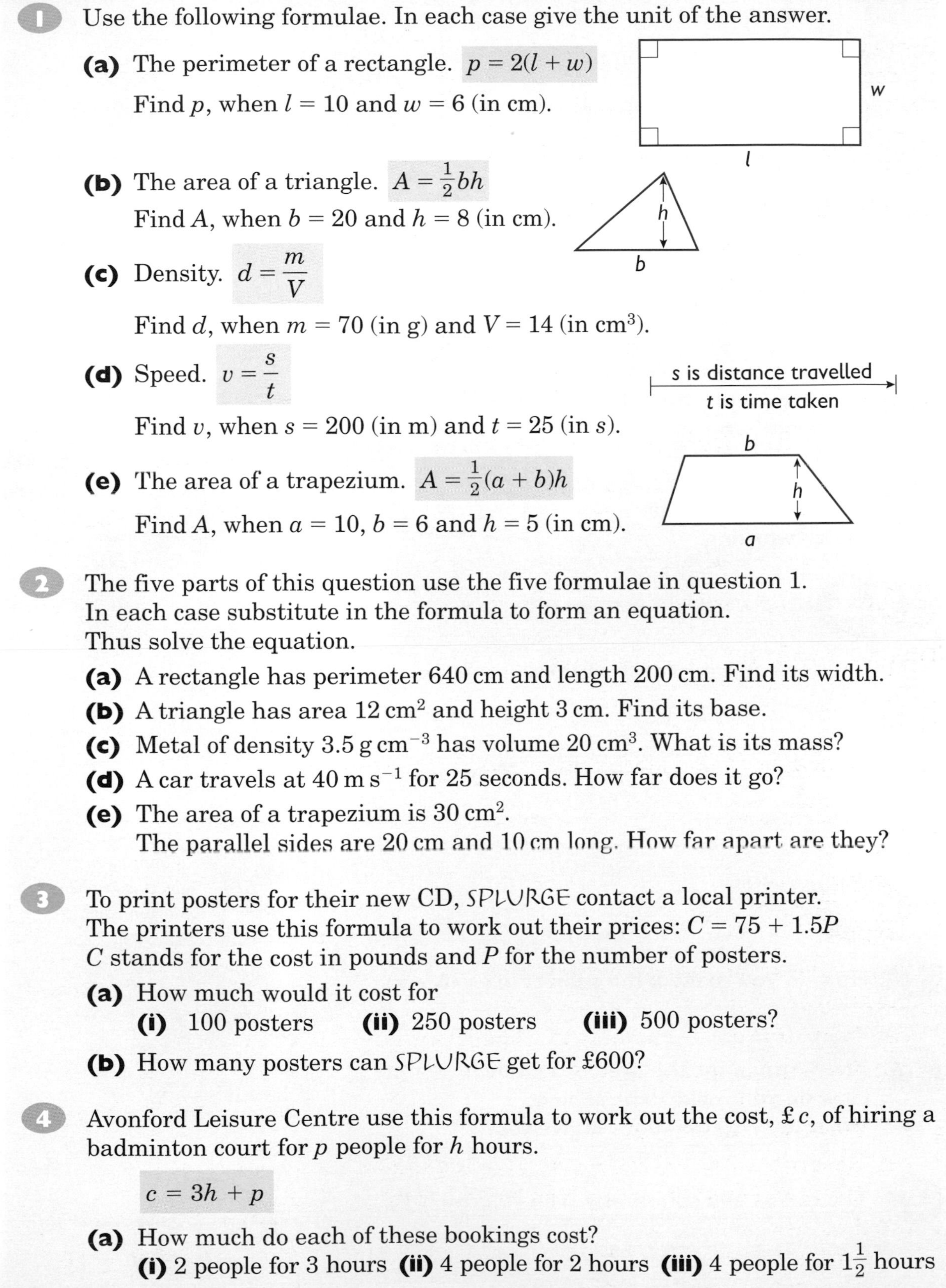

Exercise

1 Use the following formulae. In each case give the unit of the answer.

(a) The perimeter of a rectangle. $p = 2(l + w)$

Find p, when $l = 10$ and $w = 6$ (in cm).

(b) The area of a triangle. $A = \frac{1}{2}bh$

Find A, when $b = 20$ and $h = 8$ (in cm).

(c) Density. $d = \frac{m}{V}$

Find d, when $m = 70$ (in g) and $V = 14$ (in cm^3).

(d) Speed. $v = \frac{s}{t}$

Find v, when $s = 200$ (in m) and $t = 25$ (in s).

(e) The area of a trapezium. $A = \frac{1}{2}(a + b)h$

Find A, when $a = 10$, $b = 6$ and $h = 5$ (in cm).

2 The five parts of this question use the five formulae in question 1.
In each case substitute in the formula to form an equation.
Thus solve the equation.

(a) A rectangle has perimeter 640 cm and length 200 cm. Find its width.

(b) A triangle has area 12 cm^2 and height 3 cm. Find its base.

(c) Metal of density 3.5 g cm^{-3} has volume 20 cm^3. What is its mass?

(d) A car travels at 40 m s^{-1} for 25 seconds. How far does it go?

(e) The area of a trapezium is 30 cm^2.
The parallel sides are 20 cm and 10 cm long. How far apart are they?

3 To print posters for their new CD, SPLURGE contact a local printer.
The printers use this formula to work out their prices: $C = 75 + 1.5P$
C stands for the cost in pounds and P for the number of posters.

(a) How much would it cost for
(i) 100 posters **(ii)** 250 posters **(iii)** 500 posters?

(b) How many posters can SPLURGE get for £600?

4 Avonford Leisure Centre use this formula to work out the cost, £c, of hiring a badminton court for p people for h hours.

$$c = 3h + p$$

(a) How much do each of these bookings cost?
(i) 2 people for 3 hours **(ii)** 4 people for 2 hours **(iii)** 4 people for $1\frac{1}{2}$ hours

(b) A group of 4 people pay £10. How long do they have the court for?

Rearranging formulae

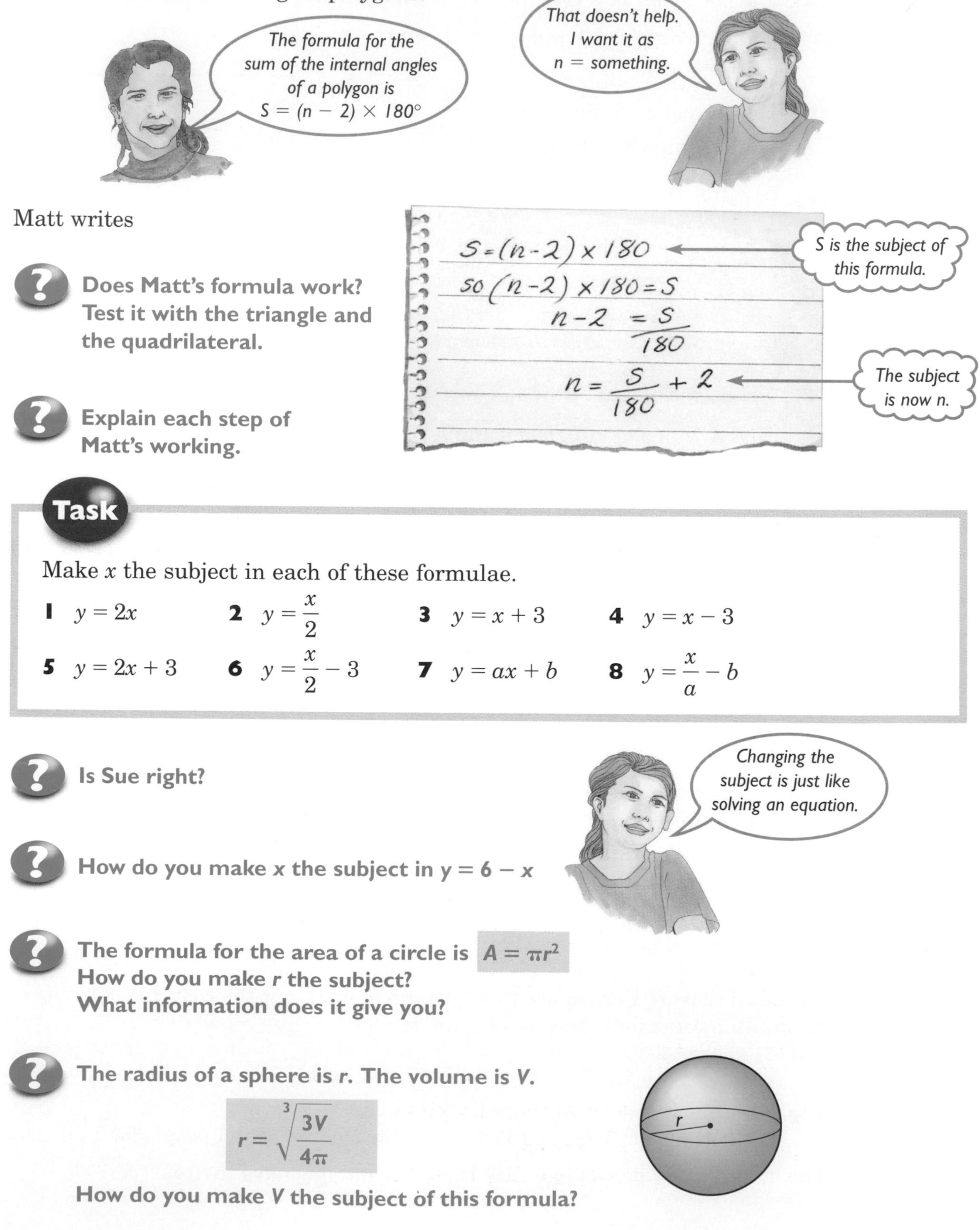

Matt and Sue are working on polygons.

Matt writes

Does Matt's formula work? Test it with the triangle and the quadrilateral.

Explain each step of Matt's working.

Task

Make x the subject in each of these formulae.

1 $y = 2x$ **2** $y = \frac{x}{2}$ **3** $y = x + 3$ **4** $y = x - 3$

5 $y = 2x + 3$ **6** $y = \frac{x}{2} - 3$ **7** $y = ax + b$ **8** $y = \frac{x}{a} - b$

Is Sue right?

How do you make x the subject in $y = 6 - x$

The formula for the area of a circle is $A = \pi r^2$
How do you make r the subject?
What information does it give you?

The radius of a sphere is r. The volume is V.

$$r = \sqrt[3]{\frac{3V}{4\pi}}$$

How do you make V the subject of this formula?

Exercise

1 For each of these formulae, change the subject to the red letter.

(a) $p = 2(l + w)$
(b) $A = \frac{1}{2}bh$
(c) $A = 180 - B - C$
(d) $T = 30W + 70$
(e) $d = \frac{m}{V}$
(f) $v = \frac{s}{t}$
(g) $V = IR$
(h) $m = \frac{a + b + c + d}{4}$
(i) $c = 2\pi r$
(j) $l = \sqrt{A}$
(k) $A = \frac{1}{2}(a + b)h$
(l) $V = \pi r^2 h$

2 Avonford Gas Supply Company uses this formula to calculate gas bills.

$$C = 1.5U + 2500$$

Standing charge.

C = cost (pence).
U = number of units used.

(a) Jill uses 1500 units.
What is the cost in pounds and pence?

(b) Rearrange the formula to make U the subject.

(c) How many units are used when the bill is

(i) £43 **(ii)** £52.75 **(iii)** £56.50?

3 Zeynah is having a party at the Royal Avonford Hotel.

Royal Avonford Hotel

Superior function room £150
plus £5 per person

DJ and Disco
£2.00 per person extra

(a) Look at their prices.
Write down the formula for the cost £C for p people (including the disco).

(b) Change the formula to make p the subject.

(c) Zeynah has £400 to spend.
How many people can she invite?

Finishing off

Now that you have finished this chapter you should be able to:

- construct a formula from given information
- substitute values into a formula.
- rearrange a formula

Review exercise

1 The interior angle $x°$ of a regular polygon depends on the number of sides of the polygon.

Interior angle.

The formula is

$$x = 180 - \frac{360}{n}$$

A regular pentagon

What is the interior angle of

(a) a regular pentagon **(b)** a regular octagon?

2 Rearrange each of these formulae to make the red letter the subject of the formula.

(a) $y = 5 + x$ **(b)** $y = 5 - x$ **(c)** $v + f = e + 2$ **(d)** $r + f = e + 2$

(e) $a = 3b$ **(f)** $c = \frac{d}{4}$ **(g)** $y = 3x + 6$ **(h)** $v = u + at$

(i) $s = qt - 4$ **(j)** $s = ut$ **(k)** $x = y + z$ **(l)** $a = 3b + c$

3 An electrician charges £15 per hour plus a call out charge of £25.

(a) Write a formula for his total charge, £c, for a call out lasting h hours.

(b) Calculate c when **(i)** $h = 2$ **(ii)** $h = 3\frac{1}{2}$.

(c) Rearrange your formula to make h the subject.

(d) How many hours did he work if his total charge is

(i) £115 **(ii)** £145 **(iii)** £77.50.

4 The pentagon shown has been split into triangles by drawing diagonals from one corner.
In this case there are 5 outside sides and 3 triangles.

Experiment with other polygons and find a formula that connects the number of outside sides, n, with the number of triangles, t.

(a) Write the formula

(i) with t as the subject **(ii)** with n as the subject.

(b) How many triangles do you get from a 10-sided polygon (a decagon)?

(c) How many sides does a polygon have if it can be split into 10 triangles?

5 Avonford Caterers are advertising their Home Buffet service.
Their rates are shown in their brochure.

(a) Write down a formula to work out the cost of a standard buffet with wine and china plates.
Let c stand for the cost in pounds and p for the number of people.

(b) Rearrange this formula to make p the subject.

(c) How many people can you invite for £425?

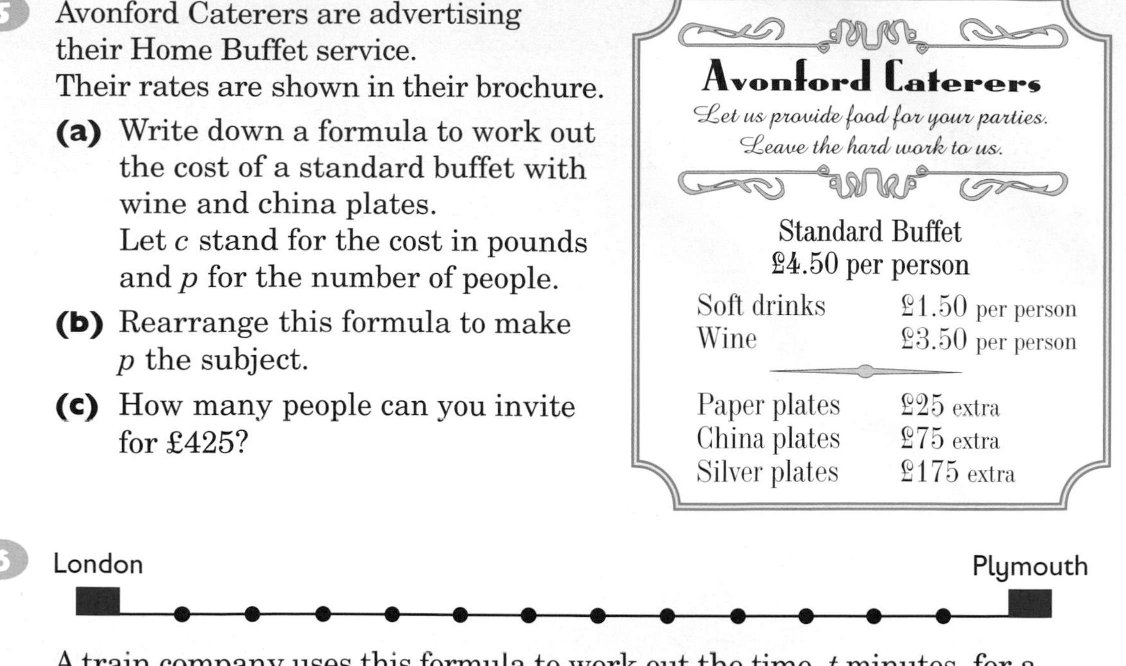

6 London ——— Plymouth

A train company uses this formula to work out the time, t minutes, for a rail journey from London to Plymouth.
The number of station stops along the way is s.

$$t = 165 + 5s$$

(a) Find the times of trains stopping at
(i) all 12 stations along the way
(ii) no stations between London and Plymouth
(iii) Reading, Taunton and Exeter only.

(b) A train takes 3 hours and 10 minutes on the journey.
How many stations does it stop at along the way?

Investigation

This diagram shows a network with nodes, arcs and regions.
Make up some more networks on your own.

(a) The table has been filled in for this network.
Copy and complete the table for your networks.

Nodes (N)	Arcs (A)	Regions (R)
4	6	4

Arc.
Node.
Region.
The outside is also a region.

(b) Find a formula giving R in terms of N and A.

(c) Rearrange your formula to make the subject **(i)** N **(ii)** A.

18 Accuracy

Robert works out the circumference, C, of 3 different circles.
He rounds all his answers to the nearest whole number.

(a) 340 cm

$C = \pi \times 340$ cm
$= 1068.141502$ cm
$= 1068$ cm (nearest whole number)

(b) 4·2 cm

$C = \pi \times 4.2$ cm
$= 13.19468915$ cm
$= 13$ cm (nearest whole number)

(c) 0·04 m

$C = \pi \times 0.04$ m
$= 0.125663706$ m
$= 0$ m (nearest whole number)

Round each of Robert's answers to a more suitable degree of accuracy.

Significant figures

Numbers can also be rounded to a number of **significant figures**.
The first (or most) significant figure is underlined in each of the following numbers.

$\underline{4}136$, $\underline{1}2.8$, $0.\underline{5}23$, $0.000\,0\underline{6}3\,2$, $\underline{4}\,200\,000$

Is the first significant figure always the first figure?

In each of the following numbers the **3rd** significant figure is underlined.

$43\,\underline{2}16$, $0.65\underline{4}\,2$, $10\underline{2}.3$, $0.000\,56\underline{3}\,72$, $86\underline{0}.32$, $64\underline{5}\,100$

Look carefully at the 0s in the numbers above.
When are the 0s not counted as significant figures? When are they counted?

Task

Write a sentence like Robert's to describe the significance of the figure 6 in each of these numbers.

6103, 32.6, 86.12, 106.2, 0.634, 0.010 61, 1 630 000

In which of the numbers is 6 the third significant figure.
Does it have the same value in each number?

Look at these two examples of rounding.
Explain what has happened.

679 456 = 679 000 (correct to 3 sig. figs.)
0.002 7 = 0.003 (1 s.f.)

Sometimes you have to be careful. In the next two examples, the answers don't look quite right at first glance.

3.796 = 3.80 (3 s.f.)

Why has the 0 been added?

4973 = 5000 (2 s.f.)

The first 0 is significant. Why?

Exercise

1 Round the following numbers to the nearest 10.

(a) 254 **(b)** 1785 **(c)** 21.35 **(d)** 103.67 **(e)** 458 361

2 Write the following numbers correct to 1 decimal place.

(a) 34.43 **(b)** 12.372 **(c)** 1.567 **(d)** 0.432 **(e)** 0.256

3 Copy the following numbers and underline the 2nd significant figure in each number.

(a) 45.3 **(b)** 127 **(c)** 0.635 1 **(d)** 0.000 845 **(e)** 587 210

4 Write each of the numbers in question 3 correct to 2 significant figures.

5 Write the following numbers correct to the number of significant figures given in the bracket.

(a) 18.34 (2) **(b)** 41.359 (3) **(c)** 1246 (3) **(d)** 0.015 (1)
(e) 460 (1) **(f)** 4986 (2) **(g)** 0.020 4 (2) **(h)** 0.109 6 (3)

6 All the statements below contain a mistake.
In each case
(i) rewrite the statement correctly **(ii)** explain the mistake.

(a) 354 = 35 correct to 2 significant figures.
(b) 4 is the second significant figure in the number 20 453.
(c) 2135 = 2130 correct to 3 significant figures.
(d) 0.419 6 = 0.42 correct to 3 significant figures.
(e) 0.23 is exactly the same as 0.230.

7 Use your calculator to do the following calculations.
Give your answers correct to 3 significant figures.

(a) $\sqrt{(4.9 + 18.73)}$ **(b)** $56 \div (23 + 14)$ **(c)** $\dfrac{3 + 4 + 2.5 + 5.5 + 6 + 1.5}{6}$

(d) $\pi \times 4.7$ **(e)** $\pi \times 3.5^2$ **(f)** $\sqrt{(3.4^2 + 1.6^2)}$

Activity

1 750 is accurate to 2 significant figures.
(a) Explain why 746 = 750 correct to 2 significant figures.
(b) Write down a smaller number that will give 750 when rounded to 2 significant figures.

? **What is the smallest answer you could have given in (b)?**

(c) Write down the largest number that will give 750 when rounded to 2 significant figures.

2 3560 is accurate to 3 significant figures.
(a) What is the smallest number that this could be?
(b) What is the largest number this could be?

Range of values

Ranjit measures the diameter of a penny.
He says 'It is 21 mm to the nearest mm'.

Avonford High School
CHARITY DAY
Help us to lay a kilometre of pennies.
Add your spare pennies to the end of the line.

How many pennies are needed to make a line 1 kilometre long?

How much money can be raised this way? (Give your answer to the nearest penny.)

Avonford High School only raised £465.11.

How is this possible?

Write down the smallest number that will give 21 mm (to the nearest mm).
Write down a measurement larger than 21.49 mm that will round to 21 mm.

Explain why the largest possible measurement can be described as <21.5 mm.

Any measurement is really a range of possible measurements.

$20.5\text{ mm} \le 21\text{ mm to the nearest mm} < 21.5\text{ mm}$

A distance of 1 kilometre was measured to the nearest metre.

$999.5\text{ m} \le 1000\text{ m } (= 1\text{ km}) < 1000.5\text{ m}$

1000.5 m = 1 000 500 mm

The calculation $1\,000\,500 \div 20.5$ **will give the largest possible number of pennies in a kilometre of pennies. Work this out.**

Work out the smallest number of pennies in a kilometre of pennies.

Task

1 **(a)** Measure the thickness of a *Formula One Maths* textbook.
(b) How accurate is your measurement?
(c) Write your measurement as a range of values and work out
  (i) the greatest possible height of a pile of 10 identical *Formula One Maths* books
  (ii) the smallest possible height of a pile of 10 identical *Formula One Maths* books.
(d) Place 10 textbooks in a pile. Measure the height of the pile. Compare your measurement with your answers in (c).

2 **(a)** Measure the height and width of your *Formula One Maths* textbook.
(b) Write each measurement as a range of values.
(c) Work out the smallest and largest possible area for its front cover.

Exercise

1 Brian weighs 67 kilograms to the nearest kilogram.

(a) Round these weights to the nearest kilogram.
67.4 kg, 66.97 kg, 66.52 kg, 67.499 9 kg

(b) Write down the smallest possible value for Brian's weight.

(c) Write down the largest possible value for Brian's weight.

(d) Write Brian's weight as a range of values.

2 Write each of the following measurements as a range of values.

(a) 45 cm to the nearest cm

(b) 124 litres to the nearest litre

(c) 58 grams to the nearest gram

(d) 4.1 cm to the nearest mm ← *Change 4.1 cm to mm first.*

(e) 2.5 km to the nearest metre. ← *Change 2.5 km to metres first.*

3 The measurements of the photograph are accurate to the nearest millimetre.

(a) Write each measurement as a range of values.

(b) Work out the smallest possible measurement for the perimeter of the photograph.

(c) Work out the largest value of the perimeter.

(d) Work out the smallest and largest values for the area of the photograph.

4 Explain why it might not be safe for the following people to travel together in the lift.
David 65 kg, Brian 92 kg, Bronwen 74 kg,
Pat 54 kg, Peter 86 kg, Bruce 95 kg,
Ahmed 89 kg, Mark 93 kg

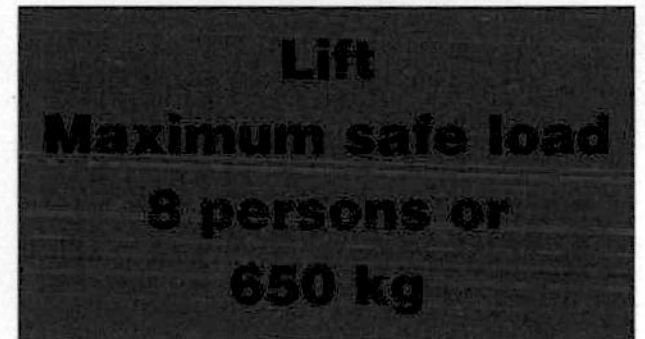

5 Mary has 64 CDs. Each CD is 1 cm wide (to the nearest mm).
She needs a shelf to store her CDs.
How long should she make her shelf?

6 **(a)** Write the following numbers correct to 2 significant figures.
1.24, 1.237, 1.165 2, 1.151, 1.249, 1.249 99

(b) The number 1.2 is given correct to 2 significant figures.
Write this as a range of numbers.

7 Write each of the following as a range of numbers.

(a) 230 (correct to the nearest 10)
(b) 5.62 (correct to 2 decimal places)
(c) 200 (correct to 1 s.f.)
(d) 200 (correct to 2 s.f.)
(e) 0.4 (correct to 1 s.f.)
(f) 0.40 (correct to 2 s.f.)

8 Explain why 0.40 is not always the same as 0.4.
Use your answers to questions 7(e) and (f) to help you.

Estimating answers

Mirna measures the trolley and a tin of dog food.
She calculates

Volume of trolley = average depth x average width x length
= 63 cm x 47 cm x 98 cm
≈ 60 x 50 x 100 cm^3
= 300 000 cm^3
Volume of a tin = area of circle x height
= $\pi \times 3.7^2 \times 11$ cm^3
≈ $3 \times 4^2 \times 10$
≈ 50 x 10
= 500 cm^3
Estimated number of tins = 300 000 ÷ 500 = 600

My guess: 573

These numbers have all been rounded to 1 significant figure.

? **Why is $3 \times 4^2 \approx 50$?**

? **Why has Mirna chosen to guess a number that is less than her estimate?**

Task

For each of the calculations below
(a) round all the numbers to 1 significant figure
(b) use the rounded values to estimate the answer to the calculation
(c) do the calculation and compare your answer with your estimated value.

1 $47 \times 123 \times 87$ **2** $2.35 \times (8.2 + 1.78)$ **3** $(735 + 209) \div 32$

4 $\pi \times 2.7^2$ **5** $10.7^2 + 4.2^2$ **6** $0.53 \times 1.87 \times 82$

Estimating square roots and cube roots

Write down the first 10 square numbers and the first 10 cube numbers.

? **Explain why $\sqrt{39} \approx 6$ and $\sqrt[3]{69} \approx 4$.**
Give approximate values for $\sqrt{97}$ and $\sqrt[3]{120}$.

? **Estimate the answer to $\sqrt{(423.194)}$.**

Exercise

1 Give an approximate answer for each of the following.

(a) 87.9×2.97 **(b)** $487 \div 5.23$ **(c)** $(67 + 32) \div 1.78$

(d) $\sqrt{35}$ **(e)** $\sqrt{119}$ **(f)** $\sqrt{340}$

(g) $\sqrt{(52 \times 1.9)}$ **(h)** $2.8^2 + 7.13^2$ **(i)** $\sqrt{(4^2 + 5^2)}$

(j) $\dfrac{12.7 + 108}{8.7}$ **(k)** $\pi \times 4.8$ **(l)** $\pi \times 2.7^2$

2 Use your calculator to evaluate the calculations in question 1.
Give your answers correct to 3 significant figures.

3 Select the correct answer (i–iv) for each of the calculations below.
Explain your choice.

(a) $385^2 =$ **(i)** 164 025 **(ii)** 148 225 **(iii)** 1425 **(iv)** 1615

(b) $45.6 \times 8.1 =$ **(i)** 412.36 **(ii)** 305.26 **(iii)** 369.36 **(iv)** 402.46

(c) $3.9^2 + 2.1^2 =$ **(i)** 17.52 **(ii)** 36.3 **(iii)** 12.44 **(iv)** 19.62

(d) $\dfrac{12.3 + 18.9}{7.2} =$ **(i)** 4.33 **(ii)** 6.33 **(iii)** 15.33 **(iv)** 20.33

4 **(a)** A company makes bedsheets that are 2.3 m long.
They buy the material in 50 m lengths costing £40 each.
Estimate the cost of the material for one sheet.

(b) The distance from London to Aberdeen is about 480 miles.
Estimate the journey time at a speed of between 55 and 60 miles per hour.

(c) Estimate the number of seconds you sleep in one day.

(d) John counts the number of turns his bicycle wheel makes in a distance of 100 m. The answer is 46 (to the nearest whole number).
Estimate the diameter of the wheel.

5 Estimate the number of jelly men in the jar.
Show all your working.

A jelly man is 1 cm thick.

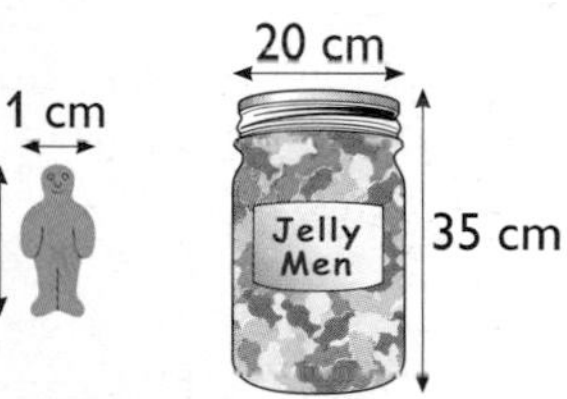

Activity

1 Write all the prices on this bill correct to the nearest pound (£).

2 Use your answers to estimate the total bill.

? **Why does this not give a very good estimate?**

A better way to estimate the bill is to
(i) add up all the complete pounds, then
(ii) look at the pence and group these in amounts that are roughly equal to £1 e.g. 79p + 24p ≈ £1
(iii) add any complete pounds to the total from (i).

Corn flakes	£1.49
Milk	£0.79
Bacon	£2.24
Apples	£1.14
Bananas	£1.39
Frozen peas	£2.35
Coffee	£4.49
Chicken	£1.99

3 Use this method to find a better estimate for the total bill.

4 Find some old shopping bills.
See how quickly you can estimate the total of each bill.

Finishing off

Now that you have finished this chapter you should be able to:

- understand significant figures
- round numbers and measurements to a given number of decimal places or significant figures, or to the nearest unit
- understand that any measurement is really a range of values
- provide efficient estimates for many calculations.

Review exercise

1 Write the following numbers correct to the number of decimal places given in the brackets.

(a) 12.347 (1) **(b)** 4.843 5 (2) **(c)** 0.135 6 (3) **(d)** 0.025 3 (2)
(e) 0.0478 (3) **(f)** 4.999 (1) **(g)** 3.999 9 (2) **(h)** 0.999 9 (3)

2 Write the following numbers to the number of significant figures stated in the brackets.

(a) 12.1 (1) **(b)** 1.35 (2) **(c)** 103.6 (3) **(d)** 0.509 (2)
(e) 0.352 (2) **(f)** 0.604 5 (3) **(g)** 0.001 7 (1) **(h)** 999 (1)

3 Find a rough estimate for each of the following calculations.
Then use your calculator to give the answer correct to 3 significant figures.

(a) 4.77×23.8 **(b)** $82.8 \div 14.6$ **(c)** 56.3×0.573 **(d)** $7.65 \div 6.345$
(e) 675×256 **(f)** $45.6 \div 14.76$ **(g)** 0.32×0.127 **(h)** $537.8 \div 34.7$
(i) 3.45^2 **(j)** 127.5^2 **(k)** $\sqrt{(34.5 + 18.23)}$ **(l)** $\sqrt{(12 \times 2.34)}$
(m) $\dfrac{3.793 \times 0.458}{4.89}$ **(n)** $\dfrac{43.8 \times 3.62}{4.83}$ **(o)** $\dfrac{975 \times 0.838}{41.9}$ **(p)** $\dfrac{57.8}{5.45 \times 11.35}$

4 The 1995 populations of a number of countries are given in the table.
The figures are accurate to the nearest 1000.
Write each number as a range of values.

Country	Population
Australia	18 107 000
Denmark	5 229 000
France	58 286 000
Japan	125 156 000
United Kingdom	58 306 000
United States	263 563 000

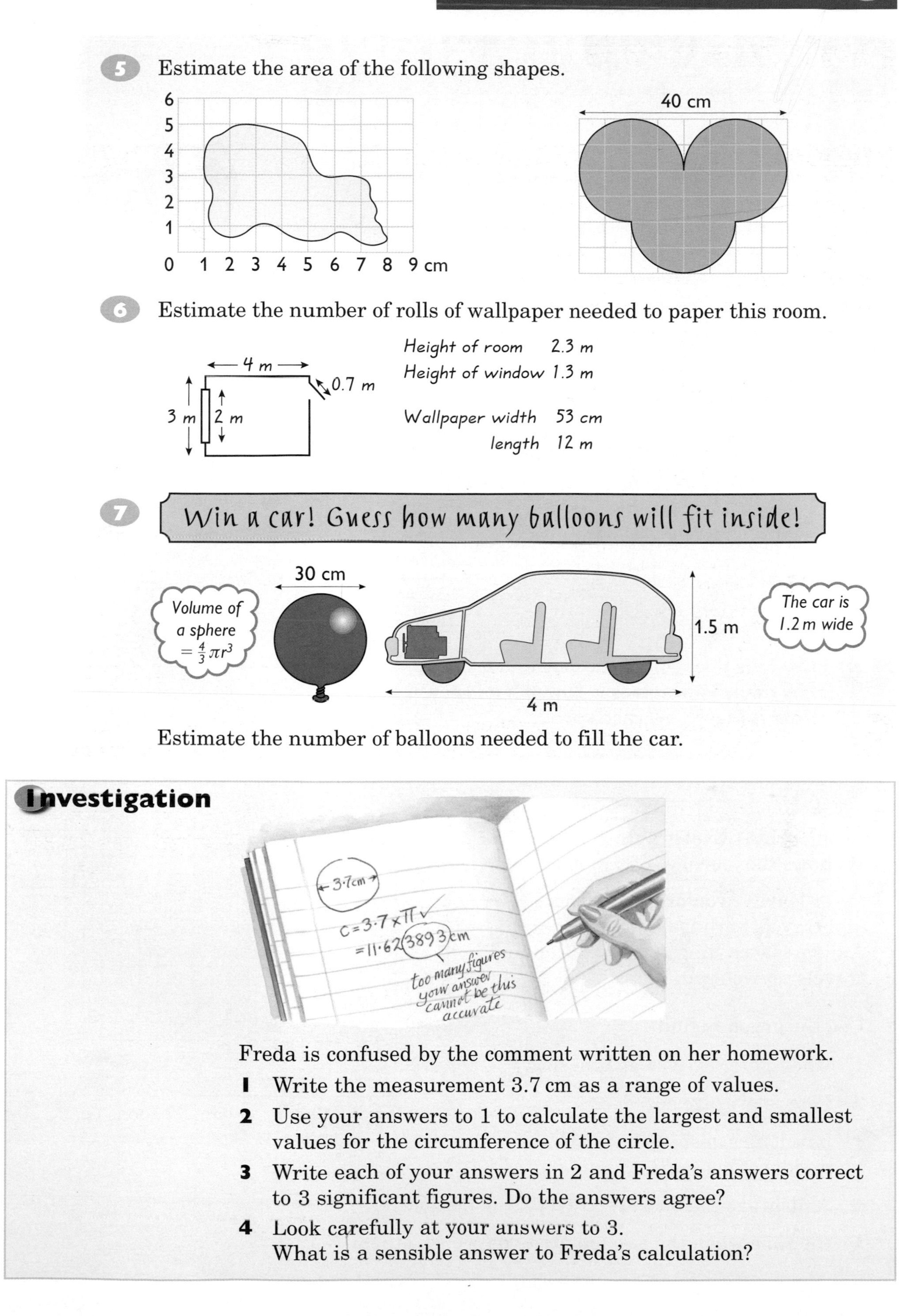

5 Estimate the area of the following shapes.

6 Estimate the number of rolls of wallpaper needed to paper this room.

7

Win a car! Guess how many balloons will fit inside!

Estimate the number of balloons needed to fill the car.

Investigation

Freda is confused by the comment written on her homework.

1 Write the measurement 3.7 cm as a range of values.

2 Use your answers to 1 to calculate the largest and smallest values for the circumference of the circle.

3 Write each of your answers in 2 and Freda's answers correct to 3 significant figures. Do the answers agree?

4 Look carefully at your answers to 3.
What is a sensible answer to Freda's calculation?

19 Real life graphs

John always walks to school.

Nnena takes the bus.

Mary comes by car.

Can you identify which line on this graph represents each person?

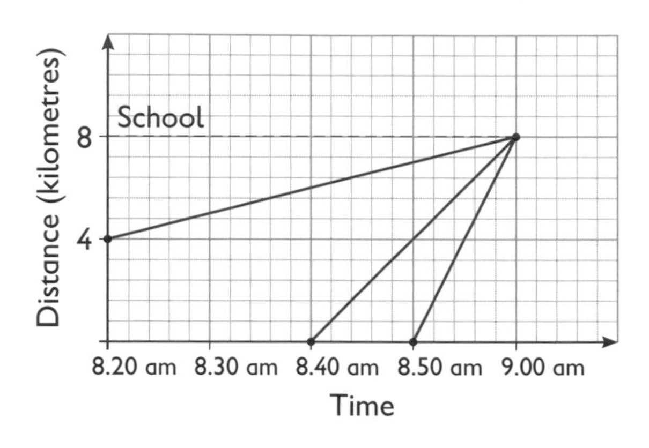

How long does John's journey take?
How many kilometres is Nnena's journey?
What is Mary's average speed?

Task

Look at this travel graph.
It shows the journeys of two motorists.

Peter leaves Avonford at 9.00 am and travels to Bingsley.
Amishi leaves Bingsley at 9.30 am and travels to Avonford.

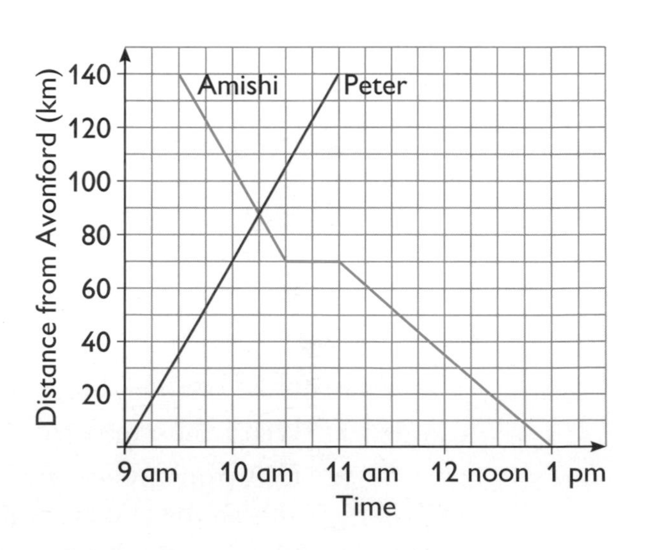

Use the graph to find

1 when Peter arrives at Bingsley

2 Peter's average speed

3 how long Amishi rests in the middle of her journey

4 Amishi's speed on the second part of her journey

5 the time when the two motorists pass each other.

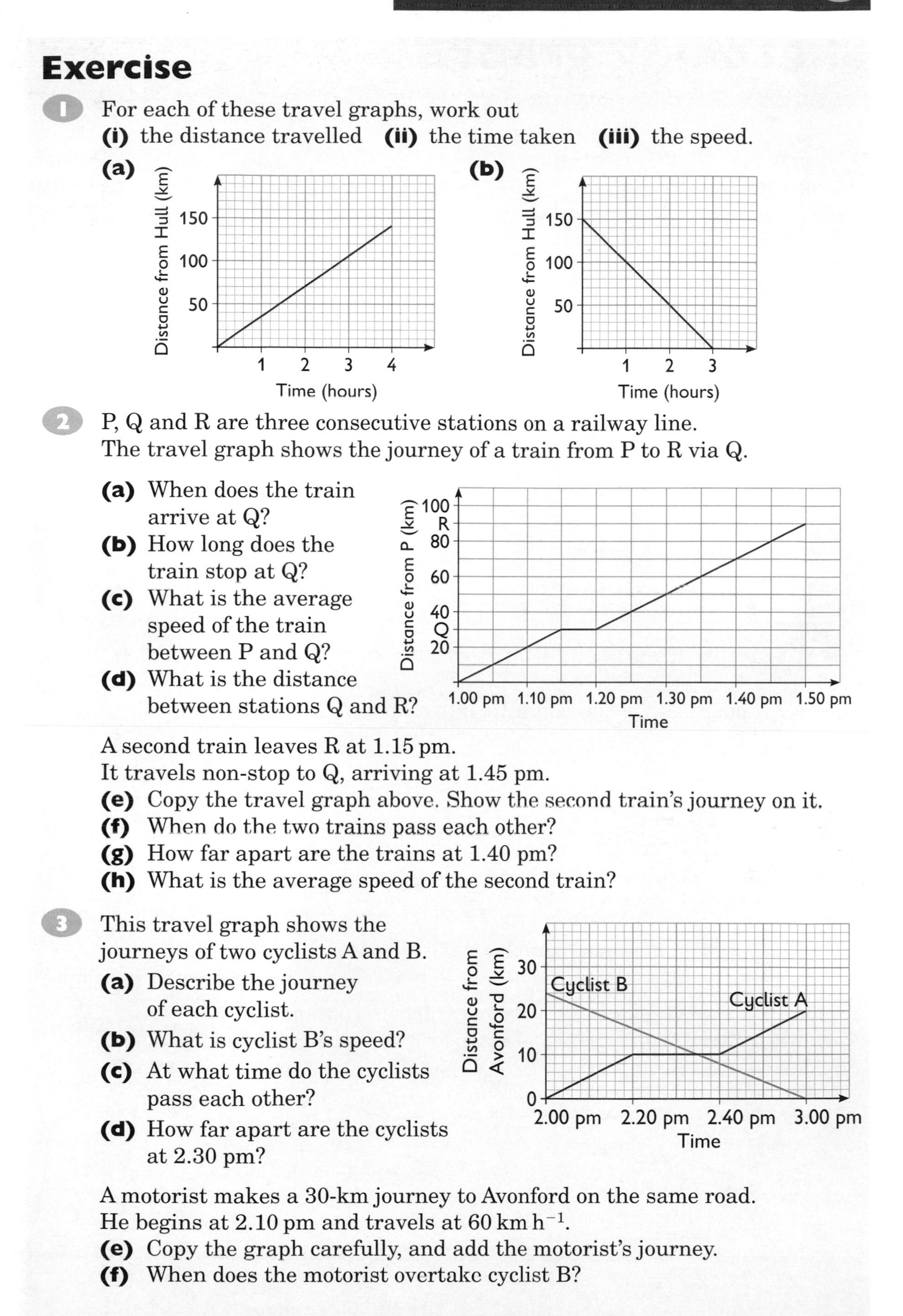

Exercise

1 For each of these travel graphs, work out
(i) the distance travelled **(ii)** the time taken **(iii)** the speed.

(a) **(b)**

2 P, Q and R are three consecutive stations on a railway line.
The travel graph shows the journey of a train from P to R via Q.

(a) When does the train arrive at Q?
(b) How long does the train stop at Q?
(c) What is the average speed of the train between P and Q?
(d) What is the distance between stations Q and R?

A second train leaves R at 1.15 pm.
It travels non-stop to Q, arriving at 1.45 pm.
(e) Copy the travel graph above. Show the second train's journey on it.
(f) When do the two trains pass each other?
(g) How far apart are the trains at 1.40 pm?
(h) What is the average speed of the second train?

3 This travel graph shows the journeys of two cyclists A and B.
(a) Describe the journey of each cyclist.
(b) What is cyclist B's speed?
(c) At what time do the cyclists pass each other?
(d) How far apart are the cyclists at 2.30 pm?

A motorist makes a 30-km journey to Avonford on the same road.
He begins at 2.10 pm and travels at 60 km h^{-1}.
(e) Copy the graph carefully, and add the motorist's journey.
(f) When does the motorist overtake cyclist B?

Sketching graphs

Dr Beckham is Quality Control Manager at Avonford Dairy. She wants to know how the milk level varies with time while the bottle is being filled.

Dr Beckham draws this sketch graph.

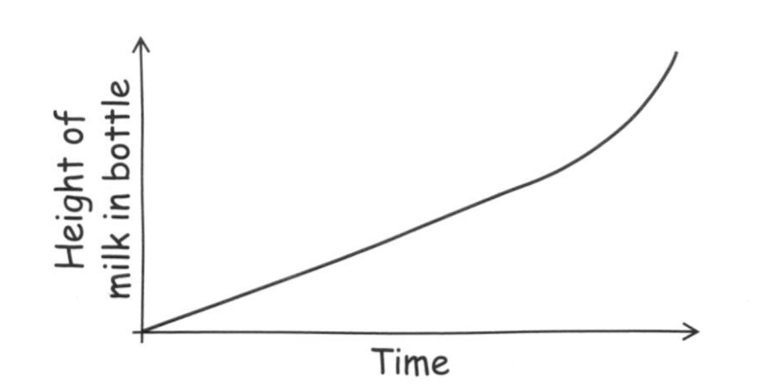

Task

Water is poured, at a constant rate, into each of these containers:

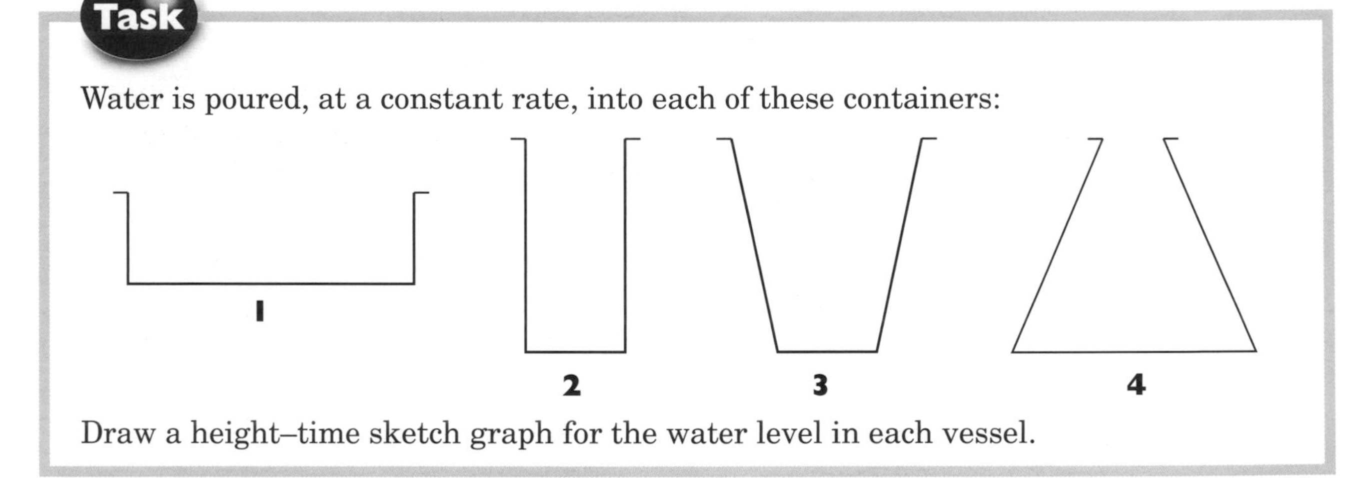

Draw a height–time sketch graph for the water level in each vessel.

Avonford Dairy also sell orange juice in cardboard containers. When being filled, the height–time curve looks like the graph below.

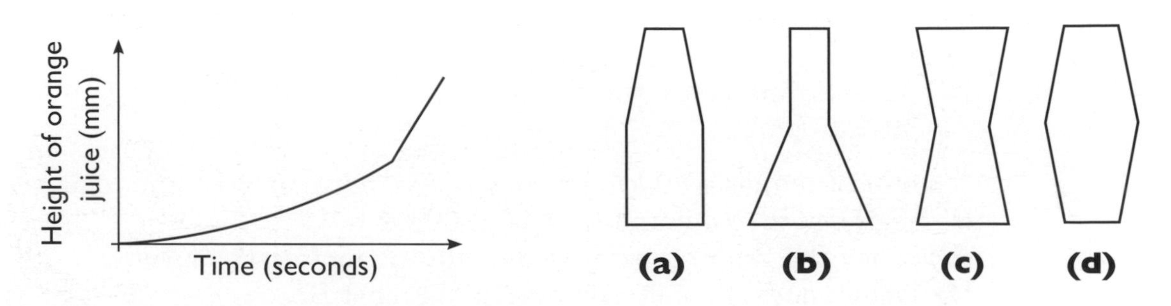

Which diagram on the right shows the container shape?

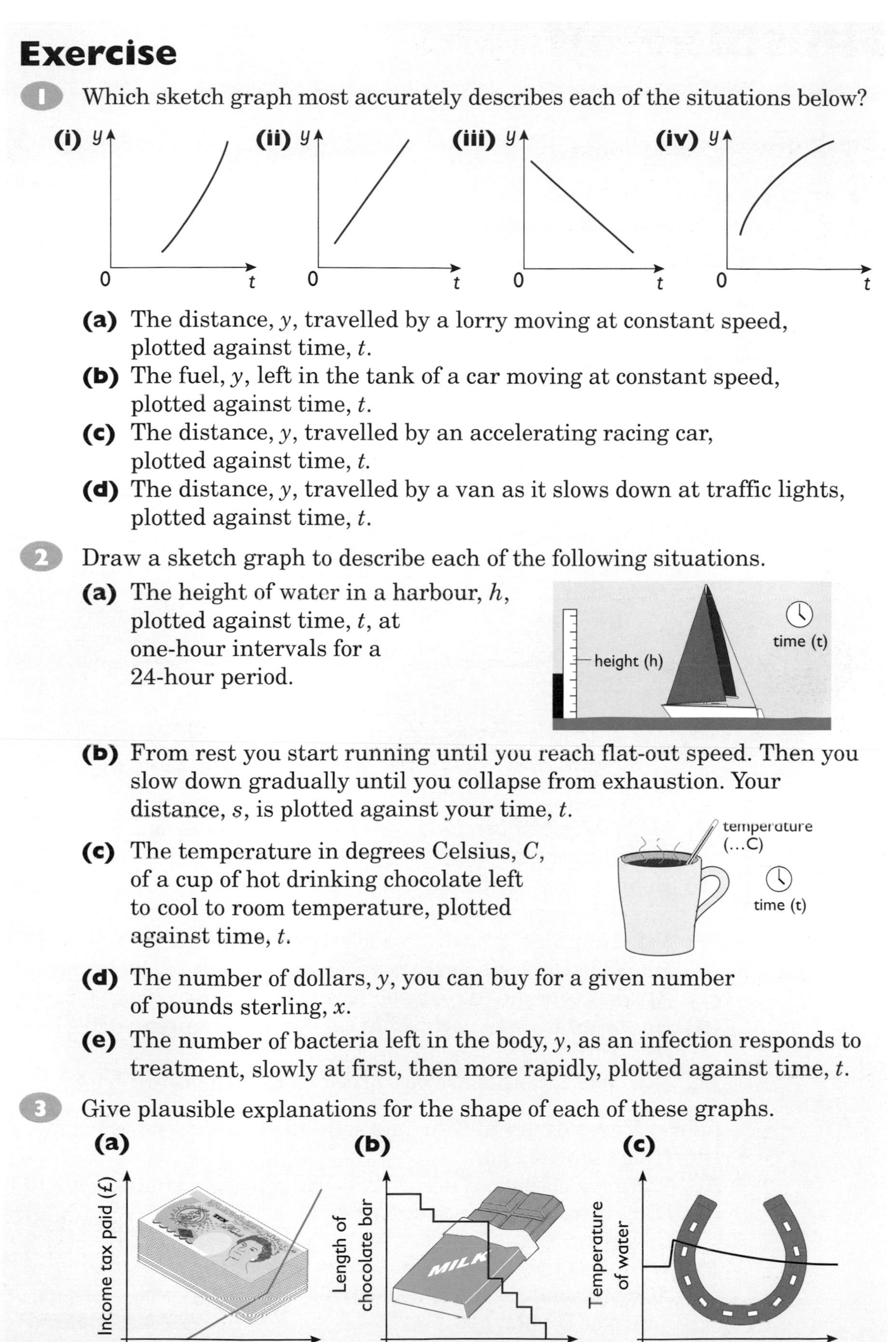

Exercise

1 Which sketch graph most accurately describes each of the situations below?

(a) The distance, y, travelled by a lorry moving at constant speed, plotted against time, t.

(b) The fuel, y, left in the tank of a car moving at constant speed, plotted against time, t.

(c) The distance, y, travelled by an accelerating racing car, plotted against time, t.

(d) The distance, y, travelled by a van as it slows down at traffic lights, plotted against time, t.

2 Draw a sketch graph to describe each of the following situations.

(a) The height of water in a harbour, h, plotted against time, t, at one-hour intervals for a 24-hour period.

(b) From rest you start running until you reach flat-out speed. Then you slow down gradually until you collapse from exhaustion. Your distance, s, is plotted against your time, t.

(c) The temperature in degrees Celsius, C, of a cup of hot drinking chocolate left to cool to room temperature, plotted against time, t.

(d) The number of dollars, y, you can buy for a given number of pounds sterling, x.

(e) The number of bacteria left in the body, y, as an infection responds to treatment, slowly at first, then more rapidly, plotted against time, t.

3 Give plausible explanations for the shape of each of these graphs.

Finishing off

Now that you have finished this chapter you should be able to:

- draw a travel graph
- obtain information from a given travel graph
- draw sketch graphs to illustrate real situations
- obtain information from given sketch graphs of real situations.

Review exercise

1 The graph represents Mr Watson's journey from London to Stockport.

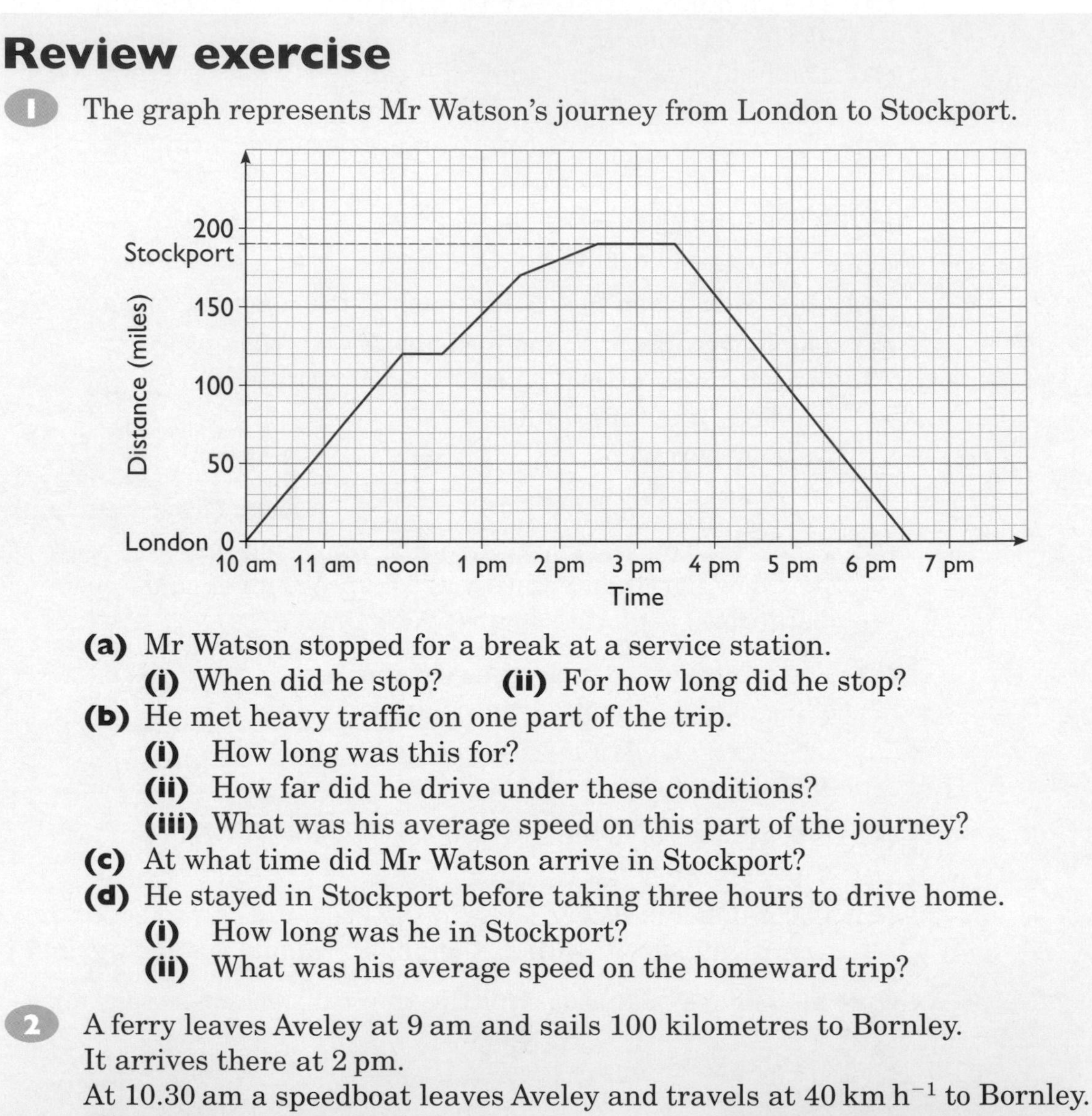

(a) Mr Watson stopped for a break at a service station.
 (i) When did he stop? **(ii)** For how long did he stop?

(b) He met heavy traffic on one part of the trip.
 (i) How long was this for?
 (ii) How far did he drive under these conditions?
 (iii) What was his average speed on this part of the journey?

(c) At what time did Mr Watson arrive in Stockport?

(d) He stayed in Stockport before taking three hours to drive home.
 (i) How long was he in Stockport?
 (ii) What was his average speed on the homeward trip?

2 A ferry leaves Aveley at 9 am and sails 100 kilometres to Bornley.
It arrives there at 2 pm.
At 10.30 am a speedboat leaves Aveley and travels at 40 km h^{-1} to Bornley.

(a) Draw a travel graph to describe these journeys.

(b) What was the speed of the ferry?

(c) At what time did the speedboat arrive at Bornley?

(d) When did the vessels meet?

(e) How far apart were the vessels at 11 am?

3 The graph shows the price of Oldcastle United plc shares at the close of business each week day (Monday to Friday) over a 10-day period.

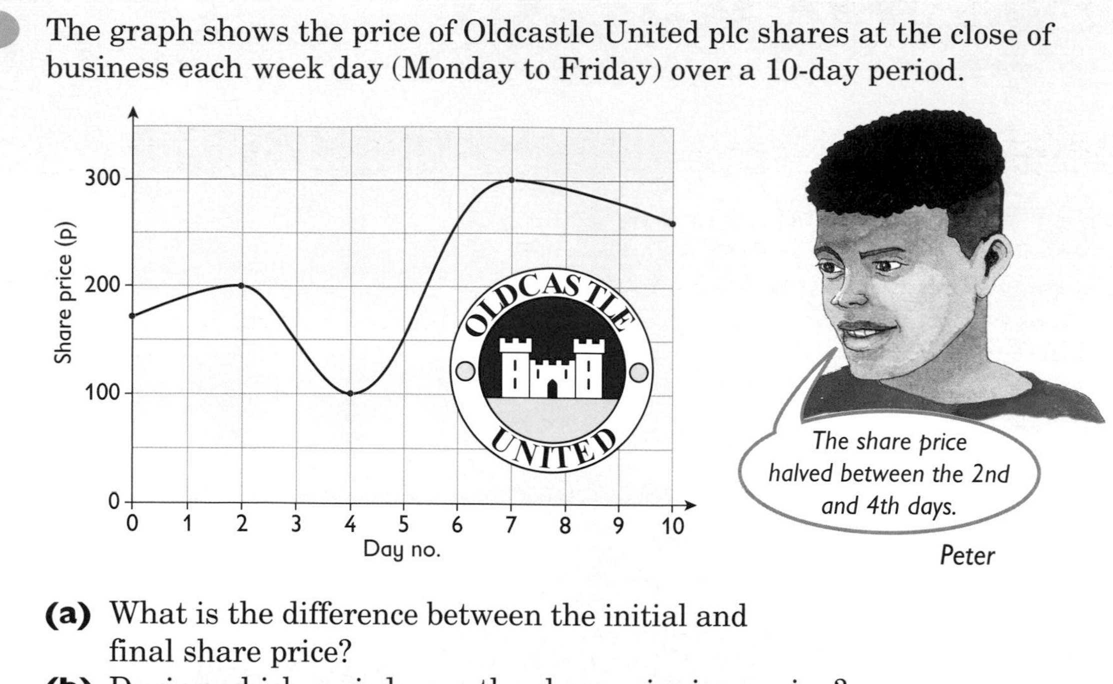

(a) What is the difference between the initial and final share price?
(b) During which periods was the share price increasing?
(c) When is the best time to buy shares?
(d) When is the best time to sell shares?
(e) Is Peter's statement true or false?

4 The diagram shows the shape of Jane's bath. Water runs in at a constant rate. Which sketch graph below best describes the height–time graph? Justify your answer.

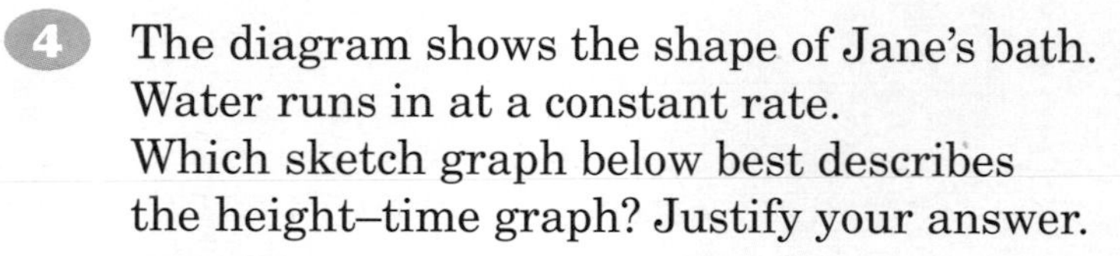

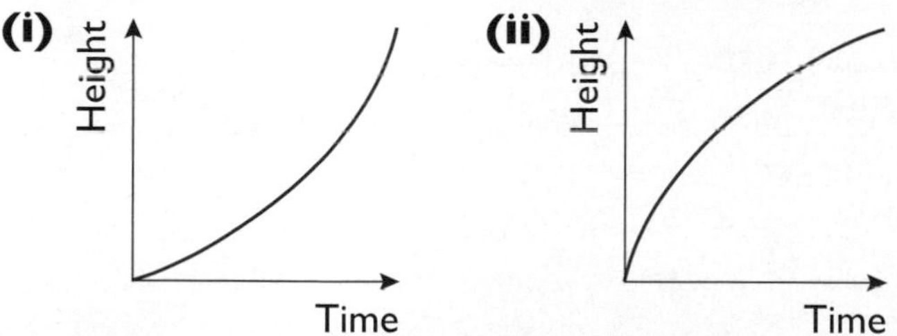

5 The graph opposite shows the height of perfume in a bottle varying with time when the bottle is filled at a constant rate.

Which of these shapes is the correct shape of the bottle?

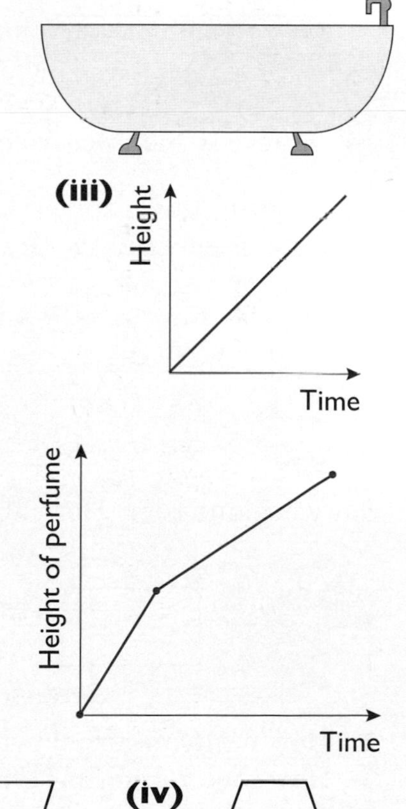

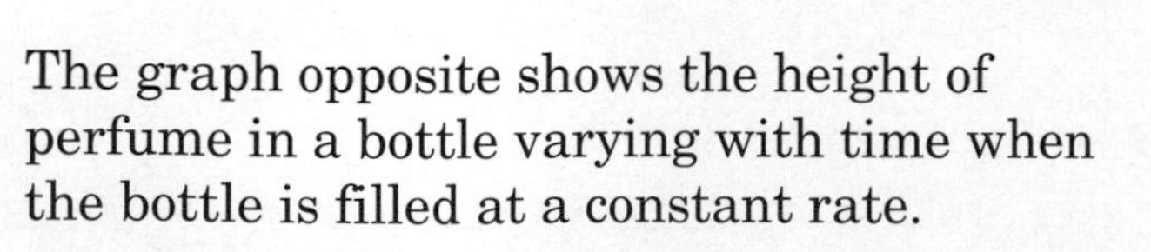

20 Transformations

Translation, reflection and rotation

What is a transformation?

Here are three types of transformation.

The **translation** which maps A to B can be described as 5 in the x direction and -2 in the y direction.

It is written $\begin{pmatrix} 5 \\ -2 \end{pmatrix}$.

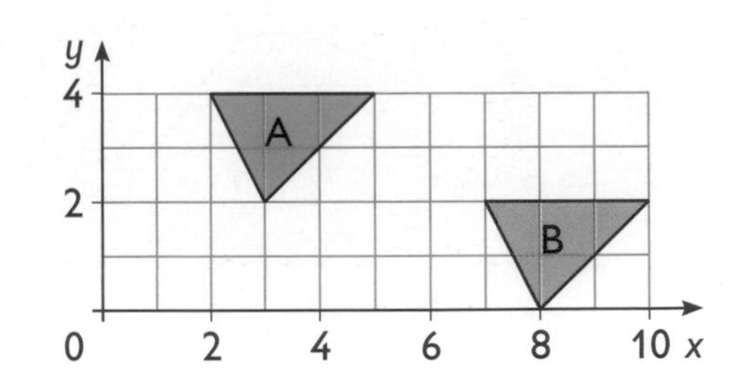

Describe the translation which maps B to A.

Here are some **reflections**.

(a)

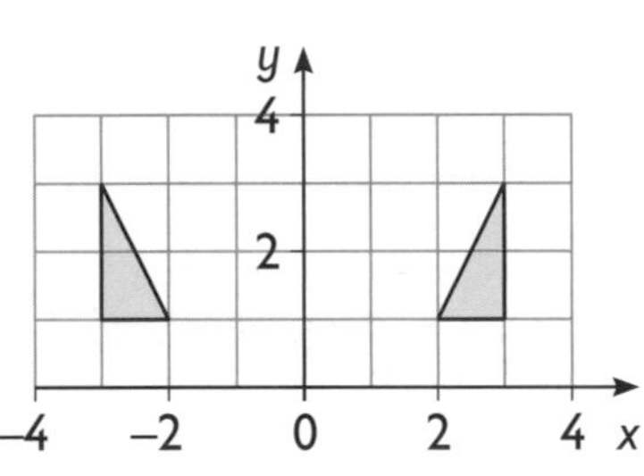

(b)

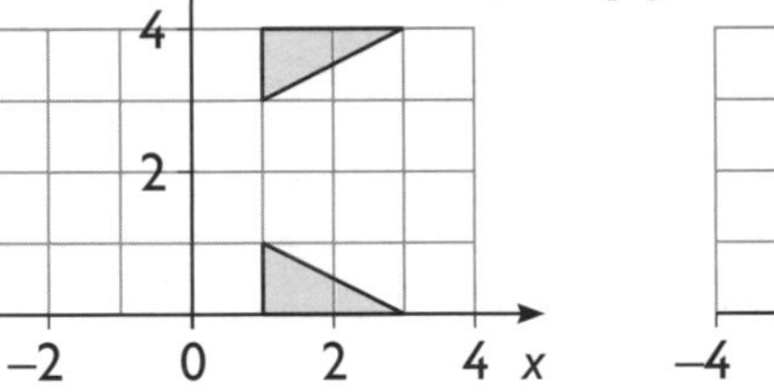

(c)

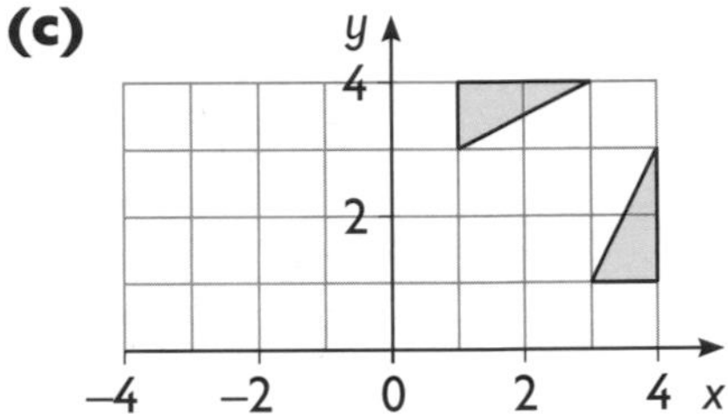

Where is the mirror line in each of these reflections?

In the diagram the orange triangle (**object**) is **rotated** to form the green triangle (**image**).

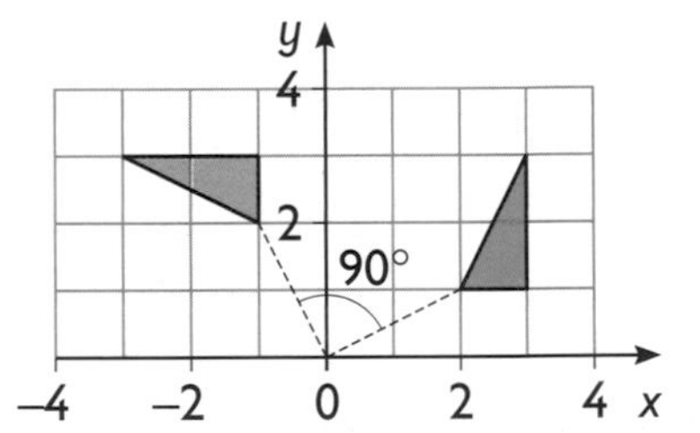

Describe this rotation fully.

Sally designs tiles. Here are some of her designs.

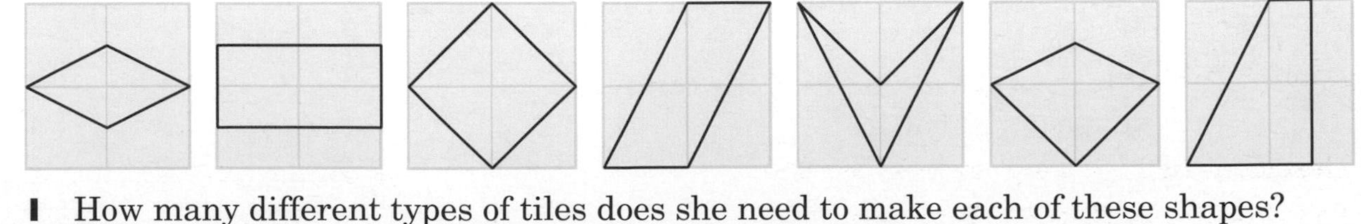

1 How many different types of tiles does she need to make each of these shapes?
2 How many different types does she need in total?

In translations, reflections and rotations the object and image are **congruent**.

What does congruent mean?

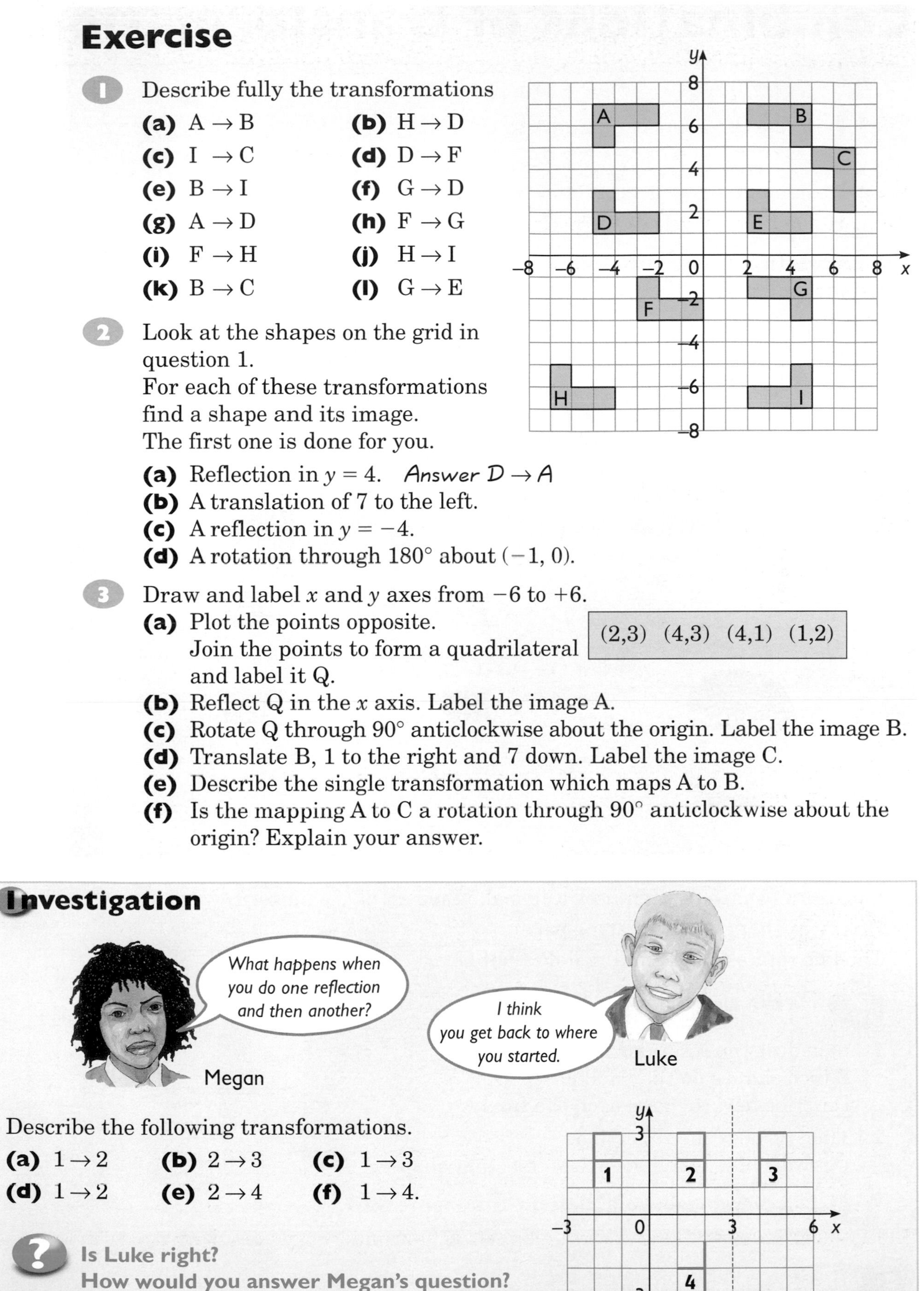

Exercise

1 Describe fully the transformations

(a) A → B **(b)** H → D
(c) I → C **(d)** D → F
(e) B → I **(f)** G → D
(g) A → D **(h)** F → G
(i) F → H **(j)** H → I
(k) B → C **(l)** G → E

2 Look at the shapes on the grid in question 1.
For each of these transformations find a shape and its image.
The first one is done for you.

(a) Reflection in $y = 4$. *Answer D → A*
(b) A translation of 7 to the left.
(c) A reflection in $y = -4$.
(d) A rotation through 180° about $(-1, 0)$.

3 Draw and label x and y axes from -6 to $+6$.

(a) Plot the points opposite.
Join the points to form a quadrilateral and label it Q.

(2,3) (4,3) (4,1) (1,2)

(b) Reflect Q in the x axis. Label the image A.
(c) Rotate Q through 90° anticlockwise about the origin. Label the image B.
(d) Translate B, 1 to the right and 7 down. Label the image C.
(e) Describe the single transformation which maps A to B.
(f) Is the mapping A to C a rotation through 90° anticlockwise about the origin? Explain your answer.

Investigation

Describe the following transformations.

(a) 1 → 2 **(b)** 2 → 3 **(c)** 1 → 3
(d) 1 → 2 **(e)** 2 → 4 **(f)** 1 → 4.

? **Is Luke right?**
How would you answer Megan's question?

Combinations of transformations

Describe the transformation A to C.

Describe the transformation which maps
(a) A to B (b) B to C.

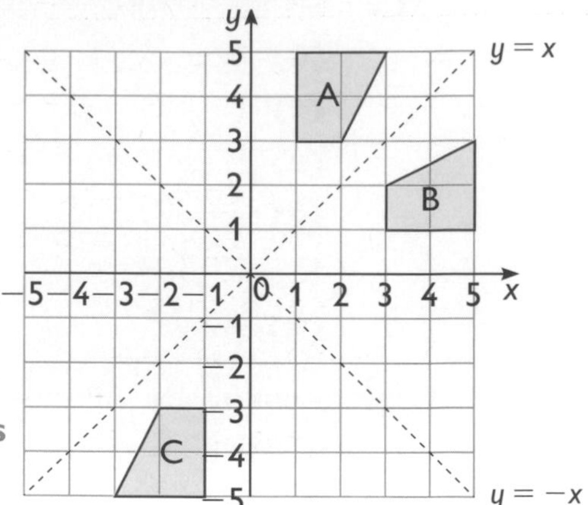

The combination of these two transformations takes A to C.

Describe another combination of transformations which takes A to C.

Task

Sam is designing a window.
This is what he has in mind.

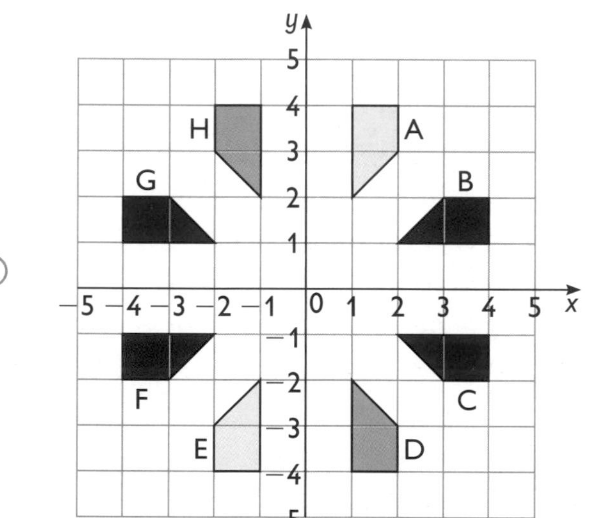

Sam draws shape A, then uses it to make the rest of the pattern.
He traces shape A onto tracing paper.
Then he rotates it by pinning it down at O.

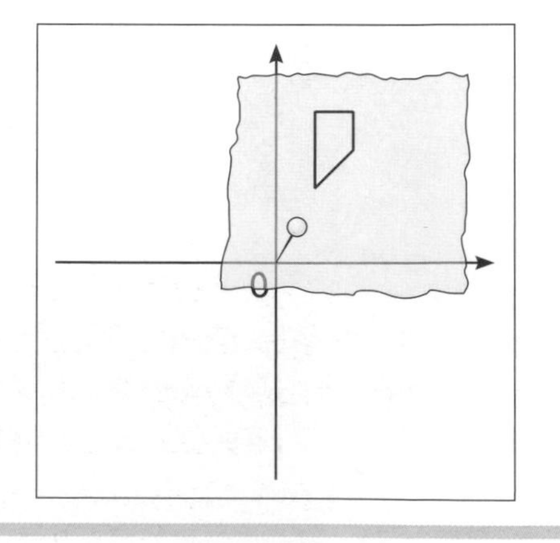

He can also reflect it. How?

1 Sam does the rotations first.
Which shapes do they make?
Describe how to make each one from A.

2 Then he does the reflections.
Describe how to make each of the other shapes.

Did everyone in your class do it the same way?

A shape is reflected three times.
Can the image be obtained by a single rotation of the object? Explain your answer.

Exercise

1 Draw and label x and y axes from -6 to $+6$.

(a) Plot these points. (1,3) (2,5) (4,6) (4,4)

Join the points to form a quadrilateral. Label it A.

(b) Rotate A through 90° clockwise about the origin. Label the image B.

(c) Reflect B in the x axis. Label the image C.

(d) Describe a combination of two transformations which maps C to A.

(e) Describe a single transformation which maps **(i)** A to C **(ii)** C to A.

2 Katie is making a pattern by reflecting triangles in mirror lines.

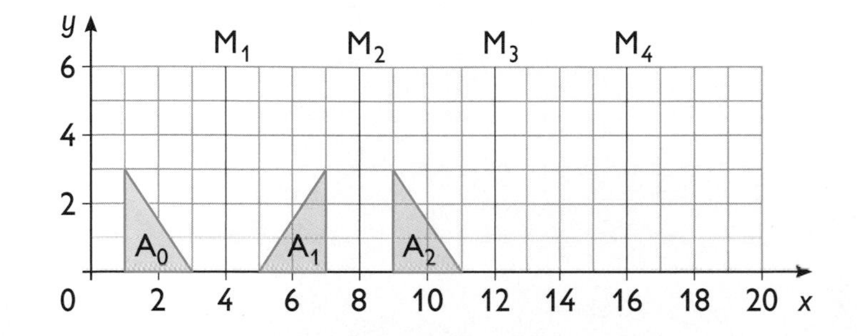

Triangle A_0 is reflected in M_1 to form A_1, triangle A_1 is reflected in M_2 to form A_2, and so on.

(a) Copy the diagram and draw triangles A_3 and A_4.

(b) Describe the transformation which maps A_0 to A_2.

(c) Write down the equations of the mirror lines
(i) M_1 **(ii)** M_2.

(d) Reflect triangle A_0 in $y = 3$ to form triangle B.

(e) Describe the single transformations which map
(i) B to A_1 **(ii)** B to A_3.

(f) Describe combinations of two transformations which map
(i) B to A_1 **(ii)** B to A_3.

3 Draw and label axes with the x axis from -6 to 10 and the y axis from 0 to 8. Plot the following points and join them to form a triangle T.

(−1,4) (−4,4) (−4,6)

(a) Reflect T in the y axis and label the image A.

(b) Translate A, 4 to the right and 3 down. Label the image B.

(c) Describe another combination of a reflection followed by a translation which maps T to B.

(d) Describe a combination of a translation followed by a reflection which maps T to B.

(e) Find two more transformations which map T to B.

Enlargement

Look at this picture. 

Which of the pictures below are enlargements of it? Explain your answer.

(a)  **(b)**  **(c)**  **(d)** 

Nicola is enlarging the picture to go on a wall of size 8 m by 6 m.
She wants to make the picture as large as possible.

How does she do it? What space is left over?

Task

Triangle A′B′C′ is an enlargement of triangle ABC.

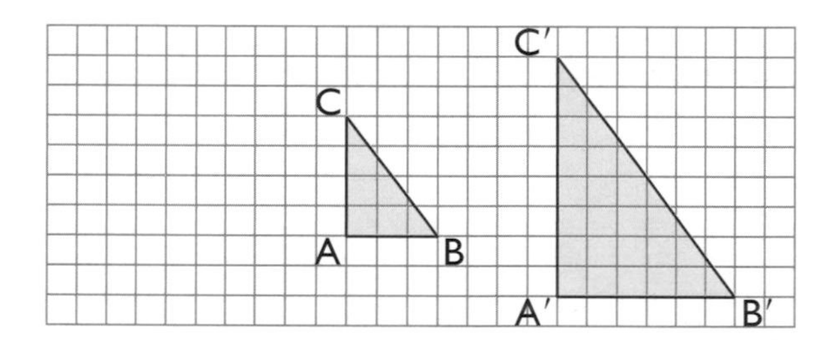

1 Copy these triangles onto a sheet of centimetre squared paper.

2 Draw and extend the lines A′A, B′B and C′C.
You should find that they meet at a point.
Label this point O. It is the **centre of enlargement**.

3 **(a)** Use your drawing to find the lengths of the sides of the two triangles.
Work out **(i)** A′B′ ÷ AB **(ii)** A′C′ ÷ AC **(iii)** B′C′ ÷ BC.
(b) What do you notice?

4 **(a)** Find **(i)** OA′ ÷ OA **(ii)** OB′ ÷ OB **(iii)** OC′ ÷ OC.
(b) What do you notice?

5 What is the scale factor of enlargement for this transformation?

What sort of scale factor gives an image smaller than the object?

Exercise

1 For each diagram below

(i) make a copy on centimetre squared paper

(ii) enlarge the shape by the scale factor given, using C as the centre of enlargement.

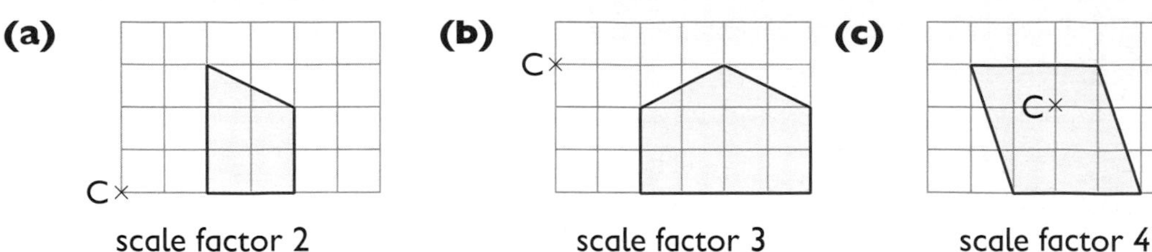

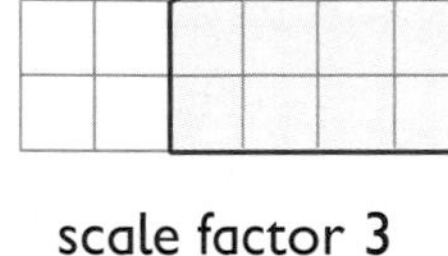
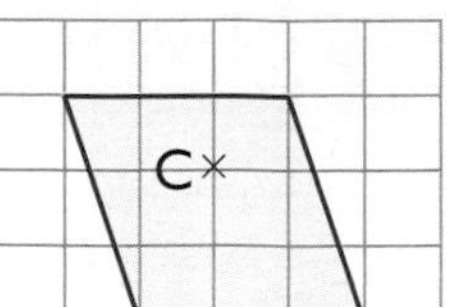

2 Look at the diagram.

O is a light source.

ABCD is the rectangular frame on a film.

A′B′C′D′ is the rectangular image on a screen.

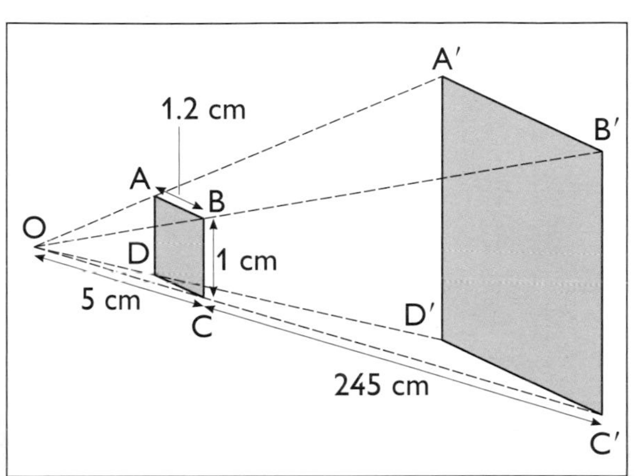

OA = OB = OC = OD = 5 cm

AA′ = BB′ = CC′ = DD′ = 245 cm

AB = 1.2 cm and BC = 1 cm

(a) Work out

(i) the length OC′

(ii) the scale factor of enlargement of the image A′B′C′D′.

(b) Work out **(i)** A′B′ **(ii)** B′C′.

(c) Work out the area of **(i)** ABCD **(ii)** A′B′C′D′.

(d) Work out the ratio Area of ABCD : Area of A′B′C′D′

Give your answer in its lowest terms.

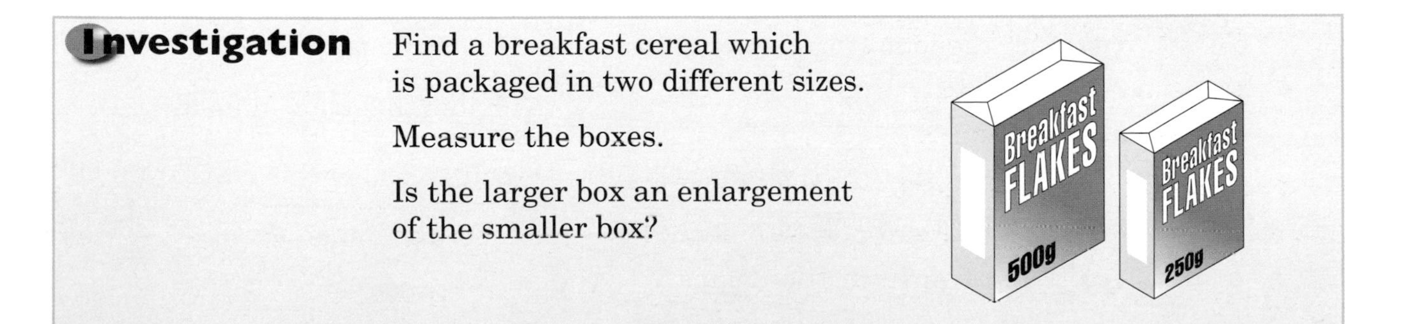

Investigation Find a breakfast cereal which is packaged in two different sizes.

Measure the boxes.

Is the larger box an enlargement of the smaller box?

Finishing off

Now that you have finished this chapter you should be able to:

- work with rotations, reflections, translations and combinations of all these transformations
- enlarge simple shapes using a centre of enlargement.

Exercise

1 Describe fully the following transformations.

(a) A $\rightarrow$ B

(b) B $\rightarrow$ C

(c) E $\rightarrow$ F

(d) C $\rightarrow$ A

(e) H $\rightarrow$ G

(f) F $\rightarrow$ G

(g) E $\rightarrow$ D

(h) H $\rightarrow$ I

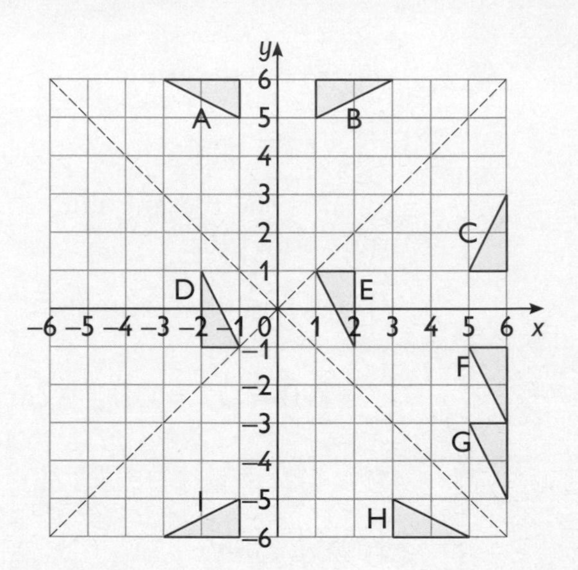

2 Look at the diagram in question 1.
The transformation C to F is a reflection in the x axis.

Find a transformation which is

(a) a rotation through 90° clockwise about O

(b) a translation $\begin{pmatrix} 4 \\ -2 \end{pmatrix}$

(c) a reflection in $y = -1$

(d) a rotation through 180° about $(2,-1)$.

3 Draw and label x and y axes from 0 to 10.

(a) Plot the following points and join them to form a trapezium.

(i) A(1,1) B(1,3) C(2,3) D(3,1) Label this trapezium T.

(ii) A′(10,10) B′(10,8) C′(9,8) D′(8,10) Label this trapezium U.

(b) Rotate T through 180° about (4,3) and label the image V.

(c) What transformation maps V to U?

4 Draw and label x and y axes from 0 to 15.

(a) Plot these points and join them to form quadrilateral Q.

(6,8) (6,11) (8,12) (9,11)

(b) Plot the following points and join them to form the quadrilateral Q′.

(9,3) (9,9) (13,11) (15,9)

(c) Explain why Q′ is an enlargement of Q.

(d) Write down the co-ordinates of the centre of enlargement.

(e) Work out the scale factor of enlargement.

(f) Explain how this enlargement of Q affects
(i) its perimeter **(ii)** its area.

5 Draw x and y axes and label the x axis from 1 to 20 and the y axis from 0 to 15.

(a) Plot points (6,10) (7,5) (10,8) and join them to form triangle T.

(b) Measure the angles of triangle T.

(c) Enlarge triangle T by scale factor 2 using (1,8) as the centre of enlargement.

(d) Measure the angles of the image. What do you notice?

Activity

You need two sheets of A4 paper, a ruler and a pair of scissors for this activity.

Copy the table and fill in your results as you work through the activity.

Size of paper	Longer side (mm)	Shorter side (mm)	Longer ÷ shorter
A4			
A5			
A6			

Measure the sides of a sheet of A4 paper.

Work out the ratio:

longer side ÷ shorter side

Cut the second sheet of A4 paper in half (see diagram).
Each half is now an A5 sheet.
Measure the sides and work out the ratio for an A5 sheet.

Cut one of the A5 sheets in half (see diagram) to get two A6 sheets.
Measure the sides and work out the ratio for an A6 sheet.

A4
A5
A6

? What do you notice about the three ratios that you have worked out?

? What do you notice when the three differently sized sheets are placed one on top of the other (see diagram)?

21 Probability

Calculating probability

The probability of picking a strawberry cream is $\frac{1}{4}$.

The probability of picking a toffee is $\frac{2}{5}$.

Alison has a bag of 20 sweets.
There are toffees, strawberry creams and mints.

How many toffees are there in the bag?
How many strawberry creams?
How many mints?

What is the probability of picking a mint?
Add up the three probabilities.
Explain your answer.

When Alison picks a sweet out of her bag, there are three possible **outcomes**, a toffee, a strawberry cream and a mint.
These are called **mutually exclusive** outcomes, because only one of them can happen.

What do you get when you add up the probabilities of all the mutually exclusive outcomes of an experiment?

Alison's friend Sarah also has a bag of the same three types of sweet.

What is the probability of picking a strawberry cream from Sarah's bag?

Sarah has 12 toffees in her bag.
How many sweets does she have altogether?
How many mints and strawberry creams does she have?

Task

Try this game with a partner.

Your partner puts 5 cubes in a bag, some red and some blue.
You take a cube out, write down its colour and put it back.
Repeat 10 times, guessing how many red and how many blue cubes are in the bag.
Your partner must not say if you are right or wrong.
Continue, and after each 10 goes make another guess.
When you are reasonably sure you know the numbers of cubes, ask your partner to tell you if you're right.

Discuss your results with the rest of the class.
How many experiments did it take to guess correctly?
When guessing, how sure could you be?

Exercise

1 Sophie is playing an arcade game. Each turn costs 50p.
Each time she plays, one of four symbols appears.
She wins £5 if she gets a star.
She loses if she gets an apple, a banana or a pear.

(a) What is the probability of getting a star?

Sophie plays the game 40 times.

(b) How much does she spend?

(c) How many times does she expect to win £5?

Probabilities:	
Apple	0.3
Banana	0.4
Pear	0.25
Star	...

2 Lynne has three bags containing red, green and blue balls.

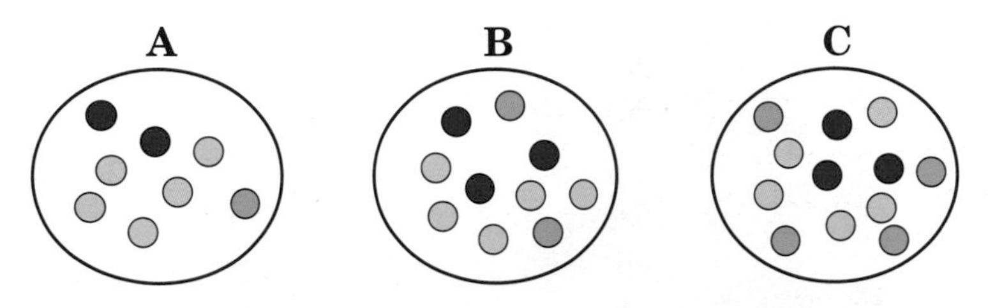

(a) Copy and complete this table.

	A	B	C
Probability of picking red			
Probability of picking green			
Probability of picking blue	$\frac{1}{8}$		

(b) Lynne wants to pick a green ball.
Which bag should she pick from?

(c) For two of the bags, the probability of picking red is the same.
Which two bags are these?

3 Simon has a bag containing red, black and white counters.
The probability of picking a red counter is $\frac{3}{8}$.
The probability of picking a black counter is $\frac{1}{2}$.

(a) What is the probability of picking a white counter?

There are 45 red counters.

(b) What is the total number of counters in the bag?

(c) How many **(i)** black **(ii)** white counters are there in the bag?

4 James has a biased die.
The probability of scoring six is twice the probability of scoring one.
The probability of scoring each of the other numbers is $\frac{1}{8}$.
Find the probability of scoring six.

Combined outcomes

Robert is using a fruit machine.
The machine has two windows.
Each window can show an orange, a banana or a pear.
Each fruit is equally likely to appear.

Robert wins 50p if he gets two bananas.
Robert wants to work out how likely he is to win.

He makes this list of possible outcomes.

Each of the outcomes on the list is equally likely.

Window 1	Window 2
orange	orange
orange	banana
orange	pear
banana	orange
banana	banana
banana	pear
pear	orange
pear	banana
pear	pear

? What is the probability of getting two bananas?

? Robert plays 9 games.
How much profit or loss would you expect him to make?

Robert writes: P(winning) $= \frac{1}{9}$

P(losing) $= \frac{8}{9}$

This is a quick way of writing "The probability of winning".

Another fruit machine has three windows.
Each window can show one of the same three fruits – orange, banana and pear.
Each fruit is equally likely to appear in each window.
You win if you get three bananas.
Make a list of all the possible outcomes and use your list to find P(winning).

? This machine also charges 10p a turn.
The owner wants it to make about the same average profit per game as Robert's machine.
How much should the prize money be?

Exercise

1 Mark throws two ordinary dice, one red and the other green, and adds the scores together.

(a) Copy and complete this table to show all the possible outcomes.

(b) Use your table to find
- **(i)** P(4)
- **(ii)** P(7)
- **(iii)** P(11)
- **(iv)** P(more than 8)
- **(v)** P(5 or less)
- **(vi)** P(prime number).

Die 1 (Red) across the top; **Die 2 (Green)** down the side.

Die 2 (Green) \ Die 1 (Red)	1	2	3	4	5	6
1	2	3	4			
2	3					
3						
4						
5						
6						

2 Helen takes 5 tops and 3 pairs of shorts on holiday.

Helen chooses a top and a pair of shorts at random one day.

(a) Make a list of all the possible combinations she could choose.

(b) What is the probability that she chooses a top and a pair of shorts which are the same colour?

3 Darren spins these two spinners.
He finds his score like this.

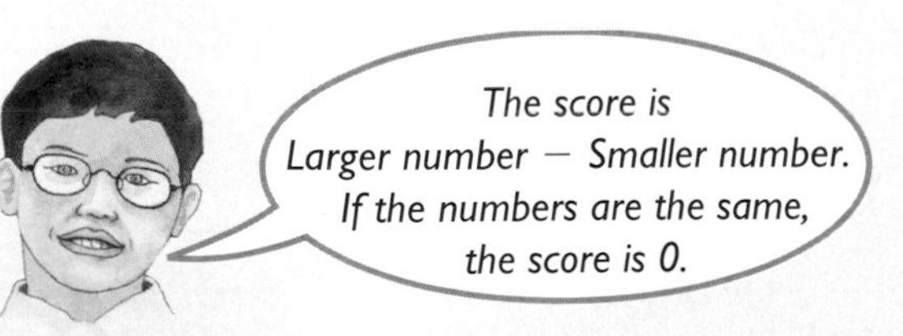

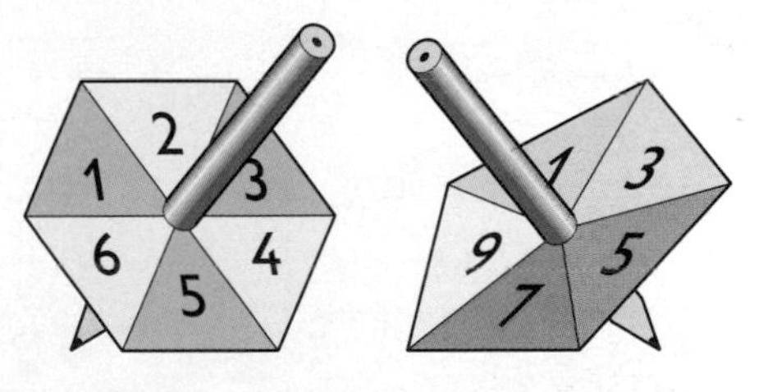

(a) Make a table or list to show all the possible outcomes.

(b) Find

(i) P(0) **(ii)** P(1) **(iii)** P(more than 4)

(iv) P(10) **(v)** P(even number).

4 **(a)** **(i)** Make a list to show all the possible outcomes of tossing 2 coins.

(ii) Find P(one head and one tail).

(b) **(i)** Make a list to show all the possible outcomes of tossing 3 coins.

(ii) Find P(two heads and one tail).

(c) **(i)** Make a list to show all the possible outcomes of tossing 4 coins.

(ii) Find P(two heads and two tails).

Estimating probability

10 hasn't come up in the National Lottery for months, so it will probably come up in the next few weeks.

I have three sons already, so my next baby will probably be a boy as well.

Tomorrow it could rain or be fine, so there is a 50% chance of rain.

None of these statements is correct. Explain why.

Justin wants to find out the probability that a person selected at random is left-handed.
He asks everyone in his class whether they are right-handed or left-handed.

Right-handed	24
Left-handed	4
Total	28

Justin can use these results to find the **relative frequency** of left-handedness.

$$\text{Relative frequency of left-handedness} = \frac{4}{28} = 0.14$$

Relative frequency can be used as an estimate of probability.

From Justin's results, P(left-handedness) = 0.14.

? Do Justin's results mean the probability a new baby will be left-handed is 0.14?

Task

Justin finds this article in a magazine.

Are you right-eyed or left-eyed?

Look at a spot on the far wall. Make your thumb and forefinger into a circle and look at the mark through it, at arm's length.

Now close your left eye – is the mark still in the circle? If it is, you are right-eyed.

Open your left eye and close your right eye. Is the mark still in the circle? If it is, you are left-eyed.

Use the instructions in the magazine article to find out whether each person in your class is right-eyed or left-eyed.

Also find out whether they are right-handed or left-handed.

Write your results in a two-way table like this.

	Right-handed	Left-handed
Right-eyed		
Left-eyed		

Use your results to estimate the probability that someone selected at random is (a) left-eyed (b) left-handed (c) both.

Exercise

1 Gemma carries out a survey one morning.
She writes down the colours of cars passing her house.
Here are her results.

Red	Blue	Green	Black	White	Other
43	32	19	14	26	16

(a) How many cars are in Gemma's survey?

(b) Estimate the probability the next car that passes Gemma's house is
(i) red **(ii)** blue **(iii)** black.

(c) There are 500 cars in a car park near Gemma's house.
How many of these cars would you expect to be
(i) red **(ii)** blue **(iii)** black?

2 Jamie is practising ten-pin bowling.
He keeps a record of the number of pins he knocks down on each turn.

Turn	0	1	2	3	4	5	6	7	8	9	10
Pins down	7	2	3	8	5	11	14	10	12	5	3

Estimate the probability that on his next turn Jamie scores
(a) 6 **(b)** 10 **(c)** 4 or less **(d)** more than 7.

3 Ian is a football fan.
He keeps a record of his team's results one season.

Win	27
Lose	14
Draw	9

(a) From Ian's results, estimate the probability that the next match the team plays results in
(i) a win **(ii)** a draw.

(b) Do you think your answers are good estimates? Explain.

Activity Jo tosses two coins.

(a) What are the theoretical probabilities of Jo getting
(i) two heads **(ii)** two tails **(iii)** one head and one tail?

Toss two coins 20 times and record your results.

(b) Use your results to estimate the probabilities of getting
(i) two heads **(ii)** two tails **(iii)** one head and one tail.

(c) Compare your answers with the theoretical probabilities.

Toss the coins 20 more times and combine your results with the first 20 results.

(d) Use these combined results to estimate the probabilities again.

(e) Compare your answers with the theoretical probabilities.

Repeat until you have 100 results.

(f) How do the estimated probabilities change as the number of trials increases?

Finishing off

Now that you have finished this chapter you should be able to:

- understand that the probabilities of all mutually exclusive outcomes add up to 1
- identify all the possible mutually exclusive outcomes of two or more events
- use relative frequency to estimate probabilities from an experiment
- compare experimental and theoretical probabilities.

Review exercise

1 Michelle keeps a record of the number of goals scored by her hockey team in each match.

Based on her results, Michelle estimates these probabilities.

P(scoring no goals) = 0.15
P(scoring 1 goal) = 0.3
P(scoring 2 goals) = 0.25
P(scoring 3 goals) = 0.2

(a) What is the probability her team will score more than 3 goals?

Michelle's records show the team scored more than 3 goals 4 times.

(b) How many matches have they played?

(c) How many times did they score no goals?

2 Parvinda takes a card from an ordinary pack.
She writes down the suit of the card – Hearts, Diamonds, Clubs or Spades.
She puts the card back, shuffles the pack and takes out another card. She writes down which suit this card is from.

(a) Make a table or list to show all the possible outcomes for the two cards.

(b) Find

(i) P(both cards are the same suit)

(ii) P(both cards are red)

(iii) P(the cards are different colours)

(iv) P(a Heart and a Spade).

3 A die is biased so that P(1) = 0.3 and P(6) = 0.1
The probabilities of getting 2, 3, 4 or 5 are equal.

(a) Find P(2).

(b) Joe throws the biased die 200 times.
How many times would you expect him to get

(i) 1 **(ii)** 6 **(iii)** 5?

4 Michael and Joe are playing a game.

Michael puts some counters in a bag.
The counters are a mixture of red, yellow, green and blue.

Joe picks a counter out of the bag, records its colour, and puts it back.
He does this 50 times and records his results.

Red	Yellow	Green	Blue
15	12	5	18

(a) From Joe's results, estimate the probability of getting
(i) a blue counter **(ii)** a green counter **(iii)** a red counter.

(b) There are 20 counters in the bag altogether.
Use Joe's results to estimate how many counters there are of each colour.

5 In a game, Sue throws a red die and a blue die, each numbered 1 to 6.

She uses this rule to find her score.

Score = 2 × number on red die − number on blue die

(a) What is Sue's score if she gets
(i) a 3 on the red die and a 2 on the blue die
(ii) a 2 on the red die and a 3 on the blue die?

(b) Copy and complete this table to show all the possible outcomes.

Red die

Blue die	1	2	3	4	5	6
1	1	3				
2	0					
3	−1					
4						
5						
6						

(c) What is the probability that Sue's score is
(i) 0 **(ii)** negative **(iii)** 4 **(iv)** −1 **(v)** −5?

Activity

1 Throw two dice 100 times and record their total scores.
Use your results to estimate the probability of getting each number.
Compare your results with the theoretical probability.

2 Collect together the results from the whole class.
Use the combined results to estimate the probability of getting each number.
Compare these probabilities with the theoretical probabilities.

22 Inequalities

Inequality signs

The small end of an inequality sign (the point) is for the smaller number. The large end is for the larger number.

The Eiffel Tower $>$ Avonford Clock Tower

The Eiffel Tower is larger than Avonford Clock Tower

£2 $<$ £20

£2 is smaller than £20

Task

Copy out the following pairs of values. Put the correct inequality sign between them.

1	2 kg	☐	2100 g	**2**	$3\frac{3}{4}$	☐	3.7	**3**	10^2	☐	1000
4	2.6 m	☐	26 cm	**5**	£7.50	☐	755p	**6**	1500 m	☐	1.8 km

There are two other signs you need to know. They are $\leqslant$ and $\geqslant$

? **What do these two signs mean?**

? **Match these words with the correct inequality shown on the right.**

(a) Less than **(b)** At least **(c)** Greater than
(d) At most **(e)** Less than or equals **(f)** Greater than or equals

$\leqslant \quad <$

$> \quad \geqslant$

Avonford Town Win

At Saturday's match police said that at least 3500 fans attended. Club officials said that there were definitely less than 3800 people present.

? **What can you say about the number of people at Saturday's Avonford Town match?**

The Club's director writes down the following inequality:

3500 ≤ number at match < 3800

? **Explain what this means. Why are the signs different? Why do both inequality signs face the same way?**

Task

Write down 3 numbers that could be x in each of these inequalities.

1	$3 \leqslant x \leqslant 10$	**2**	$0 < x < 5$	**3**	$2 < x < 4$
4	$193 < x \leqslant 200$	**5**	$27 < x < 31$	**6**	$0 \leqslant x < 3$

? **The manager of Avonford Town always has less than 20 players in the squad, but he always has more than 15 players. Write an inequality to show this.**

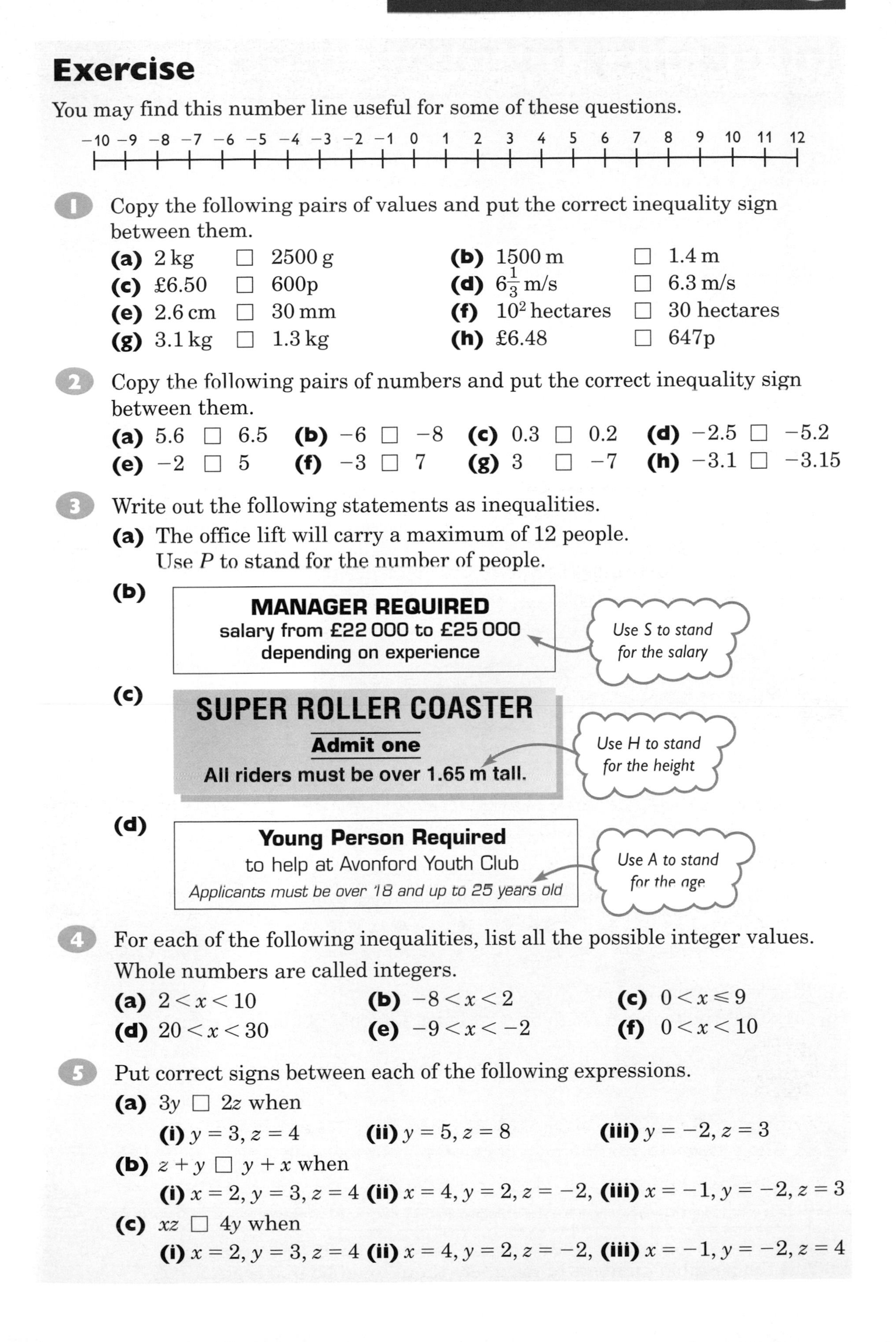

Exercise

You may find this number line useful for some of these questions.

−10 −9 −8 −7 −6 −5 −4 −3 −2 −1 0 1 2 3 4 5 6 7 8 9 10 11 12

1 Copy the following pairs of values and put the correct inequality sign between them.

(a) 2 kg □ 2500 g **(b)** 1500 m □ 1.4 m
(c) £6.50 □ 600p **(d)** $6\frac{1}{3}$ m/s □ 6.3 m/s
(e) 2.6 cm □ 30 mm **(f)** 10^2 hectares □ 30 hectares
(g) 3.1 kg □ 1.3 kg **(h)** £6.48 □ 647p

2 Copy the following pairs of numbers and put the correct inequality sign between them.

(a) 5.6 □ 6.5 **(b)** −6 □ −8 **(c)** 0.3 □ 0.2 **(d)** −2.5 □ −5.2
(e) −2 □ 5 **(f)** −3 □ 7 **(g)** 3 □ −7 **(h)** −3.1 □ −3.15

3 Write out the following statements as inequalities.

(a) The office lift will carry a maximum of 12 people.
Use P to stand for the number of people.

(b)

MANAGER REQUIRED
salary from £22 000 to £25 000
depending on experience

Use S to stand for the salary

(c)

SUPER ROLLER COASTER
Admit one
All riders must be over 1.65 m tall.

Use H to stand for the height

(d)

Young Person Required
to help at Avonford Youth Club
Applicants must be over 18 and up to 25 years old

Use A to stand for the age

4 For each of the following inequalities, list all the possible integer values. Whole numbers are called integers.

(a) $2 < x < 10$ **(b)** $-8 < x < 2$ **(c)** $0 < x \leqslant 9$
(d) $20 < x < 30$ **(e)** $-9 < x < -2$ **(f)** $0 < x < 10$

5 Put correct signs between each of the following expressions.

(a) $3y$ □ $2z$ when
(i) $y = 3, z = 4$ **(ii)** $y = 5, z = 8$ **(iii)** $y = -2, z = 3$

(b) $z + y$ □ $y + x$ when
(i) $x = 2, y = 3, z = 4$ **(ii)** $x = 4, y = 2, z = -2,$ **(iii)** $x = -1, y = -2, z = 3$

(c) xz □ $4y$ when
(i) $x = 2, y = 3, z = 4$ **(ii)** $x = 4, y = 2, z = -2,$ **(iii)** $x = -1, y = -2, z = 4$

Inequality number lines

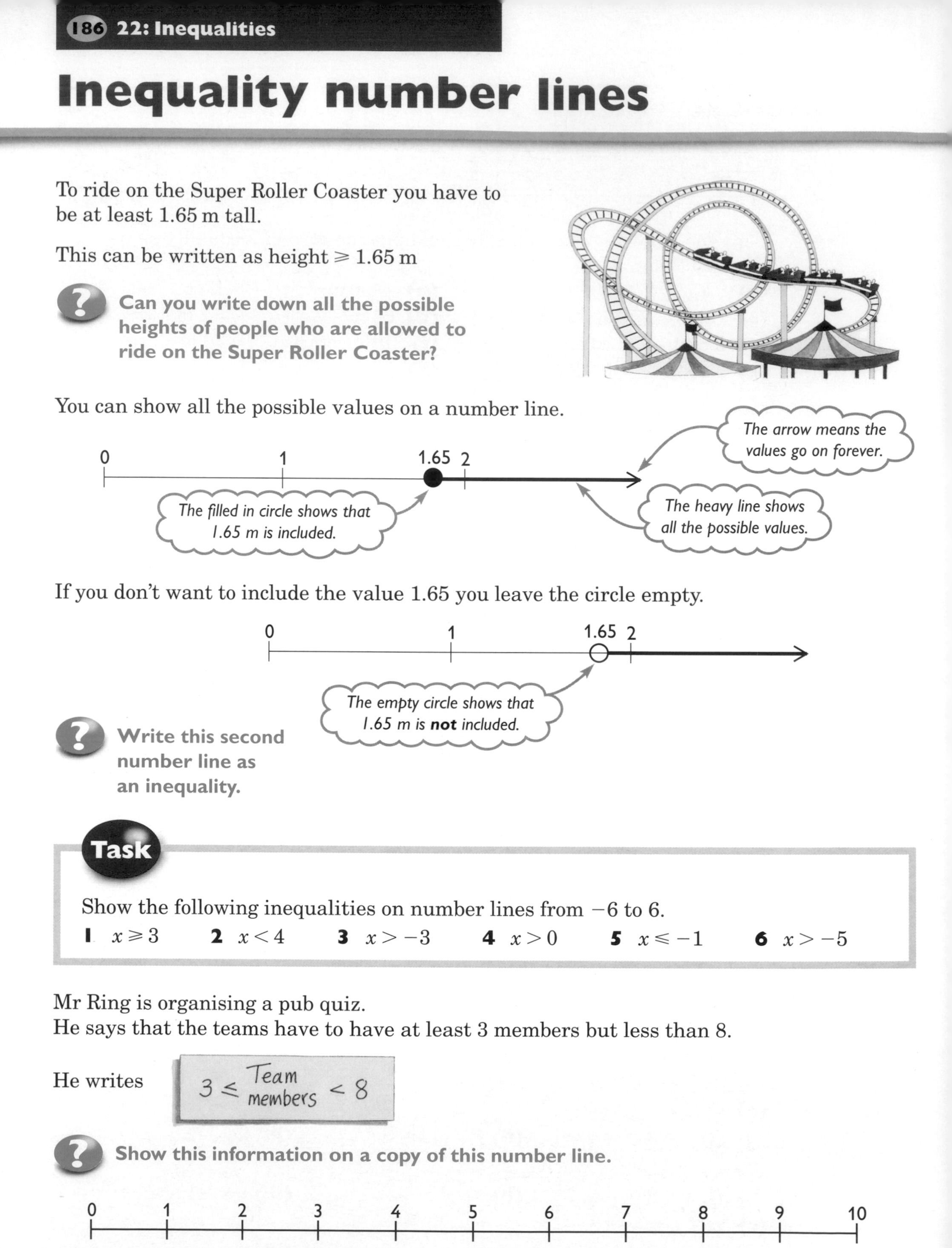

To ride on the Super Roller Coaster you have to be at least 1.65 m tall.

This can be written as height $\geqslant$ 1.65 m

Can you write down all the possible heights of people who are allowed to ride on the Super Roller Coaster?

You can show all the possible values on a number line.

If you don't want to include the value 1.65 you leave the circle empty.

Write this second number line as an inequality.

Task

Show the following inequalities on number lines from -6 to 6.

1 $x \geqslant 3$ **2** $x < 4$ **3** $x > -3$ **4** $x > 0$ **5** $x \leqslant -1$ **6** $x > -5$

Mr Ring is organising a pub quiz.
He says that the teams have to have at least 3 members but less than 8.

He writes $3 \leq$ Team members < 8

Show this information on a copy of this number line.

List all the possible numbers of members there could be in a team.

Exercise

1 To play for Avonford Youth Club table tennis team you have to be at least 12 but less than 20 years old.

(a) Write this information as an inequality and show it on a number line.

(b) Mariska plays for the team. List all her possible ages (in whole years).

2 Write down the inequalities represented by the following number lines.

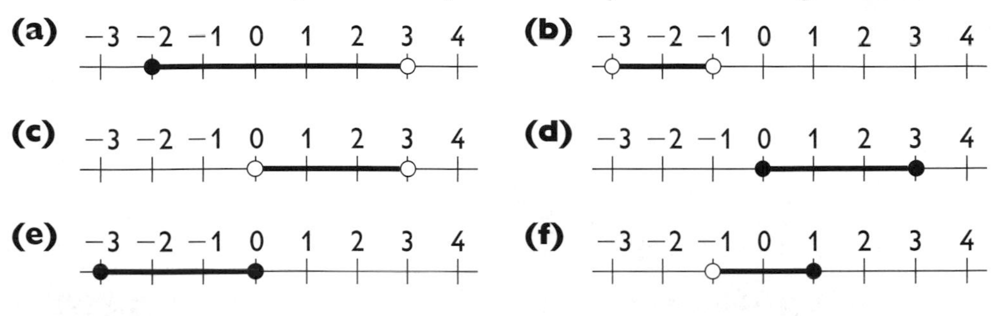

3 Draw and label number lines from -5 to 5 and show the following inequalities. Show each inequality on a separate number line.

(a) $x > 1$ **(b)** $x \leqslant 1$ **(c)** $x \geqslant -1$ **(d)** $x < -1$
(e) $x < 3$ **(f)** $-2 < x \leqslant 3$ **(g)** $-3 \leqslant x < -2$ **(h)** $-1 < x < 4$
(i) $-4 < x \leqslant -1$ **(j)** $0 \leqslant x \leqslant 3$

4 For each of the following situations, write the information in symbols using one or more inequalities.

(a) The distance x from home to school is more than 5 km.
(b) There are n people sitting on a 53-seat coach.
(c) Ian has p pence in his pocket.
He has enough money to buy a 35p bar of chocolate.
(d) Jane gets £x per week pocket money. She never gets less than £5.

Activity

This is a game for 2 players, **A** and **B**:

A thinks of a number between 1 and 100 inclusive.		**B** writes $1 \leqslant x \leqslant 100$
B guesses 60.	**A** says 'Too big.'	**B** writes $1 \leqslant x < 60$
B guesses 30.	**A** says 'Too small.'	**B** writes $30 < x < 60$
B guesses 45.	**A** says 'Too small.'	**B** writes $45 < x < 60$
B guesses 52.	**A** says 'Too small.'	**B** write $52 < x < 60$
B guesses 55.	**A** says 'That is right.'	**B** writes $x = 55$

B took 5 guesses to get **A**'s number.

Play this game with a partner. Take it in turns to think of a number. Can you guess your partner's number in less than 5 tries?

Solving inequalities

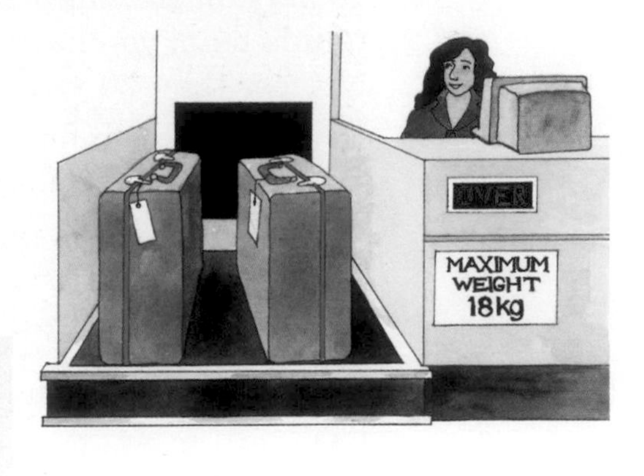

Jane is going on holiday with two suitcases.

She doesn't know how much each weighs but she does know that they are both the same weight.

She is told that she is over the weight limit of 18 kg.

She writes

Explain what the inequality tells you and what *x* stands for.

Jane then writes

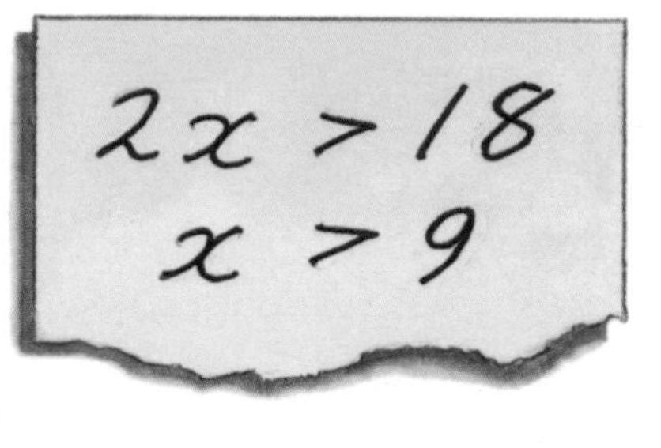

Explain what she has done.

The weight of each suitcase is still unknown but Jane does know that each one is more than 9 kg

Copy and complete the following table.
Put inequality signs between each pair of numbers.
Some have already been done.
Be careful with the negative numbers!

	Do the same to each of the numbers			
	+2	**−2**	**×2**	**÷2**
8 > 4	$10 > 6$			
−2 < 6		$-4 < 4$		$-1 < 3$
−6 < −2			$-12 < -4$	

Look at the row of inequalities.
Are all the inequalities still true? What else do you notice?

Task

Follow this example of how to solve an inequality.

Always set your working out like this.

$2x + 3 > 23$

$2x > 20$ *(subtract 3 from both sides)*

$x > 10$ *(divide both sides by 2)*

Check that some integers that satisfy $x > 10$ fit into the original inequality (make it a true statement).

Do not multiply or divide both sides of an inequality by a negative (−) number.
The Investigation opposite will help you understand why not.

Exercise

1 Solve the following inequalities.

(a) $3x \leqslant 15$ **(b)** $2x > 8$ **(c)** $x + 4 < 10$

(d) $x - 6 > 12$ **(e)** $2x + 4 > 14$ **(f)** $3x - 3 < 3$

(g) $\frac{1}{2}x - 4 < 2$ **(h)** $8 > x + 3$ **(i)** $3x + 4 \geqslant 16$

2 For each of these inequalities write down 4 possible values of x.

(a) $3x > 10$ **(b)** $x + 3 < 9$ **(c)** $x - 2 > 7$

(d) $x + \frac{1}{2} \geqslant 3$ **(e)** $3x \leqslant 27$ **(f)** $4x \geqslant 10$

(g) $\frac{x}{2} < 3$ **(h)** $2x + 1 > 6$ **(i)** $10 + 2x > 15$

3 Write each of these statements as an inequality and solve it.

(a) Jasmine buys 5 identical cakes. She receives change from £2. Use C to stand for the cost of each cake in pence.

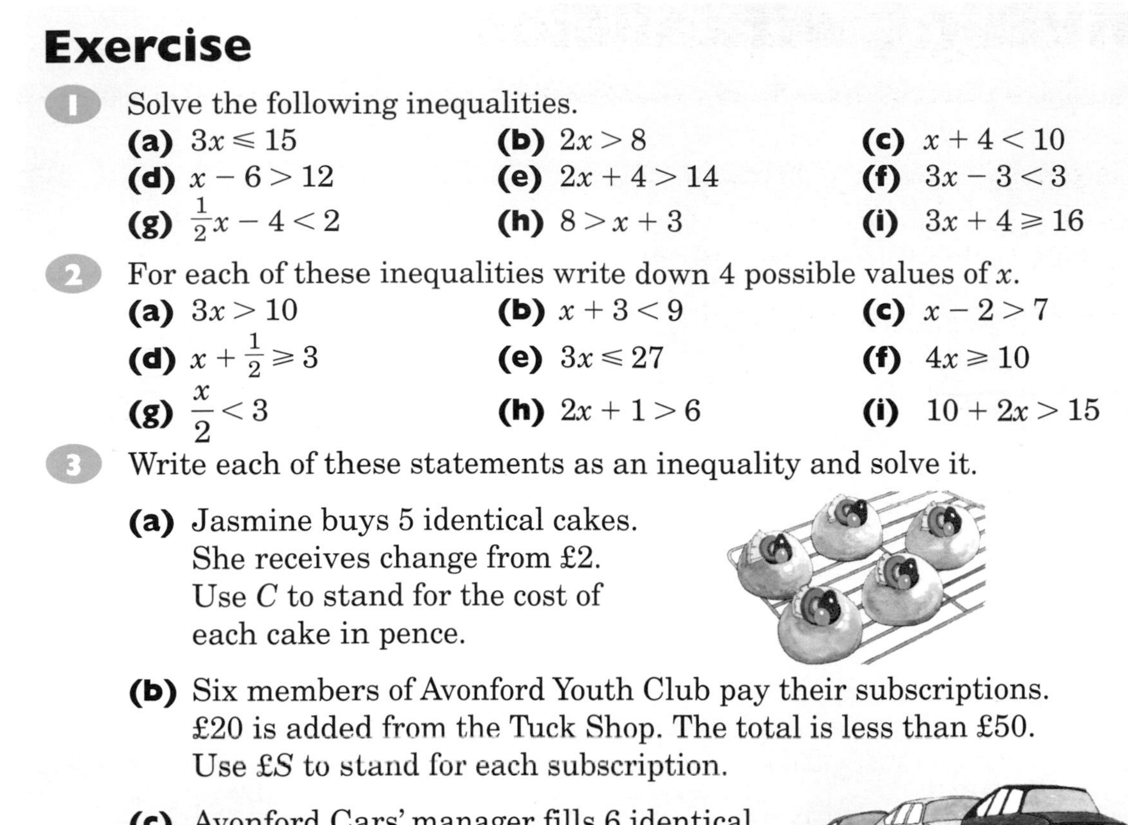

(b) Six members of Avonford Youth Club pay their subscriptions. £20 is added from the Tuck Shop. The total is less than £50. Use £S to stand for each subscription.

(c) Avonford Cars' manager fills 6 identical new cars with petrol. He also puts 90 litres into his car transporter. When he checks the pump, he sees that he has taken more than 350 litres. Use L to stand for the number of litres of petrol in each car.

Investigation

Look at the inequality $-2x < 6$

1 Which of the following numbers obey this inequality and which do not?

$$-6, -5, -4, -3, -2, -1, 0, 1, 2, 3, 4$$

2 Anya and Daniel try to solve the inequality.

Anya

$-2x < 6$

Divide by -2 $\quad -2x \div -2 < 6 \div -2$

Answer $\quad x < -3$ ✗

Daniel

Add $2x$ $\quad -2x < 6$

Subtract 6 $\quad -2x + 2x - 6 < 6 + 2x - 6$

$-6 < 2x$

Divide by 2 $\quad -6 \div 2 < 2x \div 2$

$-3 < x$

So $\quad x > -3$ ✓

They get different answers.

How do you know that Daniel's answer is right? Always use Daniel's method!

Finishing off

Now that you have finished this chapter you should be able to:

- recognise and use the inequality symbols: $<$, $>$, $\leqslant$, $\geqslant$
- write information as inequalities
- show inequalities on a number line
- solve inequalities.

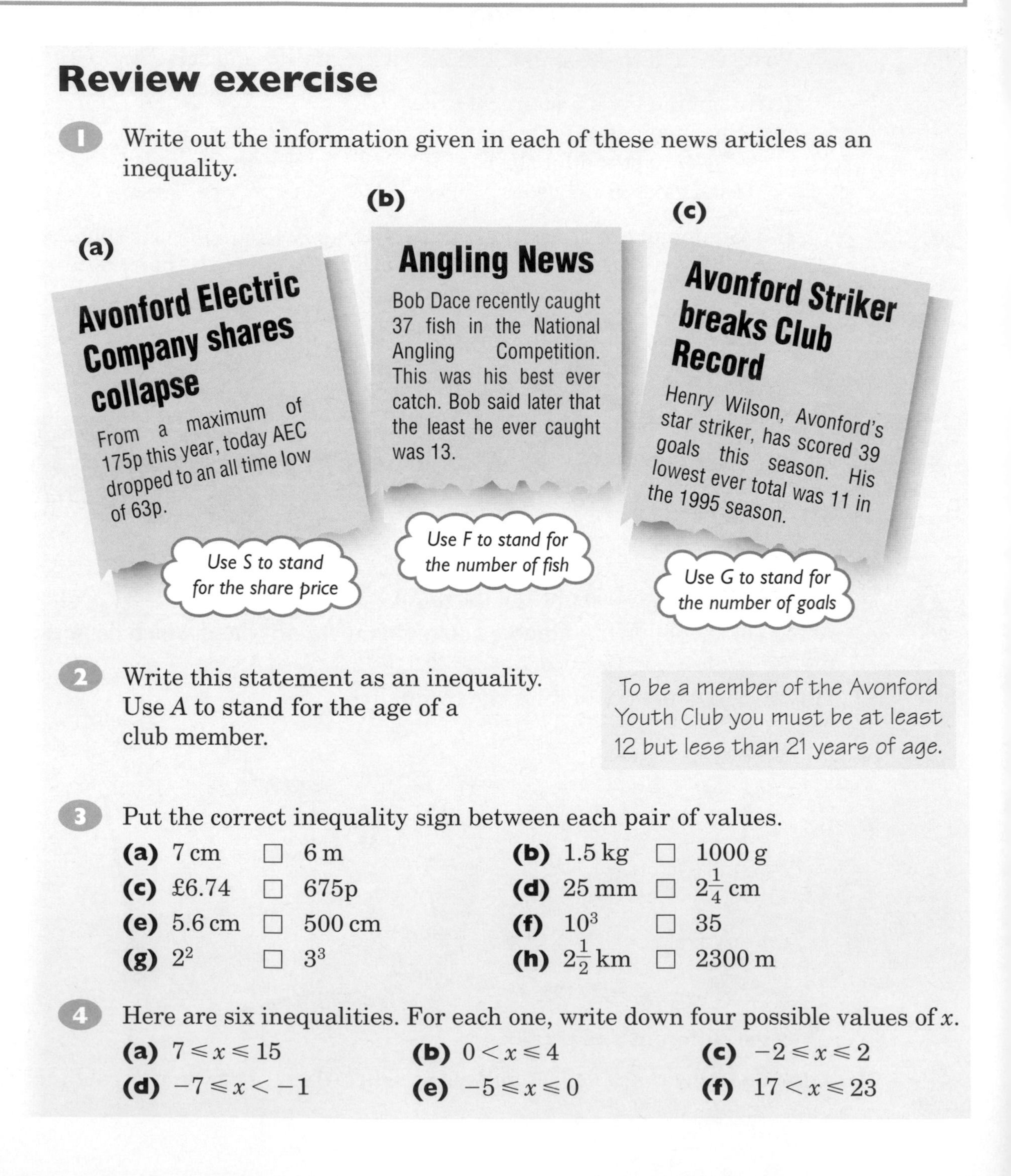

Review exercise

1 Write out the information given in each of these news articles as an inequality.

(a)

Avonford Electric Company shares collapse

From a maximum of 175p this year, today AEC dropped to an all time low of 63p.

Use S to stand for the share price

(b)

Angling News

Bob Dace recently caught 37 fish in the National Angling Competition. This was his best ever catch. Bob said later that the least he ever caught was 13.

Use F to stand for the number of fish

(c)

Avonford Striker breaks Club Record

Henry Wilson, Avonford's star striker, has scored 39 goals this season. His lowest ever total was 11 in the 1995 season.

Use G to stand for the number of goals

2 Write this statement as an inequality. Use A to stand for the age of a club member.

To be a member of the Avonford Youth Club you must be at least 12 but less than 21 years of age.

3 Put the correct inequality sign between each pair of values.

(a) 7 cm ☐ 6 m
(b) 1.5 kg ☐ 1000 g
(c) £6.74 ☐ 675p
(d) 25 mm ☐ $2\frac{1}{4}$ cm
(e) 5.6 cm ☐ 500 cm
(f) 10^3 ☐ 35
(g) 2^2 ☐ 3^3
(h) $2\frac{1}{2}$ km ☐ 2300 m

4 Here are six inequalities. For each one, write down four possible values of x.

(a) $7 \leqslant x \leqslant 15$
(b) $0 < x \leqslant 4$
(c) $-2 \leqslant x \leqslant 2$
(d) $-7 \leqslant x < -1$
(e) $-5 \leqslant x \leqslant 0$
(f) $17 < x \leqslant 23$

5 Write down what inequality each of these number lines represent.

(a) 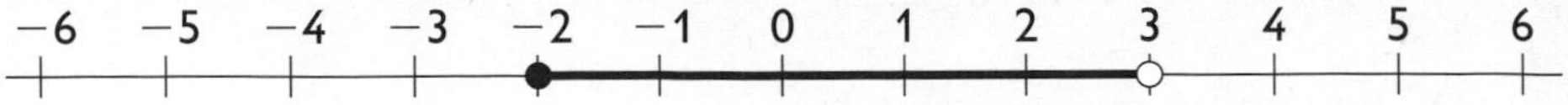 −6 −5 −4 −3 −2 −1 0 1 2 3 4 5 6

(b) 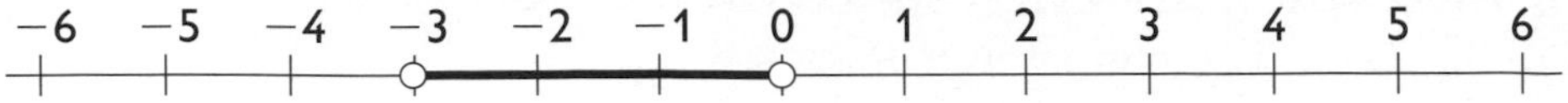 −6 −5 −4 −3 −2 −1 0 1 2 3 4 5 6

(c) 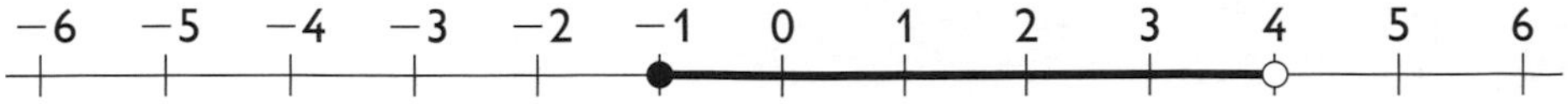 −6 −5 −4 −3 −2 −1 0 1 2 3 4 5 6

(d) 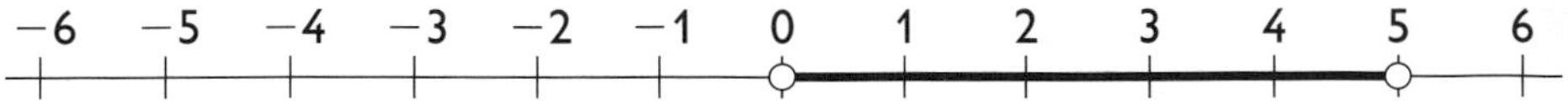 −6 −5 −4 −3 −2 −1 0 1 2 3 4 5 6

(e) 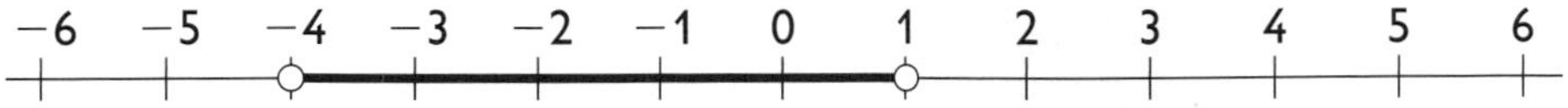 −6 −5 −4 −3 −2 −1 0 1 2 3 4 5 6

(f) 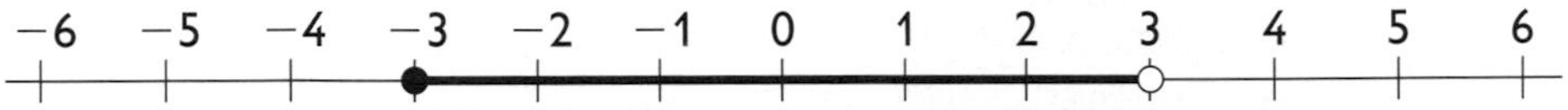 −6 −5 −4 −3 −2 −1 0 1 2 3 4 5 6

6 Draw number lines from −6 to 6 to show each of these inequalities.

(a) $-5 < x < 3$ **(b)** $0 < x < 5$

(c) $-3 \leqslant x < 0$ **(d)** $-4 \leqslant x \leqslant 4$

(e) $2\frac{1}{2} < x \leqslant 5$ **(f)** $-2 < x < 3\frac{1}{2}$

(g) $-4 \leqslant x \leqslant 6$ **(h)** $0 < x \leqslant 5$

7 Solve these inequalities. Show all your working clearly.

(a) $4x - 3 > 29$ **(b)** $3x + 4 \geqslant 16$

(c) $3x - 2 < 10$ **(d)** $2x - 7 > 8$

(e) $5x + 7 < 32$ **(f)** $8x + 6 \geqslant 38$

(g) $3(2x - 2) > 6$ **(h)** $4(3x + 2) < 44$

8 Put the correct sign between each of the following expressions.

(a) $(2a + \mathrm{b})$ ☐ $(c + 2b)$ when

(i) $a = 2, b = 3, c = 4$ **(ii)** $a = 2, b = 0, c = 1$ **(iii)** $a = 5, b = -1, c = 5$

(b) $c - a$ ☐ b when

(i) $a = 2, b = 3, c = 4$ **(ii)** $a = 2, b = -3, c = 0$ **(iii)** $a = 3, b = 7, c = 4$

(c) ab ☐ bc when

(i) $a = 2, b = 3, c = 4$ **(ii)** $a = 1, b = 1, c = 0$ **(iii)** $a = 5, b = 2, c = 1$

(d) abc ☐ $10b$ when

(i) $a = 2, b = 3, c = 4$ **(ii)** $a = 1, b = 1, c = 1$ **(iii)** $a = 2, b = 1, c = 0$

(e) $3a^2$ ☐ b^2 when

(i) $a = 2, b = 3$ **(ii)** $a = 3, b = 5$ **(iii)** $a = -3, b = -5$

23 Pythagoras

Pythagoras' rule

Look at these special triangles. They are all *right-angled*. The side opposite the right angle is called the **hypotenuse**.

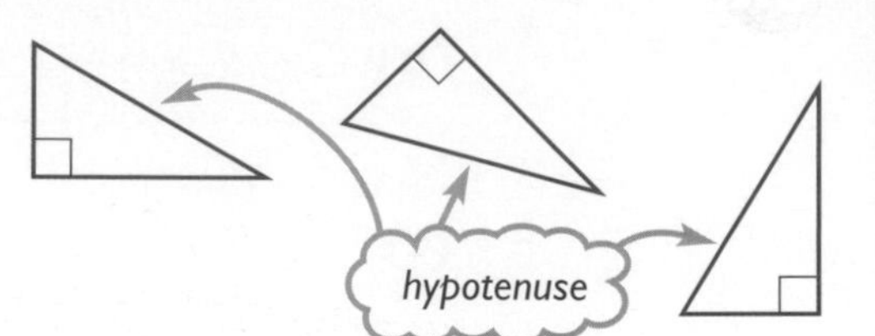

Why is the hypotenuse always the longest side?

Task

You will need centimetre squared paper, scissors and glue.

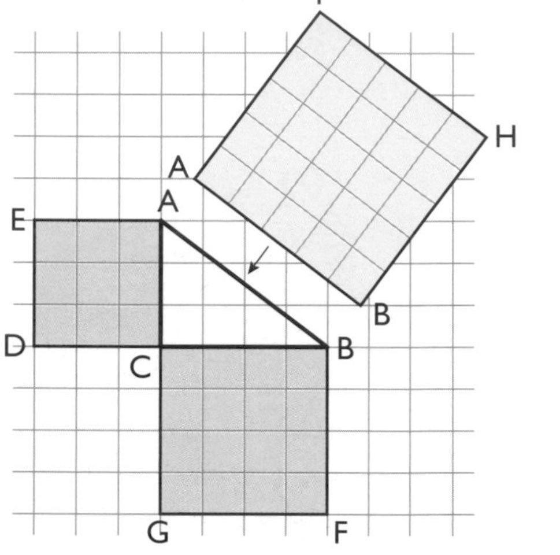

1. Draw and label triangle ABC as shown. $\hat{ACB}$ is a right angle. AC = 3 cm and CB = 4 cm.
2. Draw squares ACDE and CBFG.
3. Measure length AB. Draw square ABHI *on a separate piece of squared paper*.
4. Glue square ABHI onto hypotenuse AB.
5. Shade the two smaller squares in one colour. Shade the largest square in a different colour.
6. Repeat steps 1 to 5 for right-angled triangles with sides AC and CB as shown in the table. Copy and complete the table.

Shorter sides		Hypotenuse	Sum of areas of smaller squares	Area of largest square
AC	CB	AB		
3 cm	4 cm		$9\text{ cm}^2 + 16\text{ cm}^2 = 25\text{ cm}^2$	
5 cm	12 cm			
2.5 cm	6 cm			

What do you notice about the areas of the squares for each triangle?

A right-angled triangle has shorter sides AC = 9 cm and CB = 40 cm. What is the length of the hypotenuse?

Pythagoras' rule: The square of the hypotenuse of a right-angled triangle is equal to the sum of the squares of the other two sides.

Write down the equation connecting *a*, *b* and *c*.

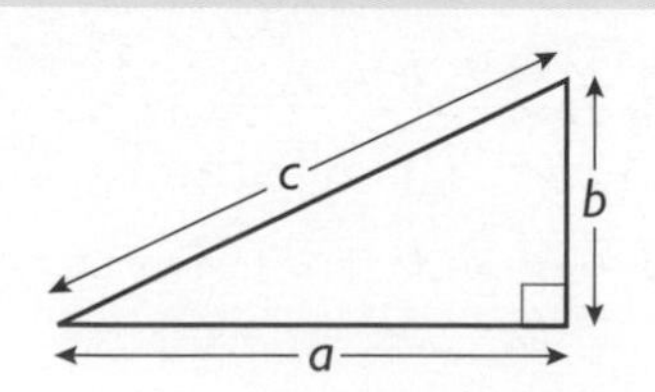

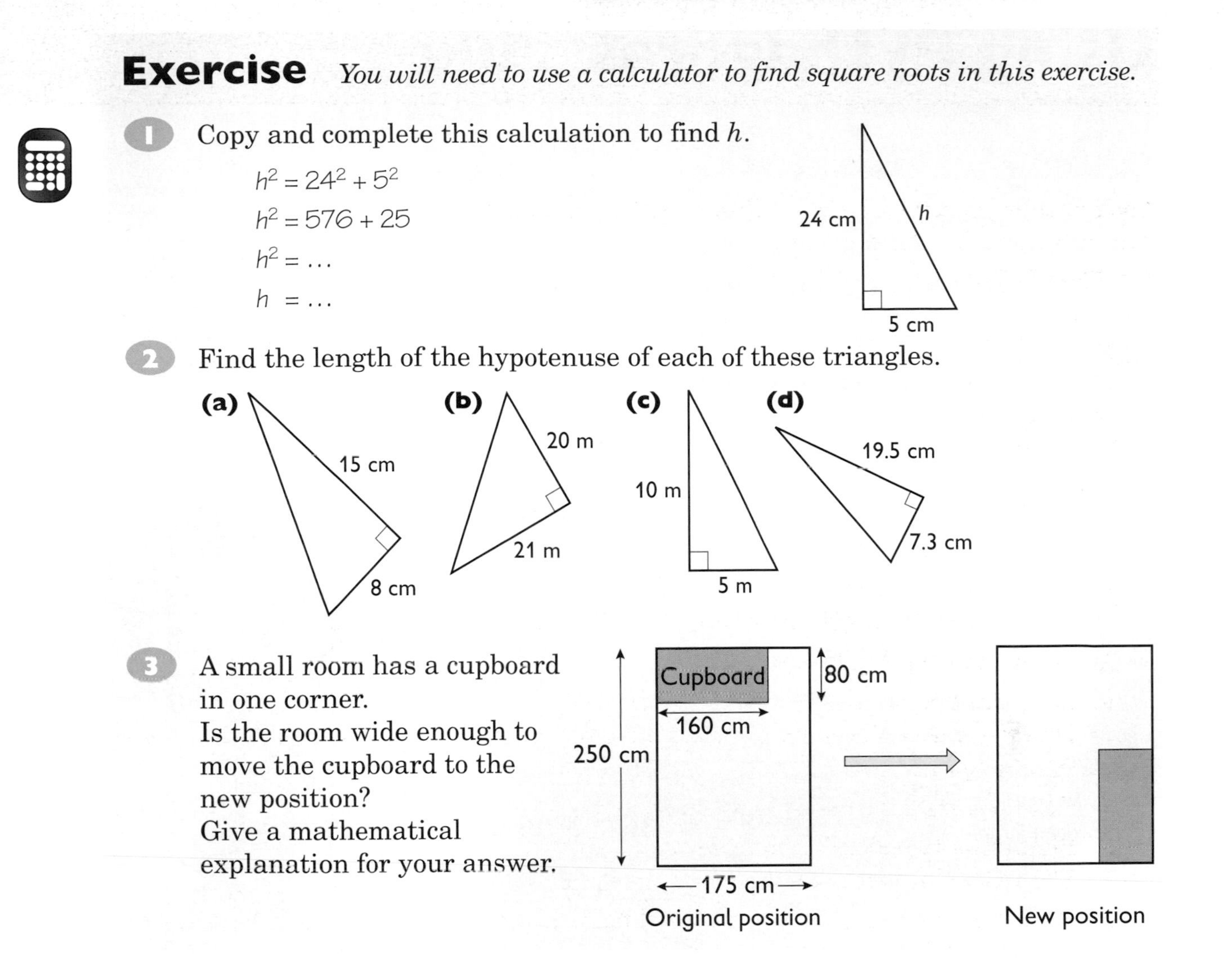

Exercise

You will need to use a calculator to find square roots in this exercise.

1. Copy and complete this calculation to find h.

   $h^2 = 24^2 + 5^2$

   $h^2 = 576 + 25$

   $h^2 = \ldots$

   $h \ = \ldots$

2. Find the length of the hypotenuse of each of these triangles.

3. A small room has a cupboard in one corner.
   Is the room wide enough to move the cupboard to the new position?
   Give a mathematical explanation for your answer.

Investigation

Look at this triangle.
The longest side is l cm and the largest angle is $X°$.
The two shorter sides are a cm and b cm.
Draw a triangle with $a = 6$, $b = 8$ and $l = 10$. Measure X.
Copy and complete the table below.

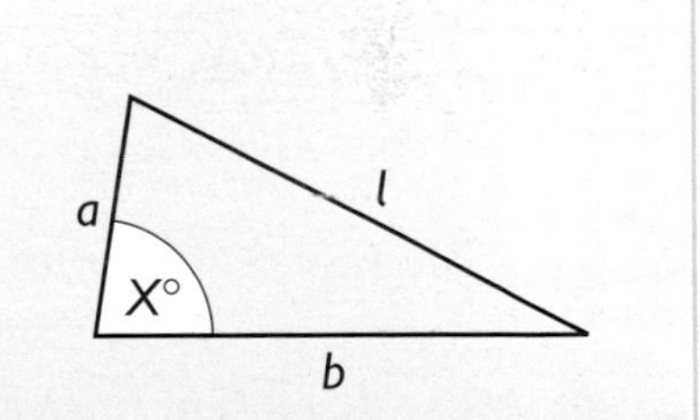

a	b	l	X	Compare l^2 to $a^2 + b^2$	Compare X to 90°
				Answer <, = or >	Answer < (acute), = (right) or > (obtuse)
6	8	10	90°	=	=
6	8	12			
6	8	9			

Now draw some triangles of your own. Enter your results in the table.
What do you notice?

Using Pythagoras' rule

Finding a shorter side

Gill is a decorator.
She uses a ladder 6 m long.
The safety instructions on the ladder say that the foot of the ladder must be at least 1.75 m from the wall on horizontal ground.
Gill calculates the maximum height her ladder will reach up a vertical wall.

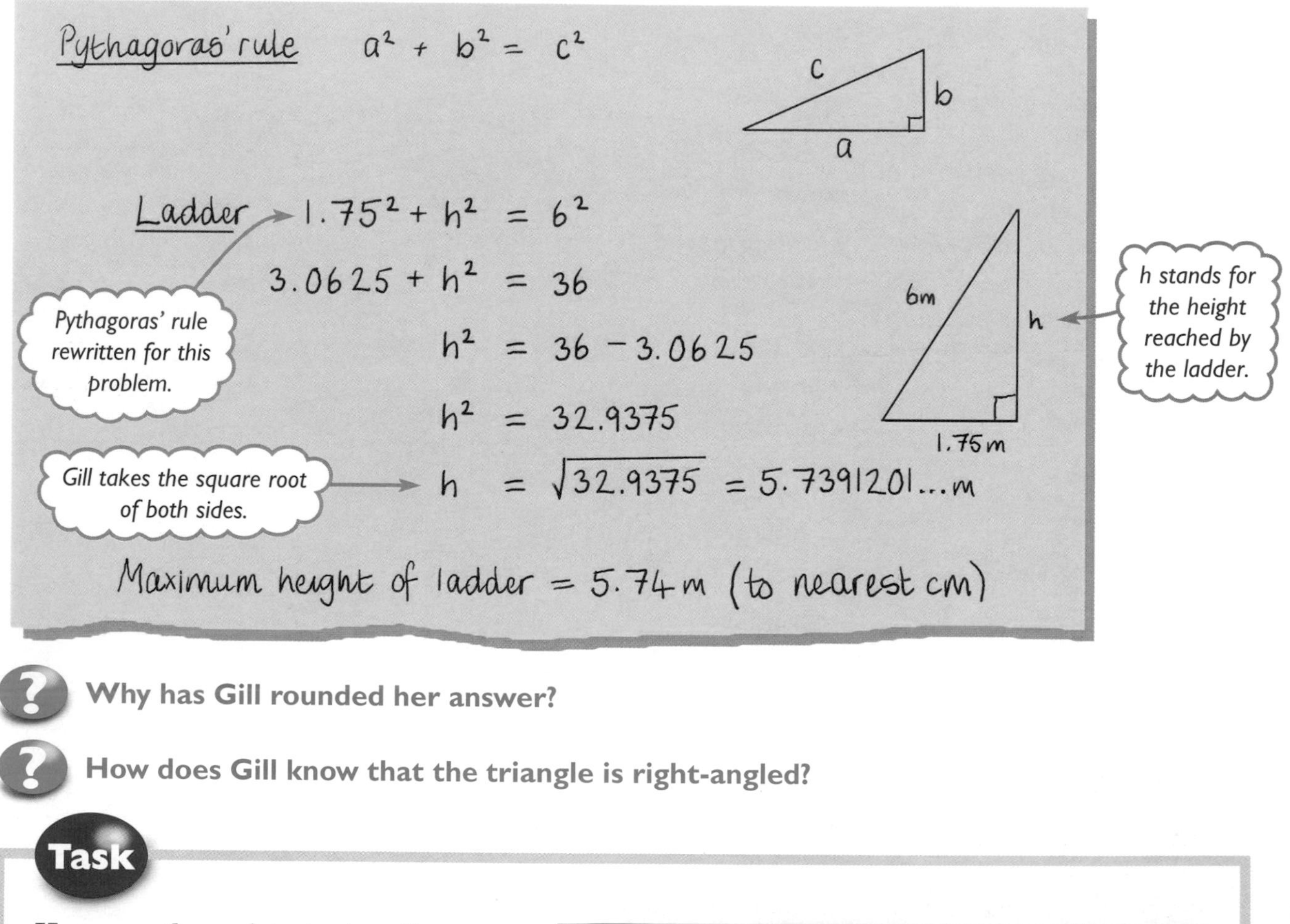

Why has Gill rounded her answer?

How does Gill know that the triangle is right-angled?

Task

Here are the safety instructions for Gill's ladder.

ATTENTION!
Foot of ladder must be between 1.75 m and 2.35 m from vertical surface.
Ground must be horizontal.

Calculate the *minimum* height the ladder can reach up the wall.

What could happen if the safety instructions are not obeyed?

Exercise

1 Find the lengths of the sides marked with a letter in each of these triangles.
Give your answers correct to one decimal place.

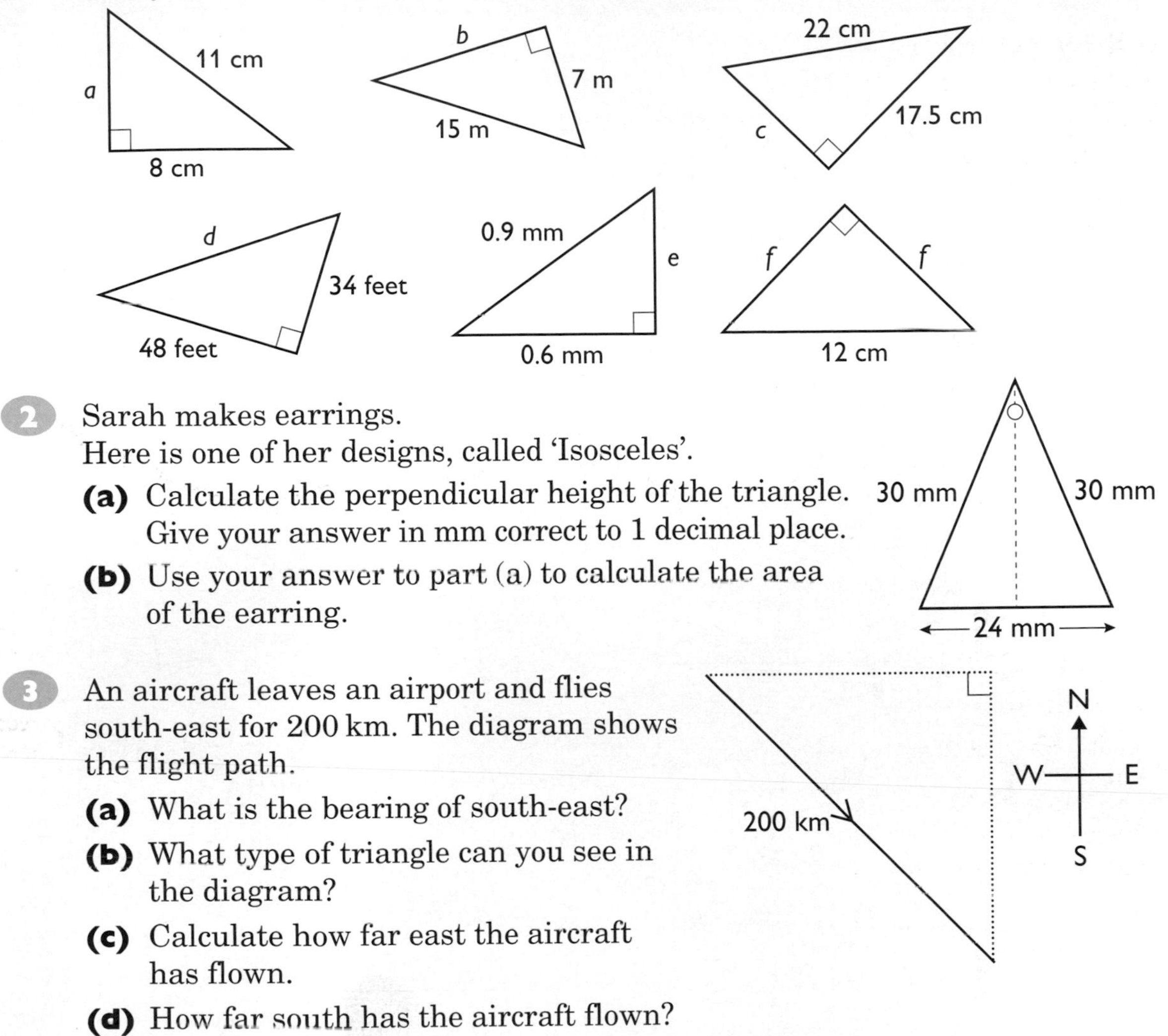

2 Sarah makes earrings.
Here is one of her designs, called 'Isosceles'.

(a) Calculate the perpendicular height of the triangle. Give your answer in mm correct to 1 decimal place.

(b) Use your answer to part (a) to calculate the area of the earring.

3 An aircraft leaves an airport and flies south-east for 200 km. The diagram shows the flight path.

(a) What is the bearing of south-east?

(b) What type of triangle can you see in the diagram?

(c) Calculate how far east the aircraft has flown.

(d) How far south has the aircraft flown?

Investigation

Builders use Pythagoras' rule to make sure that corners are 'square'.
They measure around the corner and 'three-four-five' it.

How does this ensure the corner is a right angle?
How are the measurements shown in the picture an example of 'three-four-five'?

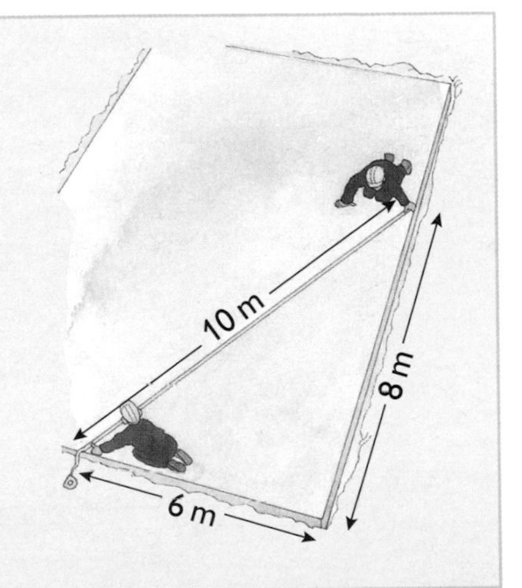

The number set 3, 4, 5 is called a **Pythagorean triple**.
Find as many different Pythagorean triples as you can.
Do not include any which are simply a multiple of one you have already found.

Finishing off

Now that you have finished this chapter you should know:

- Pythagoras' rule:

  $a^2 + b^2 = c^2$

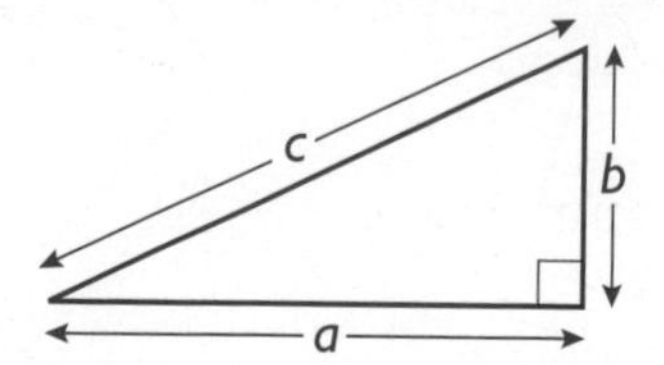

- how to use Pythagoras' rule to find the lengths of right-angled triangle sides.

Review exercise

1 Calculate the length of the hypotenuse for each of these triangles.

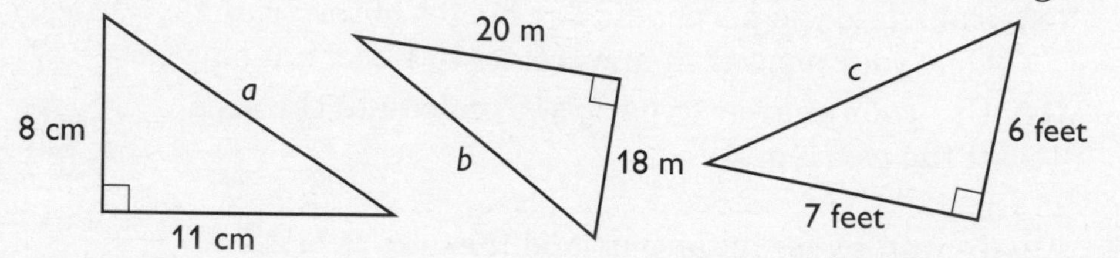

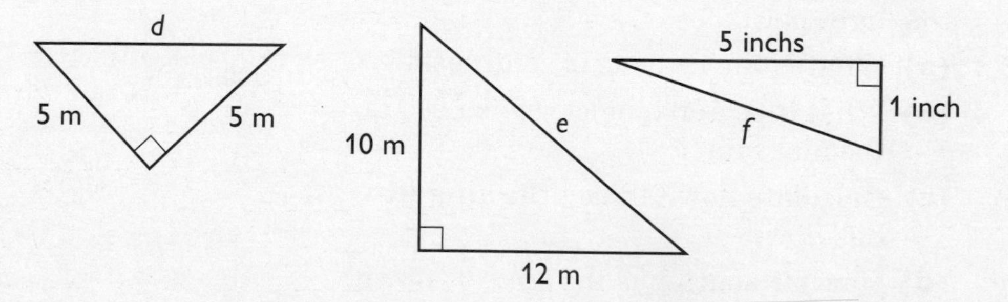

2 Calculate the length of the unknown sides in each of these triangles.

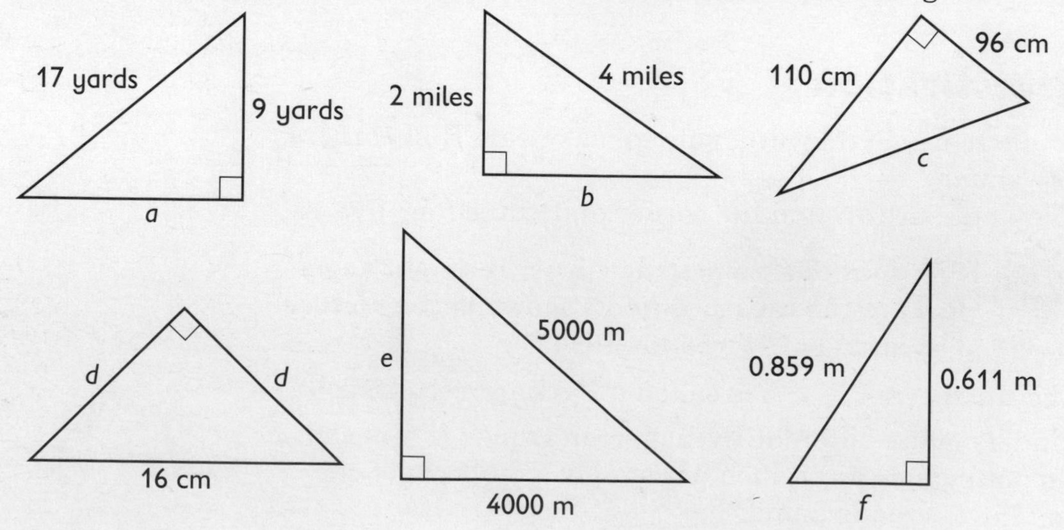

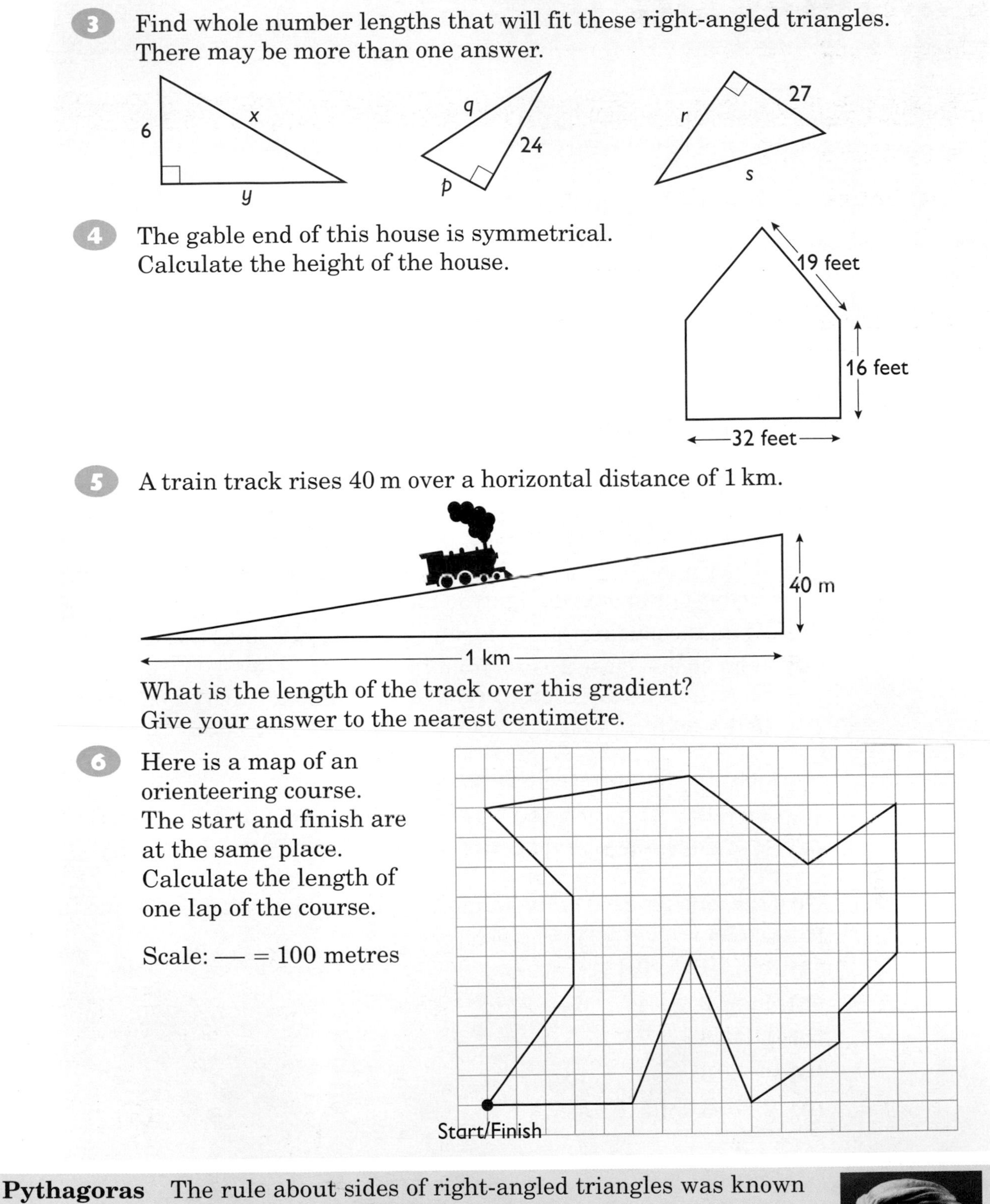

3 Find whole number lengths that will fit these right-angled triangles. There may be more than one answer.

4 The gable end of this house is symmetrical. Calculate the height of the house.

5 A train track rises 40 m over a horizontal distance of 1 km.

What is the length of the track over this gradient? Give your answer to the nearest centimetre.

6 Here is a map of an orienteering course. The start and finish are at the same place. Calculate the length of one lap of the course.

Scale: — = 100 metres

Pythagoras The rule about sides of right-angled triangles was known in Babylonia over 2500 years ago. It was first proven to be true for *all* right-angled triangles by Pythagoras. Pythagoras was a Greek philosopher who lived from 560 to 480 BC.

Pythagoras

24 Review

This chapter contains questions on work you covered in Years 7 and 8. You can use it to help you revise.

Numbers and the number system

Do not use your calculator for the questions on the next three pages.

1 Round the following numbers to the nearest 100.

(a) 23 456 **(b)** 132.6 **(c)** 6395 **(d)** 99 **(e)** 999

2 John has 5 cards. He arranges them to show the number

(a) Write this number in words.
(b) What is the largest number that John can make with these cards? Write your answer in figures and in words.
(c) Write the smallest possible number in figures and in words.
(d) John chooses just 2 cards from the five.
Show how he can choose **(i)** a prime number **(ii)** a square number **(iii)** a number which is a multiple of 5 and 3.

3 Work out the following calculations.

(a) $23 + 12 + 48$ **(b)** $204 - 98$ **(c)** 123×7 **(d)** $522 \div 6$
(e) $12.4 + 0.7 + 5.2$ **(f)** $18.3 - 3.7$ **(g)** $267 \div 8$ **(h)** 45×23

4 Give the rule for each of the number sequences below and continue the pattern for 3 more terms.

(a) 1 4 9 – – –
(b) 1 2 4 8 – – –
(c) 1 8 27 – – –
(d) 1 3 6 10 – – –
(e) 2 3 5 – – –

5

For each game, tickets are chosen from the numbers 1 to 100. Which game has the most winning tickets?

Remember, do not use your calculator for the questions on this page.

6 Write the following numbers correct to 2 decimal places.

(a) 12.423 **(b)** 5.746 **(c)** 0.3675 **(d)** 0.3446 **(e)** 0.009

7 Give the following measurements to the nearest cm.

(a) 2.3 cm **(b)** 12.7 cm **(c)** 42.16 cm **(d)** 3.634 m **(e)** 26 mm

8 Find the following fractions or percentages.

(a) $\frac{3}{5}$ of 25 **(b)** 20% of 150 **(c)** $\frac{2}{9}$ of £1.80

(d) 13% of £5 **(e)** $33\frac{1}{3}$% of £2.40 **(f)** $\frac{7}{25}$ of £42

9 Write each of these lists of numbers in order, starting with the smallest.

(a) 7, −2, 4, −1, −10

(b) 0.7, 0.32, 0.08, 0.513, 0.0138

(c) $\frac{1}{2}$, $\frac{1}{5}$, $\frac{1}{7}$, $\frac{3}{4}$, $\frac{6}{7}$

10 A photograph frame measures 35 mm by 27 mm (to the nearest mm).

(a) Write each measurement as a range of values.

(b) Calculate the largest possible value for the perimeter of the frame.

11 Work out the following multiplications and divisions.

(a) 21×100 **(b)** $1400 \div 100$ **(c)** 21.3×1000 **(d)** $12 \div 100$

(e) 13×0.1 **(f)** 540×0.01 **(g)** $6 \div 0.1$ **(h)** 200×300

(i) 1.2×400 **(j)** $600 \div 20$ **(k)** $2.4 \div 20$ **(l)** 0.5×3000

12 Write the following numbers correct to 2 significant figures.

(a) 13.62 **(b)** 123 **(c)** 45 632 **(d)** 0.025 36 **(e)** 0.002 46

13

Exchange rates

On a particular day £1 will buy: Canadian $2.09 2.3 Swiss francs 1.6 Euros US$1.37

Round the numbers to a suitable degree of accuracy, to find approximate answers to these questions.

(a) The cost in £s of

(i) a bicycle costing 150 Euros **(ii)** a pair of jeans costing US $28
(iii) an ice cream costing 2 Swiss francs
(iv) a cable car ride costing Canadian $29

(b) Jimmy has £190.
Approximately how many Euros can he buy?

Remember, do not use your calculator for the questions on this page.

14 Copy and complete this table.

Fraction	Decimal	Percentage
$\frac{1}{4}$		
	0.75	
		30%
	0.6	
$\frac{2}{5}$		
	0.04	
		120%

15 **(a)** Divide 26 in the ratio 4 : 9. **(b)** Divide 72 in the ratio 1 : 2 : 3.

(c) Sally is 6 and Jane is 8.
They receive £56 and divide it between them in the ratio of their ages.
How much does Jane receive?

16 These are the ingredients needed to make 18 scotch pancakes:

100 grams self raising flour, 30 grams caster sugar, 1 egg and 125 ml milk.

(a) List the ingredients needed to make 72 pancakes.

(b) Mrs Brown only has 75 grams of flour.
She has plenty of the other ingredients.
(i) How many pancakes can she make?
(ii) What fraction of an egg does she need?

17 Simplify the following ratios.

(a) 12 : 18 **(b)** 4 : 8 : 12 **(c)** $\frac{1}{2} : 3$ **(d)** $\frac{1}{2} : \frac{3}{4} : \frac{5}{8}$

18 Find the highest common factor of

(a) 12 and 18 **(b)** 4, 8 and 12.

(c) How have you used these answers in question 18?

19 Find the lowest common multiple of

(a) 2 and 5 **(b)** 4 and 6 **(c)** 12 and 15 **(d)** 3, 4 and 5.

20 Work out the following. (Your answers to question 19 will help you.)

(a) $\frac{1}{2} + \frac{2}{5}$ **(b)** $\frac{3}{4} - \frac{1}{6}$ **(c)** $\frac{7}{15} + \frac{5}{12}$ **(d)** $\frac{1}{3} + \frac{1}{4} - \frac{1}{5}$

21 Look at these numbers: 24, 112, 216, 225, 441, 2744.

(a) Write each of them as the product of its prime factors.

(b) Which are square numbers? What are their square roots?

(c) Write down the cube roots of 216 and 2744.

Calculations

Use your calculator for the questions on this page.

1 For each of the calculations below

(i) use your calculator to find the answer

(ii) do an approximate calculation to check your answer.

(a) 234×15 **(b)** 89×2.7 **(c)** 18.5×231

(d) $59.3 \div 9.8$ **(e)** $121 \div 0.87$

2 Fiona has used her calculator to work out the calculation $\dfrac{15 \times 7.2}{2.4 + 3.1}$

She gets the answer 48.1.

(a) Explain how you can tell that Fiona's answer is wrong.

(b) Find the correct answer. **(c)** Explain Fiona's mistake.

3 Work out the following, giving your answers to a suitable degree of accuracy.

(a) 2.3^4 **(b)** $1\frac{2}{3} + 4\frac{5}{9}$ **(c)** $(-2.7) \times (-18.1) \times (+53.4)$

(d) $(36.7 + 4.9)^4$ **(e)** $\dfrac{45 + 63}{9}$ **(f)** $\dfrac{63 - 18}{8 + 5}$

(g) $\dfrac{3.7 \times 3.42}{1.2 - 0.7}$ **(h)** $\dfrac{0.7 \times 0.35}{5 \times 14}$ **(i)** $\sqrt{45}$

(j) $\sqrt{(15 + 72)}$ **(k)** $\sqrt{(6.7^2 + 4.5^2)}$ **(l)** $\sqrt[3]{52}$

4

1 kg ≈ 2.2 pounds	1 metre ≈ 39 inches	1 stone = 14 pounds
1 litre ≈ 1.8 pints	1 km ≈ 0.625 miles	1 gallon = 8 pints

(a) George weighs 63 kg. What is this in **(i)** pounds **(ii)** stones and pounds?

(b) The distance from Paris to Geneva is 537 km. How far is this in miles?

(c) A petrol tank can hold 85 litres of petrol.
How much is this in **(i)** pints **(ii)** gallons and pints?

(d) Mary is 1.65 metres tall.
What is her height in **(i)** inches **(ii)** feet and inches?

5 8.75

The calculator display shows the result of dividing 175 by 20.

(a) The calculation involved hours. Write the answer in hours and minutes.

(b) The number is in feet. Write the answer in feet and inches.

(c) Write 8.75 pounds in pounds and ounces.

6 Write the following amounts in the units shown.

(a) 5 hours 25 min in hours **(b)** 2 km and 15 m in km

(c) 3 pounds 6 ounces in pounds **(d)** 9 feet 5 inches in feet

Algebra

1 Look at these patterns.

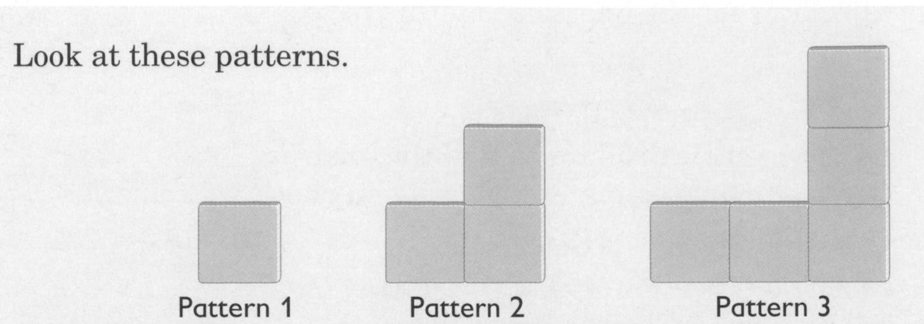

(a) How many tiles are added to make each new pattern?

(b) How many tiles are needed to make pattern number 10?

Marie uses 113 tiles to make a pattern.

(c) What is the number of the pattern that Marie makes?

Marie investigates the perimeter of each pattern.

Pattern	1	2	3	4	5	6
Perimeter	4	8	?	?	?	?

(d) Copy and complete Marie's table.

(e) What is the perimeter of pattern number 20?

(f) Which pattern has a perimeter of 112.

2 Darren organizes a trip to Avonford Adventure Park for his youth group.

Avonford Adventure Park

Coach hire £72
Entry to park £7 per person

He uses this formula to work out the cost, £C, per person.

$$C = 7 + \frac{72}{n}$$

The letter n stands for the number of people.

(a) How much does each person pay when
(i) 6 people **(ii)** 9 people **(iii)** 12 people
go on the trip?

(b) Darren works out that each person has to pay £16.
How many people go on Darren's trip?

3 Draw x and y axes, both from 0 to 7.

(a) Plot the points A (0, 5.5), B (3, 0.5) and C (6, 1.5).

These are vertices of the parallelogram ABCD. The point D is (r, s).

(b) Plot the point D on your graph. What are the values of r and s?

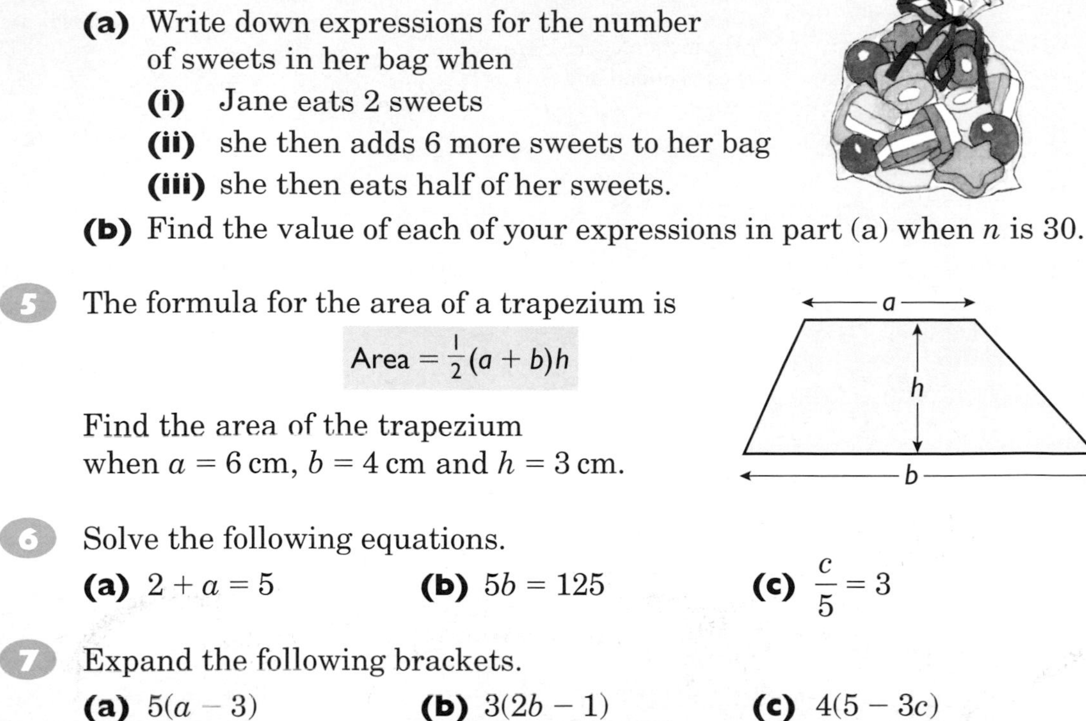

4 Jane has a bag containing n sweets.

(a) Write down expressions for the number of sweets in her bag when

(i) Jane eats 2 sweets

(ii) she then adds 6 more sweets to her bag

(iii) she then eats half of her sweets.

(b) Find the value of each of your expressions in part (a) when n is 30.

5 The formula for the area of a trapezium is

$$\text{Area} = \frac{1}{2}(a + b)h$$

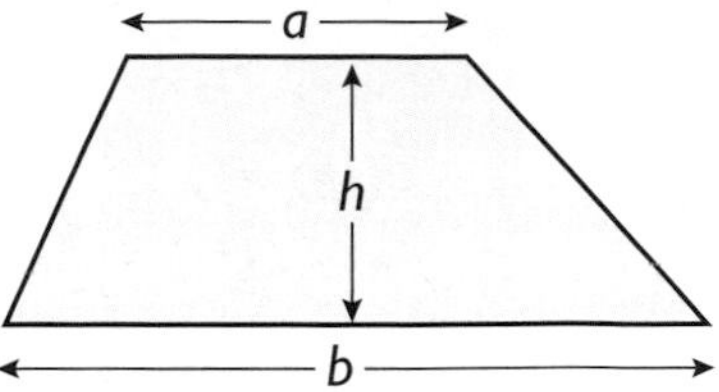

Find the area of the trapezium when $a = 6$ cm, $b = 4$ cm and $h = 3$ cm.

6 Solve the following equations.

(a) $2 + a = 5$ **(b)** $5b = 125$ **(c)** $\frac{c}{5} = 3$

7 Expand the following brackets.

(a) $5(a - 3)$ **(b)** $3(2b - 1)$ **(c)** $4(5 - 3c)$

8 Factorise the following expressions fully.

(a) $10a + 15$ **(b)** $12b - 4$ **(c)** $36 - 12c$

9 Match these sequences to the rules for the nth term.

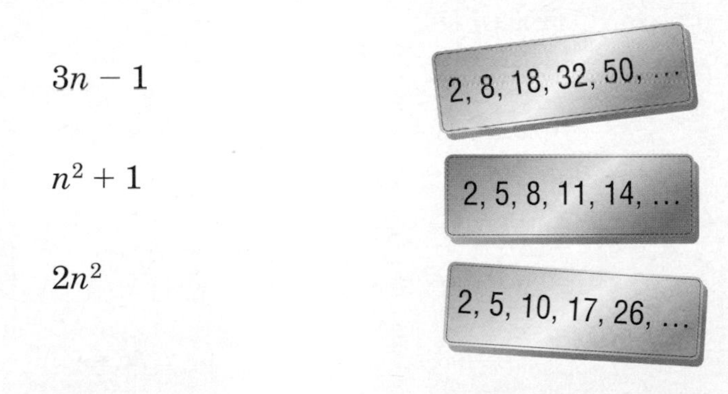

10 **(a)** Write an expression for the area of this rectangle.

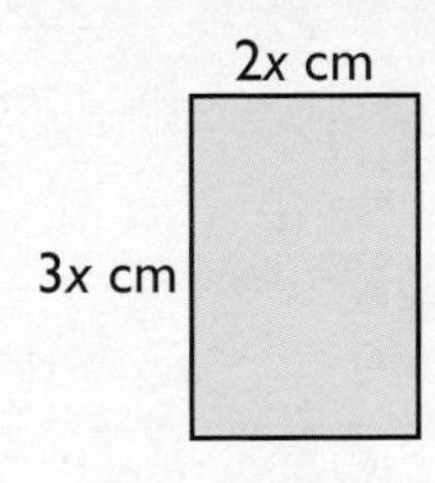

(b) The area of the rectangle is 33 cm^2.
Write an equation showing this information.

(c) Use trial and improvement to find the value of x to 1 d.p.

11 Alison and Guy are playing 'Think of a Number'.

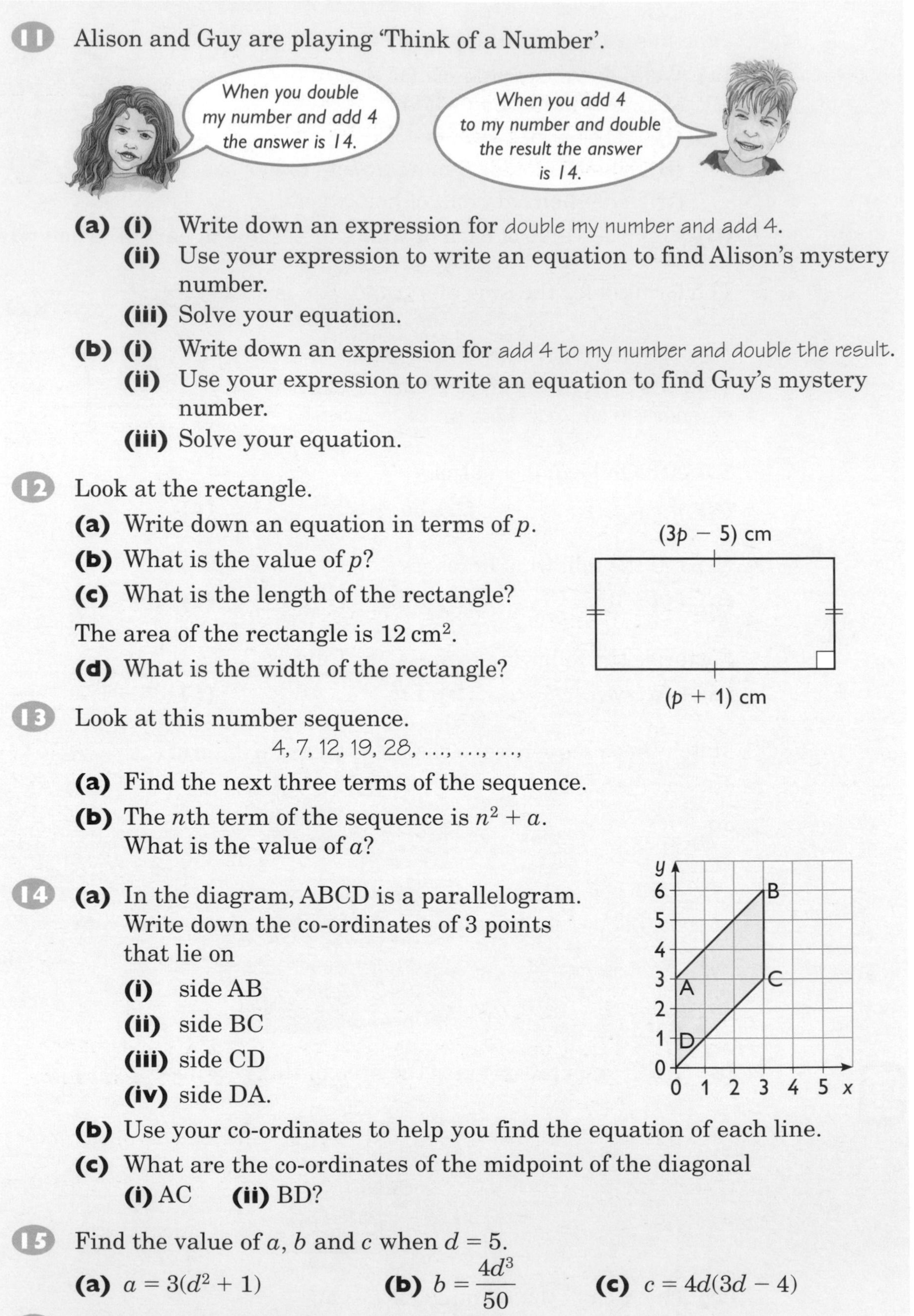

(a) (i) Write down an expression for *double my number and add 4*.

(ii) Use your expression to write an equation to find Alison's mystery number.

(iii) Solve your equation.

(b) (i) Write down an expression for *add 4 to my number and double the result*.

(ii) Use your expression to write an equation to find Guy's mystery number.

(iii) Solve your equation.

12 Look at the rectangle.

(a) Write down an equation in terms of p.

(b) What is the value of p?

(c) What is the length of the rectangle?

The area of the rectangle is 12 cm^2.

(d) What is the width of the rectangle?

13 Look at this number sequence.

4, 7, 12, 19, 28, …, …, …,

(a) Find the next three terms of the sequence.

(b) The nth term of the sequence is $n^2 + a$.
What is the value of a?

14 **(a)** In the diagram, ABCD is a parallelogram.
Write down the co-ordinates of 3 points that lie on

(i) side AB

(ii) side BC

(iii) side CD

(iv) side DA.

(b) Use your co-ordinates to help you find the equation of each line.

(c) What are the co-ordinates of the midpoint of the diagonal
(i) AC **(ii)** BD?

15 Find the value of a, b and c when $d = 5$.

(a) $a = 3(d^2 + 1)$ **(b)** $b = \dfrac{4d^3}{50}$ **(c)** $c = 4d(3d - 4)$

16 Simplify each of the following expressions.

(a) $5a + 2b - 3a + 5b$ **(b)** $6c \div 3c$ **(c)** $\dfrac{d^3}{d}$

Shape, space and measures

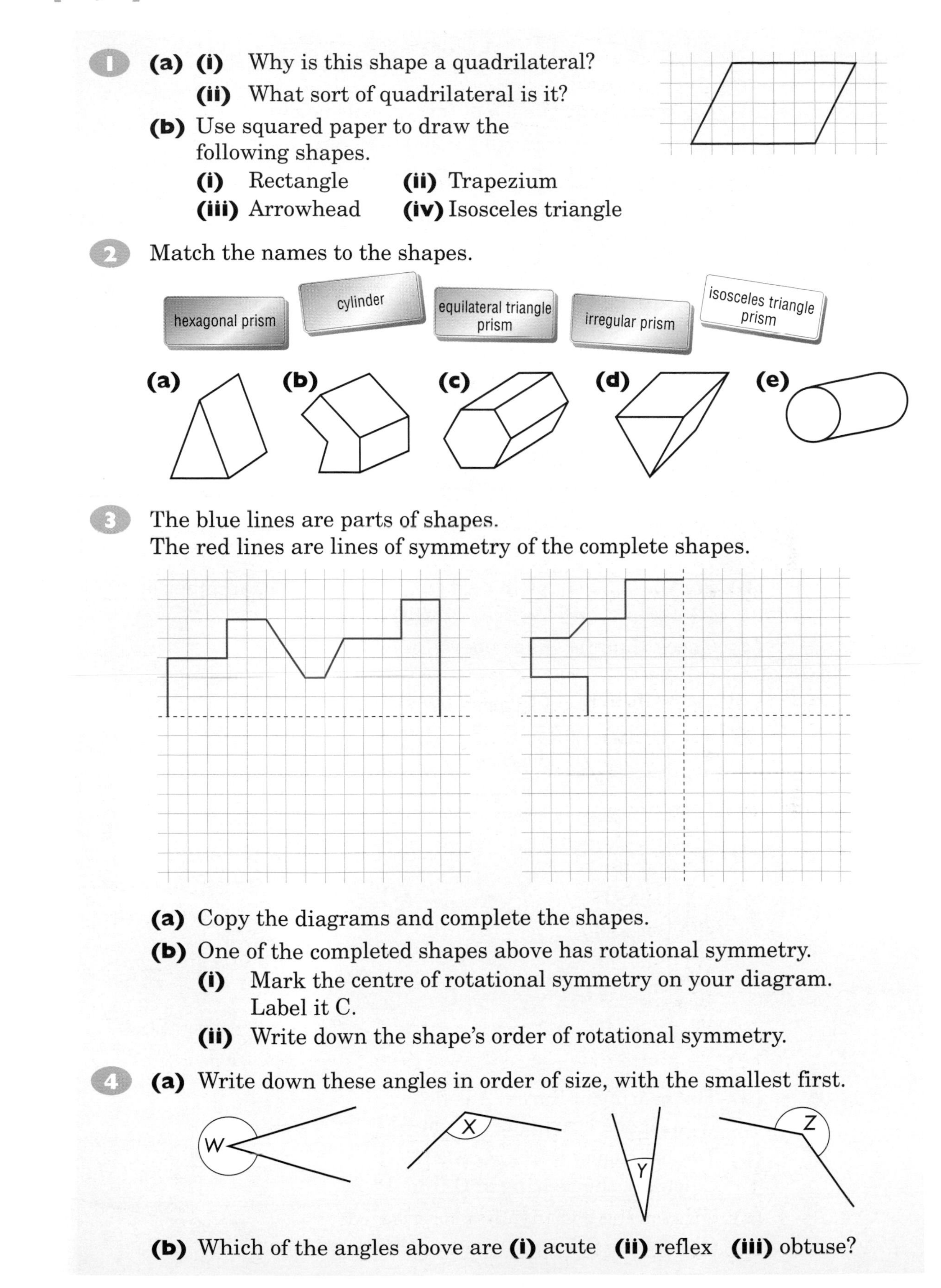

1 **(a)** **(i)** Why is this shape a quadrilateral?

(ii) What sort of quadrilateral is it?

(b) Use squared paper to draw the following shapes.

(i) Rectangle **(ii)** Trapezium

(iii) Arrowhead **(iv)** Isosceles triangle

2 Match the names to the shapes.

3 The blue lines are parts of shapes.
The red lines are lines of symmetry of the complete shapes.

(a) Copy the diagrams and complete the shapes.

(b) One of the completed shapes above has rotational symmetry.

(i) Mark the centre of rotational symmetry on your diagram. Label it C.

(ii) Write down the shape's order of rotational symmetry.

4 **(a)** Write down these angles in order of size, with the smallest first.

(b) Which of the angles above are **(i)** acute **(ii)** reflex **(iii)** obtuse?

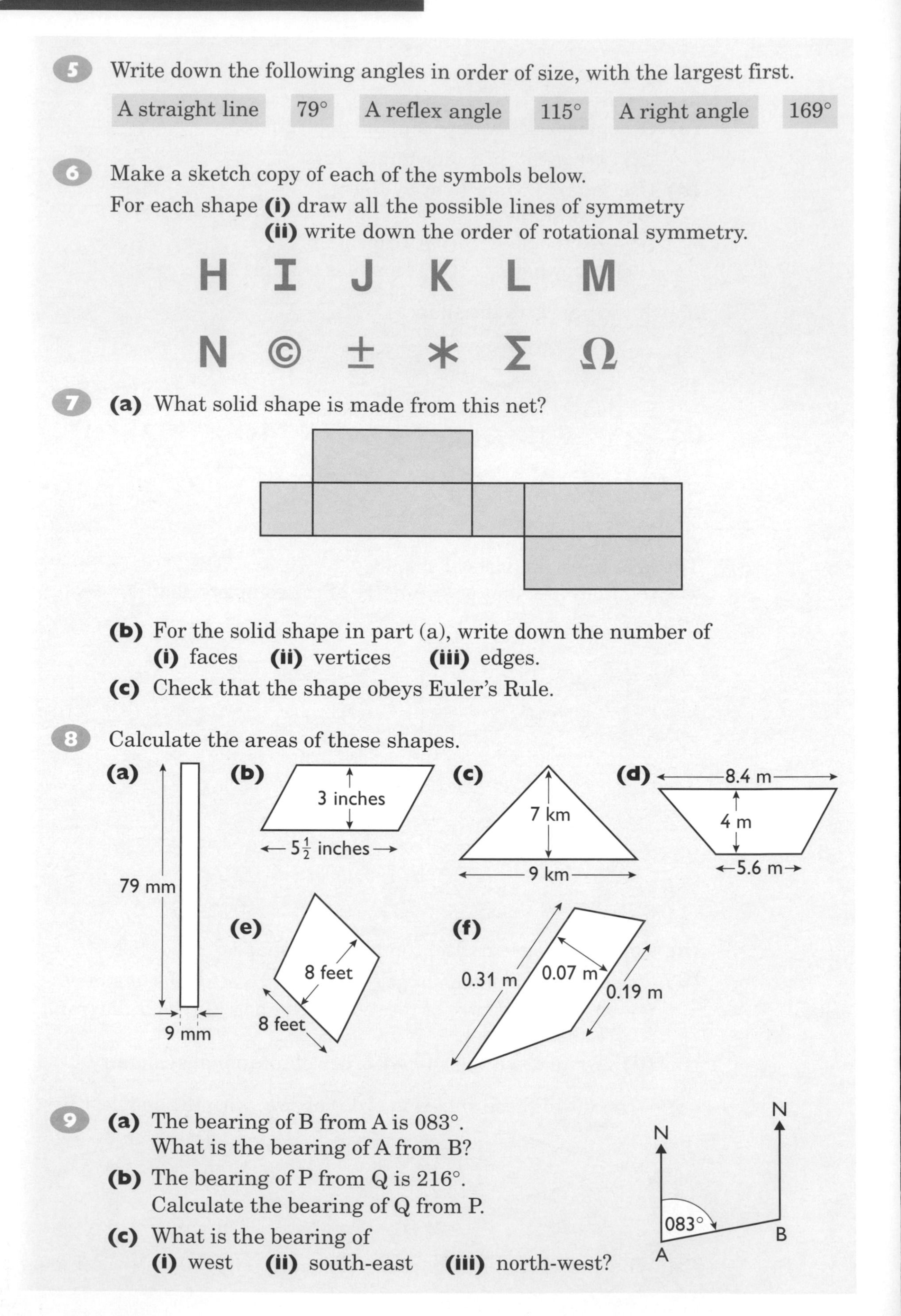

5 Write down the following angles in order of size, with the largest first.

A straight line	79°	A reflex angle	115°	A right angle	169°

6 Make a sketch copy of each of the symbols below.
For each shape **(i)** draw all the possible lines of symmetry
(ii) write down the order of rotational symmetry.

H I J K L M

N © ± ∗ Σ Ω

7 **(a)** What solid shape is made from this net?

(b) For the solid shape in part (a), write down the number of
(i) faces **(ii)** vertices **(iii)** edges.

(c) Check that the shape obeys Euler's Rule.

8 Calculate the areas of these shapes.

9 **(a)** The bearing of B from A is 083°.
What is the bearing of A from B?

(b) The bearing of P from Q is 216°.
Calculate the bearing of Q from P.

(c) What is the bearing of
(i) west **(ii)** south-east **(iii)** north-west?

10 A ship travels on a bearing of 153° for 77 sea miles, then on a bearing of 235° for a further 104 sea miles.

(a) Represent this with a scale drawing.

(b) Use your diagram to find the ship's distance and bearing from the starting point.

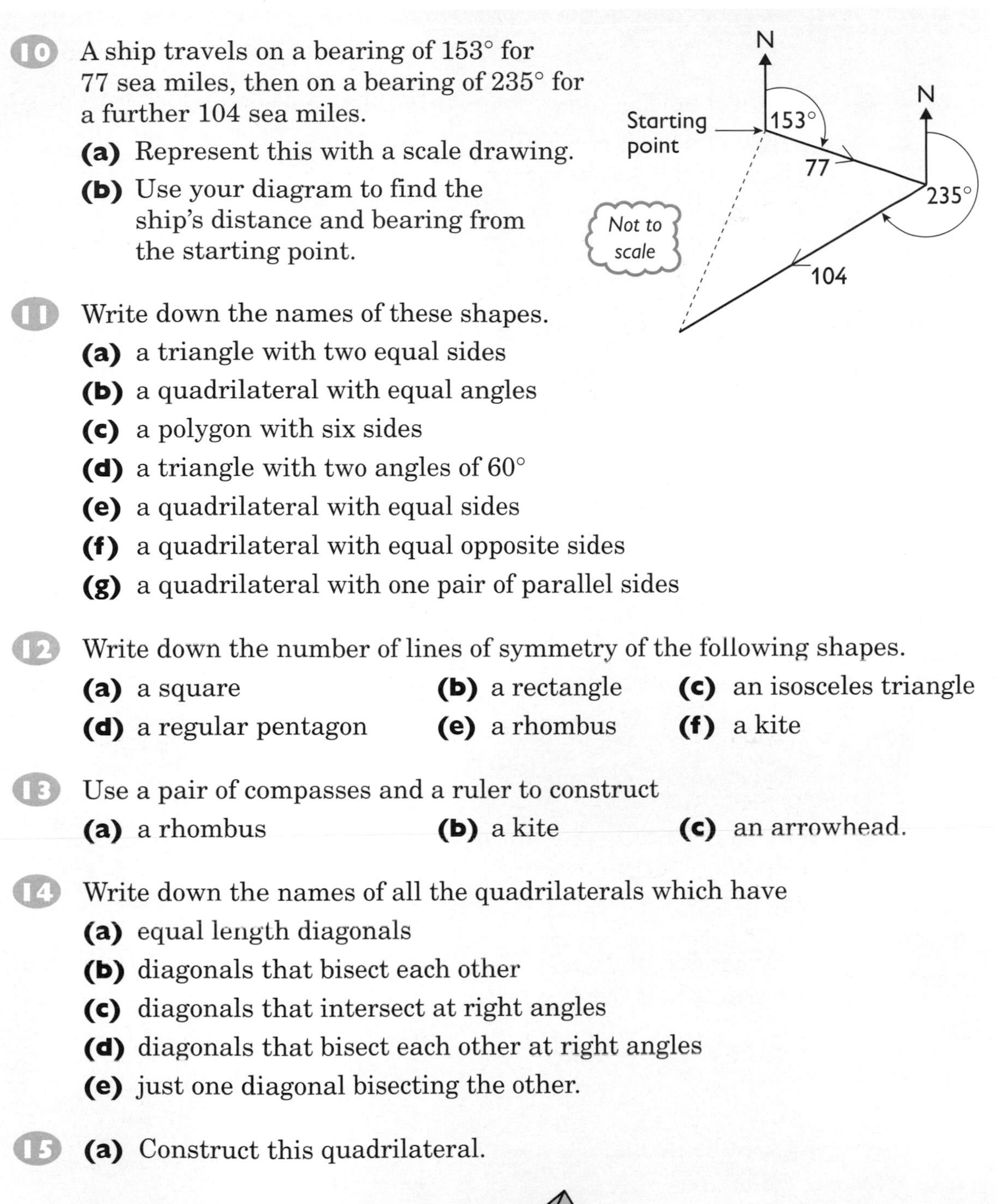

11 Write down the names of these shapes.

(a) a triangle with two equal sides

(b) a quadrilateral with equal angles

(c) a polygon with six sides

(d) a triangle with two angles of 60°

(e) a quadrilateral with equal sides

(f) a quadrilateral with equal opposite sides

(g) a quadrilateral with one pair of parallel sides

12 Write down the number of lines of symmetry of the following shapes.

(a) a square **(b)** a rectangle **(c)** an isosceles triangle

(d) a regular pentagon **(e)** a rhombus **(f)** a kite

13 Use a pair of compasses and a ruler to construct

(a) a rhombus **(b)** a kite **(c)** an arrowhead.

14 Write down the names of all the quadrilaterals which have

(a) equal length diagonals

(b) diagonals that bisect each other

(c) diagonals that intersect at right angles

(d) diagonals that bisect each other at right angles

(e) just one diagonal bisecting the other.

15 **(a)** Construct this quadrilateral.

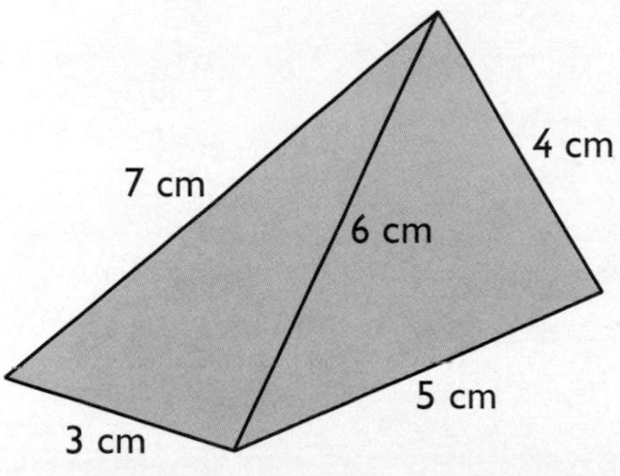

(b) Write down *two* reasons why the quadrilateral is irregular.

Handling data

1 A school holds a talent competition. There are 5 acts, A, B, C, D and E. The members of the audience vote for their favourite act.

Here are the results.

C	D	C	C	B	A	D	C	C	B	E	C
B	A	B	B	D	C	D	D	A	A	C	D
B	C	B	E	B	A	C	C	C	D	D	B
A	C	B	A	B	D	C	A	A	E	E	B
C	D	C	D	A	D	B	B	A	D	E	D

(a) Make a tally chart to show these results.
(b) How many people voted in the competition?
(c) Which act won the competition?
(d) Draw a pictogram to show these results.

2 Lisa recorded the maximum temperature, in degrees Celsius, for every day during the month of June. Here are her results.

18	21	20	21	16	17	21	20	19	19
18	20	22	21	19	20	22	21	19	20
19	18	21	21	20	21	20	17	17	18

(a) Make a tally chart to show these results.
(b) What is the mode of these data?
(c) What is the range of the temperatures?

3 40 students take a test. There are 30 questions in the test.
Here are their scores.

18	24	21	14	15	22	10	19	24	20	12	17	19	8
13	21	18	18	24	25	4	14	19	23	28	15	30	18
16	19	16	22	12	23		21	16	29	22	26	9	

(a) Copy and complete the tally chart.

Score	Tally	Frequency
1–5		
6–10		
11–15		
16–20		
21–25		
26–30		

(b) Draw a bar chart to show these results.

4 Robert did a survey to find out the favourite sports of 50 boys and 50 girls in his year group.
He drew this bar chart to show his results.

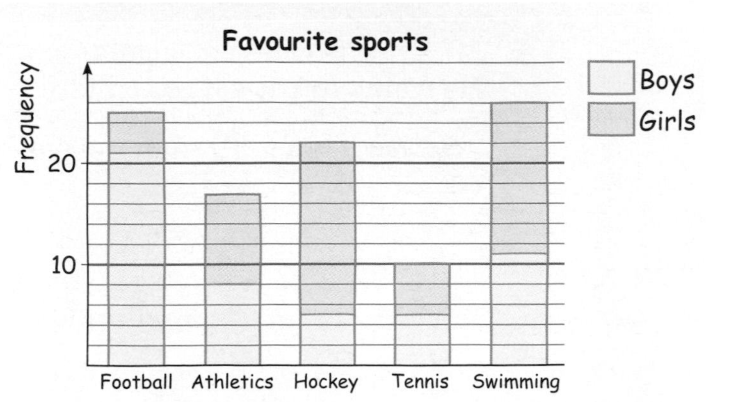

(a) State which sport

(i) is most popular overall

(ii) the boys like most

(iii) the girls like least

(iv) is preferred by an equal number of boys and girls?

(b) Copy and complete this two-way table to show the number of boys and girls who prefer each sport.

	Boys	Girls	Total
Football	21	4	25
Athletics			
Hockey			
Tennis			
Swimming			

5 This pie chart shows the items borrowed from a library one day.

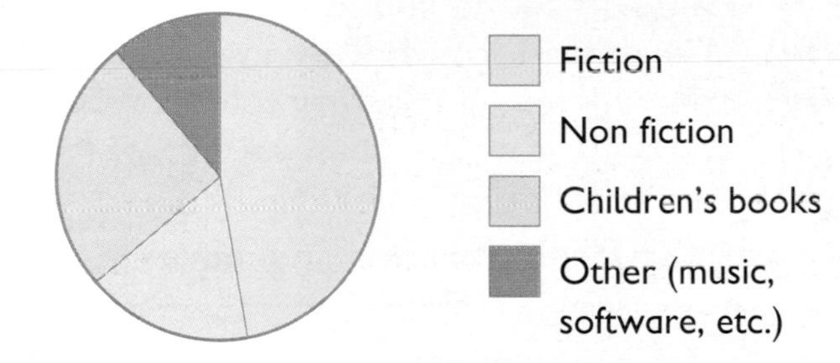

(a) What is the largest category of items borrowed?

(b) Measure the angles in the pie chart.

180 items were borrowed altogether.

(c) How many non-fiction books were borrowed?

(d) What percentage of the items borrowed were children's books?

6 A group of children do a sponsored swim.
Here are the number of lengths that each child swims.

12	25	4	10	13	11	10	16	16
22	7	12	8	10	2	30	18	15
12	11	6	9	14	15	17		

Find the mean number of lengths per child.

7 Mr Brown's class of 30 children take a mental maths test each week. He wants to know if the children's scores have improved.

Scores (out of 20) from the first week of term

12	16	8	11	15	6	5	10	14	8
16	13	10	13	18	12	3	7	14	11
16	8	9	13	11	16	14	15	10	12

Scores (out of 20) from the last week of term

10	15	17	12	13	6	11	18	8	12
19	15	16	12	20	18	7	9	11	14
13	15	7	12	8	14	16	15	18	6

(a) Find the median score for
(i) the first week of term **(ii)** the last week of term.
(b) Find the range of the scores for
(i) the first week of term **(ii)** the last week of term.
(c) Do you think the class has improved?
(d) What does the range of each set of scores tell you?

8 **(a)** Choose one of these words to describe each of the events described below.

certain | unlikely | impossible | evens | likely

(i) It will rain some time during the next week.
(ii) You will get an odd number when you throw an ordinary die.
(iii) The day after next Tuesday will be Wednesday.
(iv) You will win the lottery some day.
(v) You will score 13 when you throw two ordinary dice.
(vi) You will live to be 100 years old.
(b) Copy this probability scale and mark events (i) to (vi) on it.

0 $\frac{1}{2}$ 1

9 James is playing a game with this spinner.
(a) What is the probability that James scores
(i) 1 **(ii)** 2 **(iii)** 3?
(b) James spins the spinner 30 times.
How many times would he expect to get each number?

Sally is also using a spinner with the numbers 1, 2 and 3 on it.
Sally's spinner has 5 sides.
She spins it 50 times.
Here are her scores.

1	8 times
2	23 times
3	19 times

(c) Draw Sally's spinner.
(d) Sally spins her spinner another 50 times.
Do you think she will get the same results?

10 George works for a health and fitness club.
He wants to know what age groups use the club most. He uses members' application forms to make a list of the ages of club members.

36	32	55	28	31	36	35	22	48	51	43	37
32	27	33	40	21	29	34	56	62	43	35	26
45	51	60	38	43	25	34	72	51	30	43	29
41	41	35	31	38	41	23	48	52	44	38	32
57	45	29	51	37	32	41	22	43	38	53	27
37						26					

(a) Make a grouped tally chart to show these data.
Use the groups $20 \leqslant \text{age} < 30$, $30 \leqslant \text{age} < 40$, etc.

(b) Draw a grouped frequency chart to show the data.

(c) What is the modal class?

11 Claire, Joel and Naomi are all doing a different statistics investigation. They have each drawn a scatter diagram to show their results.

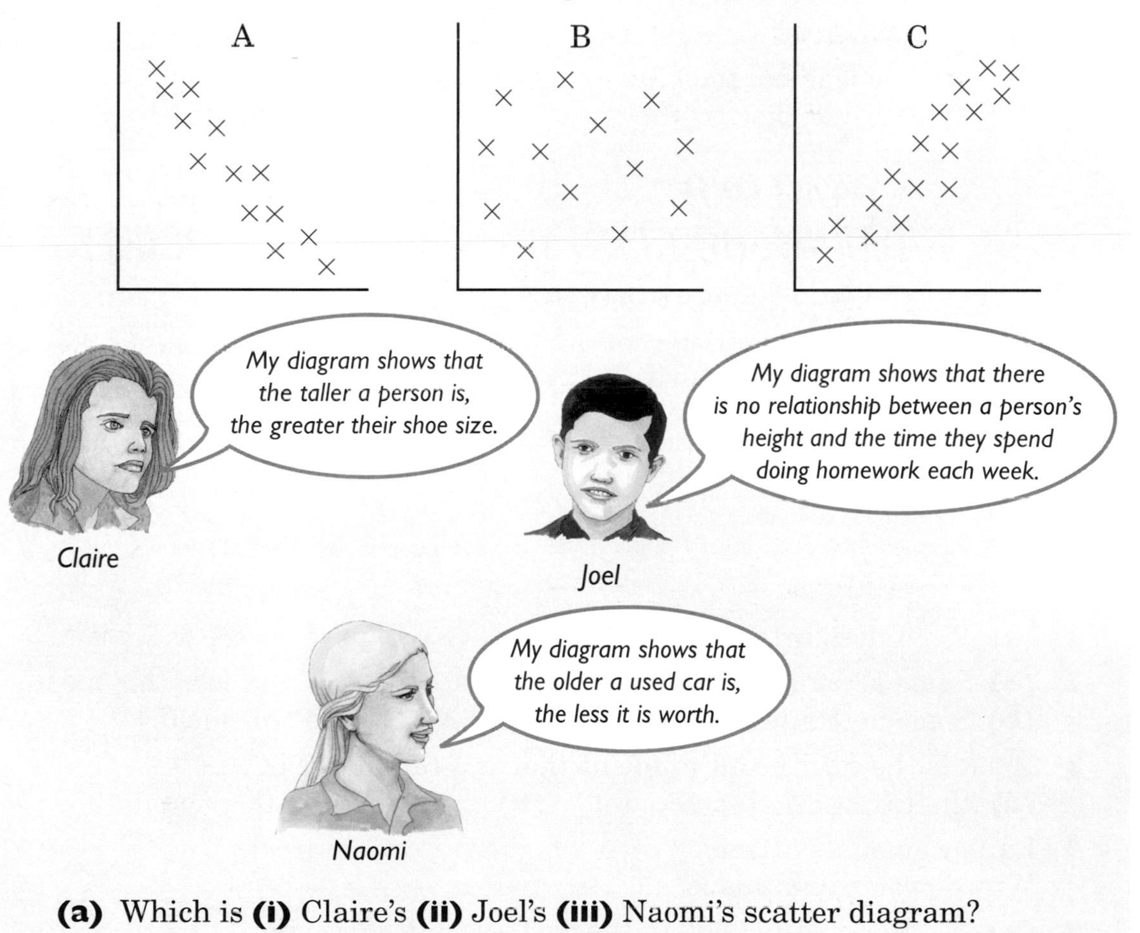

(a) Which is **(i)** Claire's **(ii)** Joel's **(iii)** Naomi's scatter diagram?

(b) Whose diagram shows
- **(i)** positive correlation
- **(ii)** negative correlation
- **(iii)** no correlation?

25 Looking after your money

Saving

Why do you save some of your wages?

Joshua

Give three reasons why Joshua saves money.

Why save money in a building society or bank account?
Why not keep it in a piggy bank at home?

You met simple interest on page 94. The formula is $I = \dfrac{PRT}{100}$.

What do *I*, *P*, *R* and *T* stand for?

Task

Read these two leaflets, then answer the questions below.
Work with a friend. Discuss the questions, then write down your best answers.

AVONFORD BUILDING SOCIETY

INSTANT ACCESS ACCOUNT

Amount	Interest rate
£1 – £499	5.75% p.a.
£500 – £1999	5.9% p.a.
£2000 +	6% p.a.

- Instant access
- Minimum investment £1
- Interest payable 31 July

AVONFORD BUILDING SOCIETY

40-DAY NOTICE ACCOUNT

Amount	Interest rate
£500 – £2499	5.95% p.a.
£2500 – £9999	6.1% p.a.
£10 000 +	6.25% p.a.

- 40 days notice for withdrawals
- Minimum investment £500
- Interest payable 31 October

1 **(a)** What does **interest rate** mean? **(b)** What does **6% p.a.** mean?

2 **(a)** Some accounts are **instant access** accounts. What does this mean?
(b) Some accounts are **notice** accounts. What does this mean?

3 What is the **minimum amount** that you can invest in
(a) the Instant Access Account **(b)** the 40-Day Notice Account?

4 Larger amounts of money earn a higher rate of interest.
Why do you think this is?

5 **(a)** Martha invests £600 in the 40-Day Notice Account.
What rate of interest does she get?
(b) Joel invests £1000 in the Instant Access Account.
What rate of interest does he get?

Exercise

1 Work out the interest earned in one year on an investment of

(a) £100 at 5% p.a. **(b)** £300 at 5% p.a. **(c)** £700 at 5% p.a.
(d) £100 at 4.5% p.a. **(e)** £500 at 4.5% p.a. **(f)** £750 at 4.5% p.a.
(g) £100 at 5.25% p.a. **(h)** £400 at 5.25% p.a. **(i)** £450 at 5.25% p.a.
(j) £100 at 4.8% p.a. **(k)** £600 at 4.8% p.a. **(l)** £950 at 4.8% p.a.

2 Harriet wants to withdraw £500 from her Avonford 40-Day Notice Account (see opposite page) to pay for her holiday.
She has to pay the £500 by 12 June at the latest.
What is the last day on which she can give notice to withdraw the money?

3 On 21 October, Liam gives notice that he wants to close his Avonford 40-Day Notice Account (see opposite page).
When will Liam be able to collect his money?

4 Here are two leaflets from South West Bank.

SOUTH WEST BANK

EASY ACCESS ACCOUNT

Amount	Interest rate
£1 – £999	5.5% p.a.
£1000 – £4999	5.8% p.a.
£5000 +	6.1% p.a.

- Minimum investment £1
- Instant access
- Interest paid 31 January

SOUTH WEST BANK

50 ACCOUNT

Amount	Interest rate
£1000 – £4999	6% p.a.
£5000 – £9999	6.05% p.a.
£10 000 +	6.15% p.a.

- Minimum investment £1000
- 50 days notice for withdrawals
- Interest paid 31 December

(a) Wai Peng invests £1200 in an Easy Access Account.
How much interest does he earn in
(i) one year **(ii)** two years?

(b) Rebecca invests £2500 in a 50 Account.
How much interest does she earn in
(i) one year **(ii)** three years?

5 Molly and Oliver look at the leaflets for Avonford Building Society (see opposite page) and South West Bank (see question 4).

(a) Molly invests £2000 in an instant access account.
Explain which account is better.

(b) Oliver invests £5000 in a notice amount.
Explain which account is better.

Part-time jobs

Many teenagers have part-time jobs or holiday jobs.

What is gained by having a job?

Why do you think Hana's mother is saying 'No'?

Task

Jason Hughes works full-time at his local supermarket. He is paid £5.20 per hour.

Jason works 5 days a week, Monday to Friday, starting at half past eight each morning.
His lunch break is 12.30 p.m. to 1.30 p.m. each day.
He finishes work at 5 p.m. except on Monday and Friday.
On Monday he finishes 30 minutes later.
On Friday he finishes an hour earlier.

Copy and complete Jason's timesheet for a normal week.

Surname		First name		Hourly rate	£
Week commencing Monday, 10 June 2002					
Day	Morning		Afternoon		Hours
	Start*	Finish*	Start*	Finish*	
Monday					
Tuesday					
Wednesday					
Thursday					
Friday					
Saturday					
Sunday					
*Use 24-hour clock times				Total hours	
e.g. 7.30 a.m. = 0730 and 5.15 p.m. = 1715				Wage	

What is the purpose of a timesheet?

One week Jason works 45 hours. How many hours overtime does he do? The overtime is paid at 'time and a half'. What does this mean? What is Jason's total wage for the week?

Exercise

1. Rachel babysits for 3 hours on Thursday, 4 hours on Friday and 4 hours on Saturday.
   She is paid £2.10 per hour.
   How much does she earn in total?

2. Mel has a holiday job.
   She works from 0945 to 1215, three days a week.
   (a) How many hours does she work in a week?
   She earns £2.30 per hour.
   (b) How much does she earn in a week?

3. Lee has a Saturday job.
   He works from 0845 to 1230 and earns £2.60 an hour.
   How much does he earn each Saturday?

4. Abi has a holiday job. The chart below shows her hours.

Day	Start	Finish	Hours
Mon	0830	1030	
Tue	0830		2
Wed	0830	1100	
Thu	0830		3
Fri	0830	1200	
Sat	0830		4

   (a) Copy the chart and fill in the missing entries.
   (b) How many hours does she work in a week?
   (c) Abi is paid £2.80 an hour.
   How much does she earn in 6 weeks?

5. Each week Ravi delivers 500 leaflets.
   (a) How much does he earn?
   Ravi saves half of what he earns.
   (b) How much does he save each week?
   (c) How many weeks must he work to save £250?

Activity Ask your classmates what jobs they do or have done.

Make a list of these jobs and for each one explain how they got the job.

What are the best paid jobs?

What are the worst paid?

My ideal room

Kim and her family are moving to a new house.

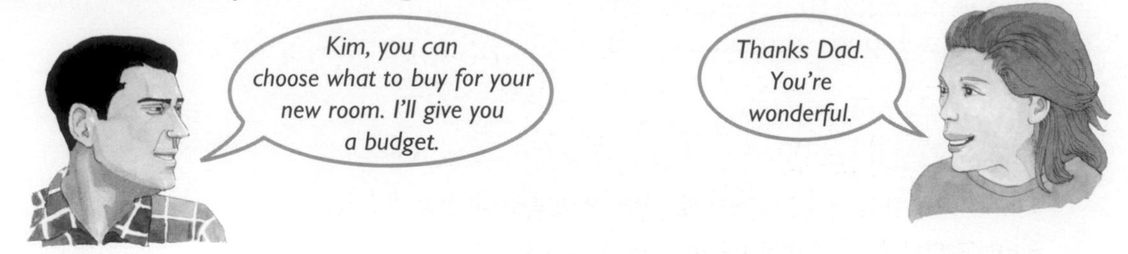

 **What would you have in your ideal room?**

Task

This is Kim's plan for her ideal room. The room measures 3 metres by 2.5 metres.

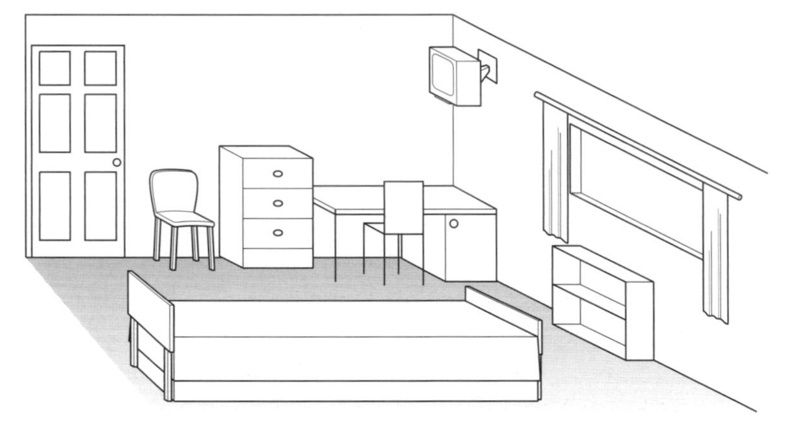

Look at the following price lists.

1 Identify which items are in Kim's room and work out the total cost of her plan.

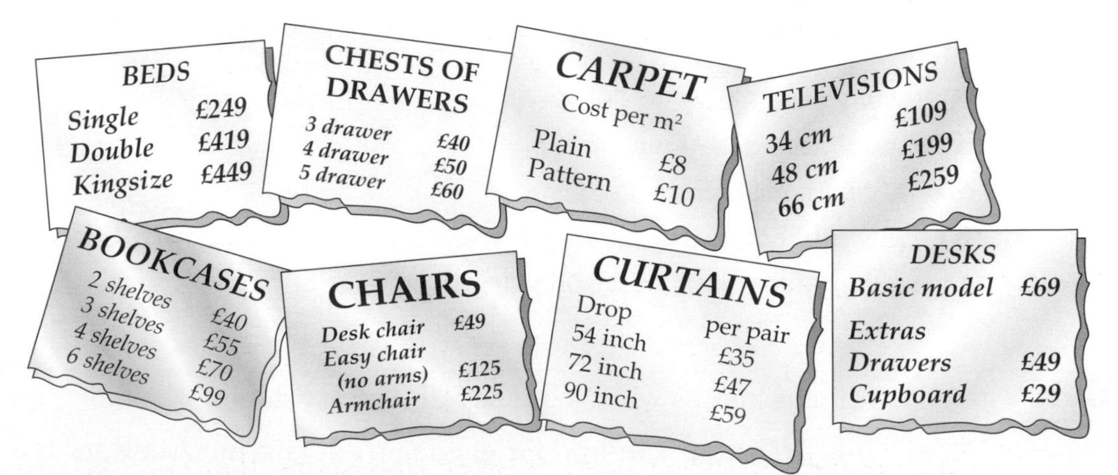

Kim's father gives her a budget of £800.

2 Is she within her budget?

Exercise

1 Tom is re-designing his bedroom.
Here are some of the calculations he does.

(a) Tom's room is 4.5 m long and 3.2 m wide.
He covers the floor with carpet costing £15 per m^2.
Work out the total cost of the new carpet.

(b) Tom's notice board is 2.8 metres wide.
His posters are each 50 cm wide.
How many posters can he fit across the board?

(c) Tom buys a quilt cover priced at £27.
He gets a 10% discount in the sale.
How much does he pay?

(d) Tom's door frame is 195 cm high by 75 cm wide.
He puts draught excluder on each post
and across the top.
How much draught excluder does Tom use?
Give your answer in metres.

(e) Tom buys a pair of curtains.
Each curtain is 120 cm wide.
Each curtain needs a curtain ring at each
end and at every 10 cm in between.
How many curtain rings does he need?

2 Cameron's room is 3 m long and 2.4 m wide.
He is going to carpet the floor.
Cameron uses square carpet tiles which measure 30 cm by 30 cm.
They cost 99p each.
Work out the total cost of the tiles.

3 Lyn's room is 3.1 m long, 2.6 m wide and 2.35 m high.
It has a door 2 m high by 0.8 m wide and a window 1.4 m by 1 m.
Lyn is painting the walls (but not the door or window).
One litre of paint covers about 12 m^2.
How much paint does she need?

Activity

Jessica's room is the same size as Lyn's (see question 3).
Jessica buys wallpaper in rolls that are 52 cm wide and 10 m long.
How many rolls does Jessica need to paper the walls of her room?

Costing a holiday

Richard, Louise and their children Jack (13 years) and Sam (11 years) are planning a holiday for 7 nights at the Palm Bay Hotel.

Look at this page from their holiday brochure. Work with a friend. Discuss the questions and then write down your best answers.

Palm Bay Hotel

	Half board			
Number of nights	**7**	**7**	**14**	**14**
Date of departure	**Adult**	**Child***	**Adult**	**Child***
1 May–23 May	419	159	629	209
24 May–30 May	499	249	749	349
31 May–13 Jun	439	169	659	229
14 Jun–27 Jun	459	179	689	239
28 Jun–11 Jul	479	189	719	259
12 Jul–18 Jul	509	229	759	319
19 Jul–22 Aug	529	249	789	349
23 Aug–5 Sep	499	249	749	349
6 Sep–27 Sep	459	179	689	239

*Child means aged under 13 years on date of departure.
Children under 2 years are free (except for insurance).

Supplements: Full board £4 per person per night (pppn)
Single occupancy £8 pppn
Flights from Gatwick £10 per person

Insurance: Adult £25, Child 0–12 years inclusive £19.

1. What does **half board** mean?
2. What are **supplements**?
3. What is **insurance**?
4. How much does insurance cost for each child?
5. On which departure dates is the holiday most expensive?

? Why is it sensible to take out holiday insurance?

? Why do many companies charge a supplement for single occupancy?

? Why are holidays more expensive in late July and August?

Exercise

Use the table opposite to work out the total cost of these bookings at Palm Bay Hotel.

1. Richard, Louise and their children Jack (13 years) and Sam (11 years) book a holiday for 7 nights, half board.
   They depart from East Midlands Airport on 8 May.
   They already have holiday insurance.
2. Ellen and Charlie book a holiday for 14 nights, full board.
   They depart from Luton on 25 May.
   They already have holiday insurance.
3. James books a holiday for himself and his two children, Dylan (7 years) and Molly (10 years).
   The holiday is for 14 nights, half board and departs from Manchester on 12 August.
   He buys holiday insurance when he books the holiday.
4. Shamicka books a holiday for 7 nights, half board, single occupancy, leaving from Gatwick on 20 July.
   She already has holiday insurance.
5. Alan, Kathy and their children Martin (4 years) and Olivia (9 months) book a holiday for 7 nights full board, departing on 27 August from Newcastle.
   They buy holiday insurance.

Investigation

Look at these flight times. The times given are local times.

Outward	Night	Day
Luton depart	0010	1125
Ibiza arrive	0325	1440

Return	Night	Day
Ibiza depart	0505	1615
Luton arrive	0620	1730

(a) How long does the flight appear to be
(i) from Luton to Ibiza **(ii)** from Ibiza to Luton?

Explain the difference in your answers.
What is the actual flight time?

The distance from Luton to Ibiza is about 1500 km.

(b) What is the average speed of the aeroplane?

Activity

The currency in Spain is the euro.

1 Find out the current exchange rate between pounds and euros.

2 To help you convert between euros and pounds
(a) copy and complete the conversion table
(b) draw a conversion graph.

3 Do you think the eight entries in the euros column are the most helpful ones for a tourist?

Euros	Pounds
1	
2	
3	
5	

Euros	Pounds
7	
10	
20	
50	

26 Investigations

In real life you often meet situations where you need to use some mathematics.
Sometimes it is not clear at the outset just what topics you will need, or the best way forward.
You have to investigate the problem first.
This chapter contains a number of questions to help you develop the skill of investigation.

As well as solving the questions, you should also explain your work clearly.

Someone else must be able to follow what you have done.

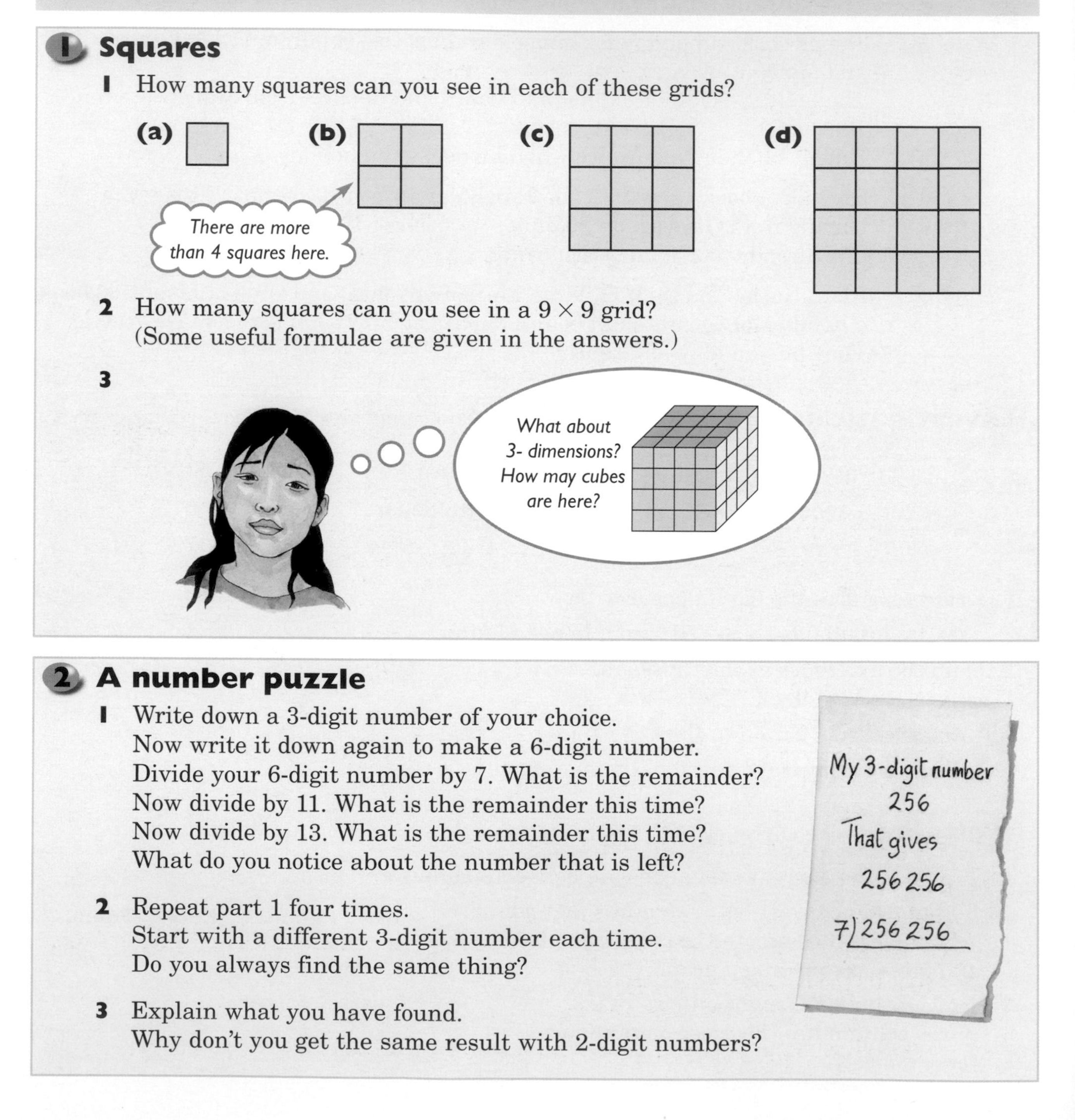

1 Squares

1 How many squares can you see in each of these grids?

(a) **(b)** **(c)** **(d)**

2 How many squares can you see in a 9×9 grid?
(Some useful formulae are given in the answers.)

3

2 A number puzzle

1 Write down a 3-digit number of your choice.
Now write it down again to make a 6-digit number.
Divide your 6-digit number by 7. What is the remainder?
Now divide by 11. What is the remainder this time?
Now divide by 13. What is the remainder this time?
What do you notice about the number that is left?

2 Repeat part 1 four times.
Start with a different 3-digit number each time.
Do you always find the same thing?

3 Explain what you have found.
Why don't you get the same result with 2-digit numbers?

3 Decoding

1 When you write English, which 5 letters do you use most often?

2 This message was recently received by Space Control. Decode it.

UNL LQLDZ TRGJL AFLLU XHFF GUUGJP
QLWU XLLP XL QLLK NLFR RFLGTL TLQK
DBCL TUGCTNHRT XHUN FGTLC MEQT
GQK G KLJBKLC

4 Square roots

Before calculators were available, calculating square roots was more difficult. One method was known as **'divide and average'**.

This is shown in the flow chart opposite.

1 Read through the flow chart. Use the flow chart to complete the table below.

n	a	b	Yes/No
11	3	3.333 333	No
	3.333 333	3.316 666	No
	3.316 666		

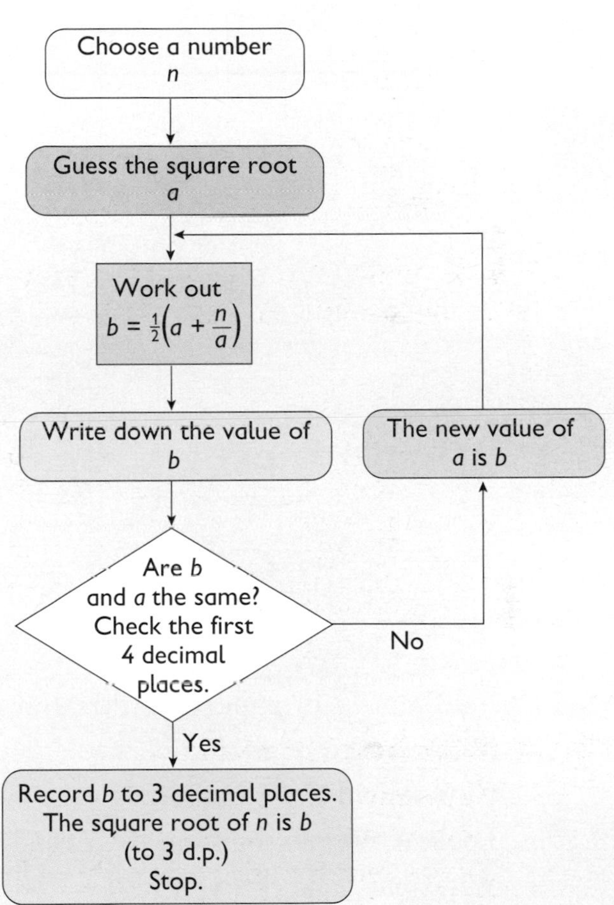

2 Investigate the following questions.

(a) What happens if the value of a is above the square root of the number?

(b) Question 1 took 3 cycles through the flow chart.
Choose a number n and try different values of a.
How does the starting point affect the number of cycles it takes to find the answer?

(c) Can you find the square root of 10 stating with $a = 1000$?

(d) Find the square root of 17 with $a = 4$.
How many extra cycles are required to improve the accuracy to 5 d.p.? to 6 d.p.?

(e) Repeat question 4 with $a = 3$.
Do you get a different answer?

(f) What happens if you start with a negative value of a?

5 Page numbering

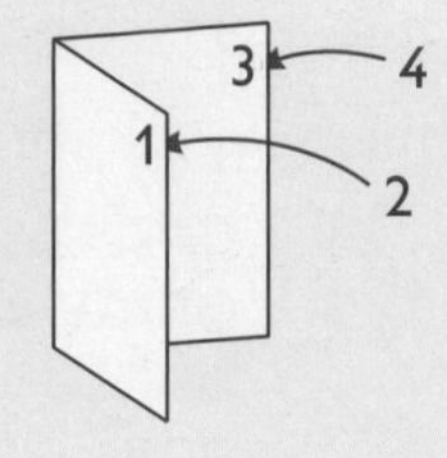

You will need some sheets of paper. A4 is a suitable size.

You can make a 4-page booklet by folding a sheet of paper along its middle.

The diagrams below show you the page numbers on the two sides of the sheet.

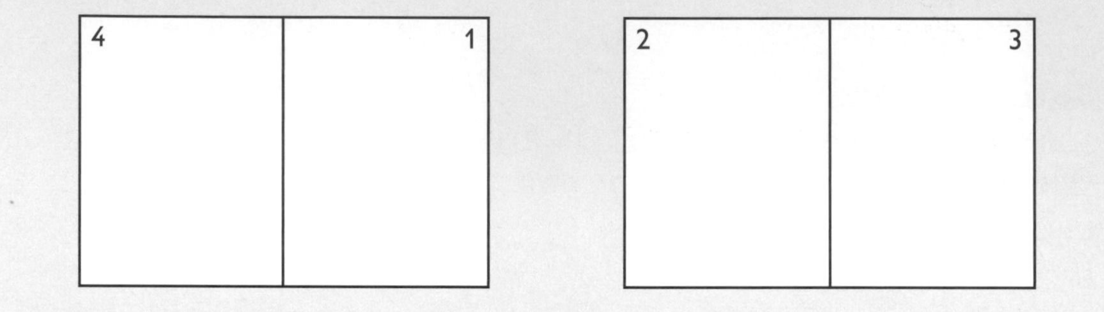

To make an 8-page booklet from a single sheet of paper, you need to make two folds and some cuts, like this.

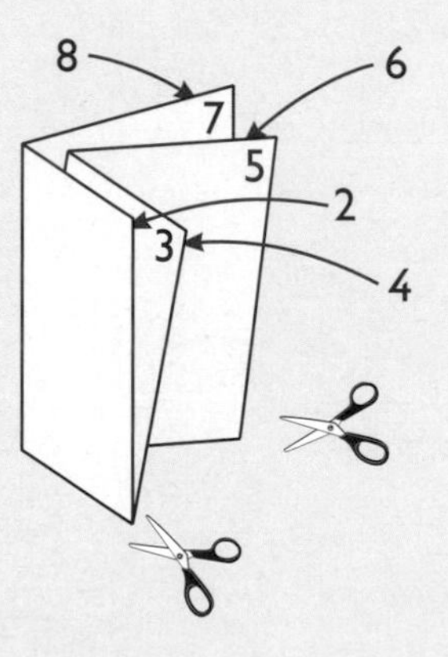

1 Fold a piece of paper like this, but don't actually make the cuts.
Mark on the page numbers.
Now unfold the paper.
Look at the numbers on the two sides of the sheet.
What patterns can you see?
How can you see where to cut?

2 Extend your investigation to 16- and 32-page booklets.

3 A 64-page book is to be made up from 16 sheets of paper.
Each sheet is folded down the middle and the sheets are stapled together along the fold-line.
How can you check that each page will end up in the right place?

6 Magic squares

1 Play this game with a partner.

Take it in turns to choose a number between 1 and 9.
Write your numbers down.
A number can only be chosen once.
The aim is to get any three of your numbers to total 15.

In this example Akosua wins.

$5 + 8 + 2 = 15$

Kofi	Akosua
6	5
4	9
1	8
7	2

Play this game a few times. Keep a record of your games.

2 In a 3×3 magic square, the numbers in

- all three rows
- all three columns
- both diagonals

add up to the same total.

8	1	6
		2

Copy and complete this magic square.
When it is finished it should contain each of the numbers 1, 2, 3, 4, 5, 6, 7, 8 and 9.

3 Take the game that Kofi and Akosua played (question one).

Mark an X over each of Kofi's numbers on the magic square, and an O over each of Akosua's numbers.

Kofi and Akosua did not know it but they were really playing noughts and crosses.

Now go through your games and turn them into games of noughts and crosses.
Which is easier, the number game or noughts and crosses?

4 Copy and complete this 4×4 magic square. Use the numbers 1 to 16

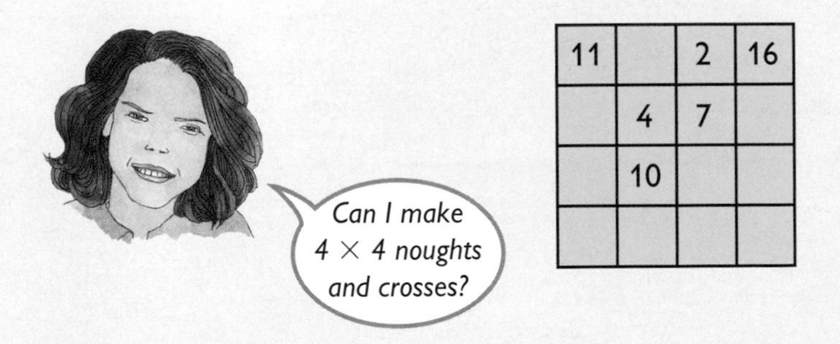

11		2	16
	4	7	
	10		

Gemma tries to use this square to make up a number game linked to a new form of noughts and crosses.

Describe her game.
Does it always work?

Rich Aunt

Deborah's kind, rich aunt writes her a letter.

Dear Deborah,

I am getting on a bit. I want to give you some of my money. You can choose from the four schemes below. All of the schemes will run for eleven years.

Scheme 1
You get £100 this year, £90 next year, £80 the year after, and so on. (£10 less each year.)

Scheme 2
You get £10 this year, £20 next year, £30 the year after and so on. (£10 more each year.)

Scheme 3
You get £10 this year, 1.5 times as much next year (so you get £15), 1.5 times as much the year after (£15 x 1.5 = £22.50), and so on.

Scheme 4
You get £1 this year. The amount of money you get doubles each year.

I look forward to hearing which of the schemes you prefer. Please tell me what you think of each scheme.

Best wishes

Aunt Moneybags

Use a spreadsheet to investigate the four schemes.

1 For each scheme copy and complete a table like the one below. Go up to eleven years.

Scheme 1 Year	Amount for Year	Running Total
1	£100	£100
2	£90	£190
3	£80	£270
4	£70	£340

2 Draw a graph with Time (in years) along the x axis and the Running Total on the y axis. Plot the four schemes on the same graph.

3 Write a letter to Aunt Moneybags to tell her what you think of each scheme. Remember to tell her which scheme you prefer.

8 The Fibonacci Sequence

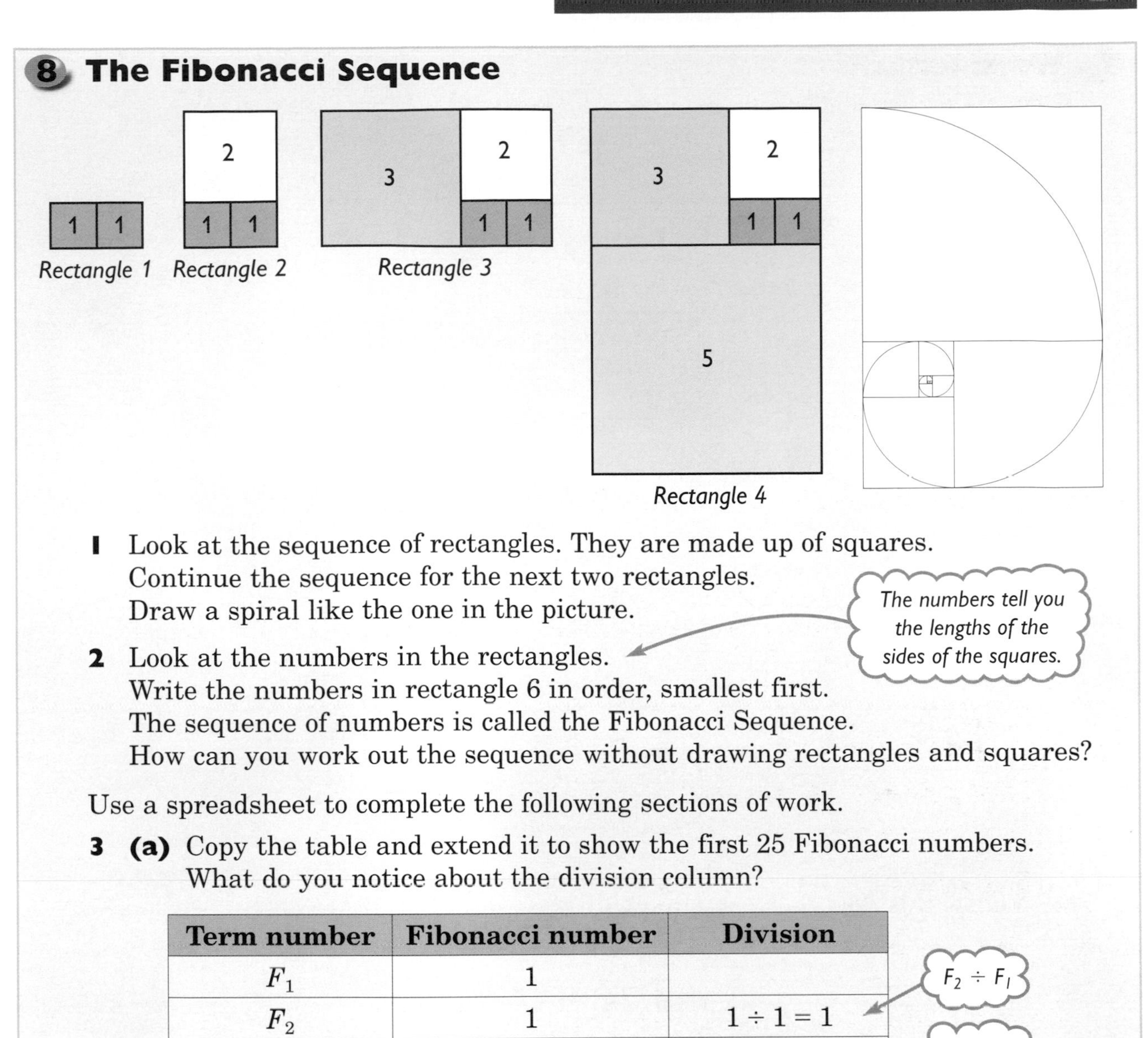

1 Look at the sequence of rectangles. They are made up of squares.
Continue the sequence for the next two rectangles.
Draw a spiral like the one in the picture.

2 Look at the numbers in the rectangles.
Write the numbers in rectangle 6 in order, smallest first.
The sequence of numbers is called the Fibonacci Sequence.
How can you work out the sequence without drawing rectangles and squares?

The numbers tell you the lengths of the sides of the squares.

Use a spreadsheet to complete the following sections of work.

3 (a) Copy the table and extend it to show the first 25 Fibonacci numbers.
What do you notice about the division column?

Term number	Fibonacci number	Division
F_1	1	
F_2	1	$1 \div 1 = 1$
F_3	2	$2 \div 1 = 2$

$F_2 \div F_1$

$F_3 \div F_2$

(b) Write down another Fibonacci-type sequence by starting with any two numbers instead of 1, 1.
Repeat part (a).

If you start with 1 and 5 the sequence is 1, 5, 6, 11, 17, ...

4 Copy the table and extend it to show the first 25 Fibonacci numbers.
Look at the sum of the numbers. What pattern do you notice?
Try this investigation again for another Fibonacci-type sequence.

Term number	Fibonacci number	Sum of numbers
F_1	1	
F_2	1	$1 + 1 = 2$
F_3	2	$2 + 2 = 4$

5 Make a list of the Fibonacci numbers that are multiples of 3.
Write down their term numbers. What do you notice?
Repeat for multiples of 5 and multiples of 8.

9 Word lengths

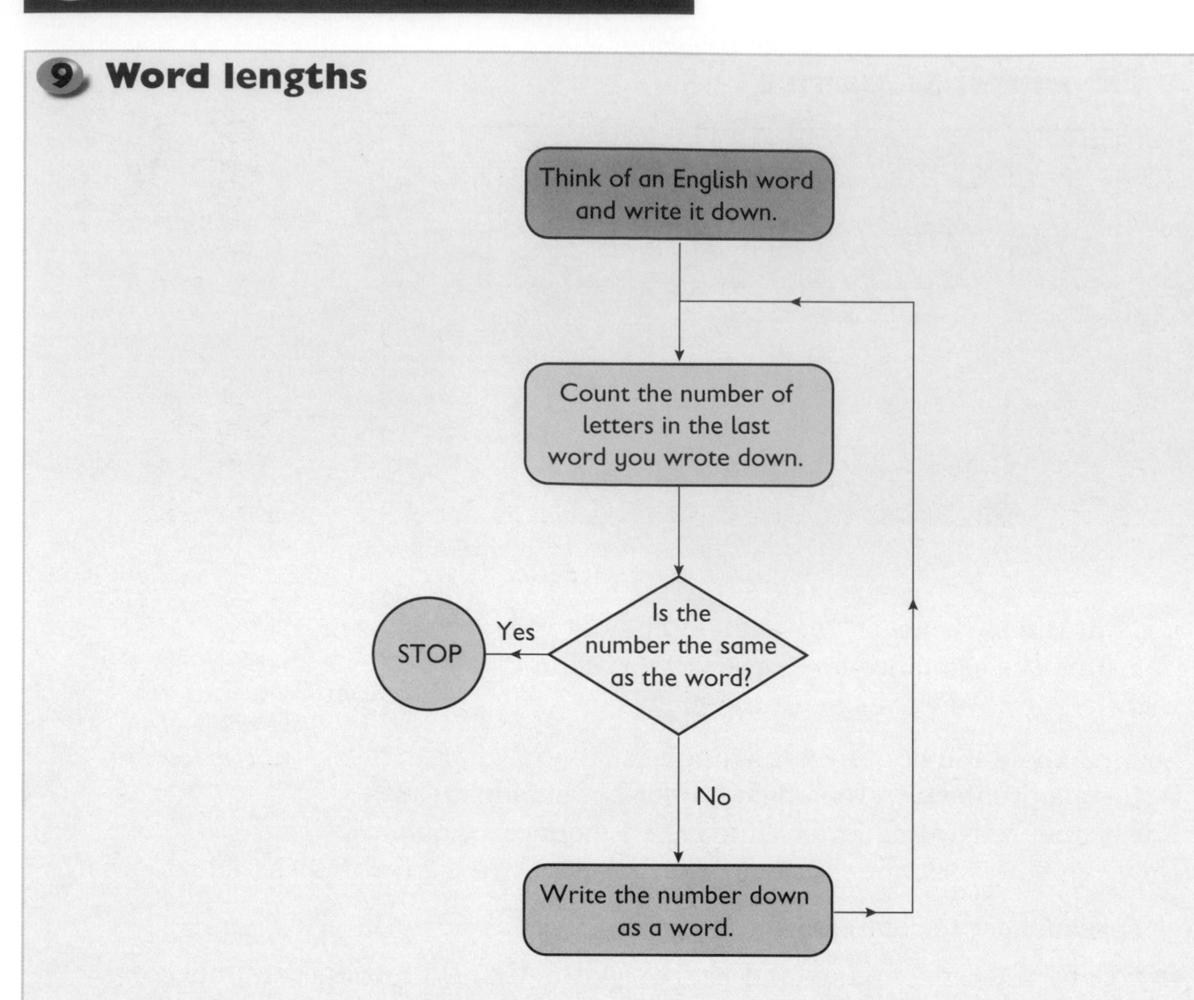

1 Follow this flow chart, starting with the word MATHEMATICS.

MATHEMATICS
ELEVEN
SIX
...

2 Work through the flow chart with four other starting words.

3 Write down the longest English word you can think of and use it as the starting word.

4 What is the greatest number of words you can write down before the flow chart tells you to stop?

5 Write down a theory about your answers.

6 Now change from English to French.
Write down four French starting words and work with French numbers.
What happens?

7 Now try Welsh, Spanish and Malay.

	Welsh	Spanish	Malay
1	un	uno	satu
2	dau	dos	dua
3	tri	tres	tiga
4	pedwar	cuatro	empat
5	pump	cinco	lima
6	chwech	seis	enam
7	saith	siete	tujuh
8	wyth	ocho	lapan
9	naw	nueve	sembilan
10	deg	diez	sepuluh

Here are some possible starting words.

Welsh

Draig (dragon)
Ar (on)
Cath (cat)
Disgleirio (to shine)
Pentref (village)

Spanish

Dragón (dragon)
Gato (cat)
En (on)
Porcion (share)
Negro (black)

Malay

Naga (dragon)
Kucing (cat)
Kampung (village)
Saham (share)
Yu (shark)

What happens with each language?

8 Find out the words for numbers 1 to 10 in some other languages and try them.

9 Make a poster showing the possible outcomes for different languages.

Explain how you can decide what will happen for any particular language by looking at its numbers written down.

Answers

1 Units (pages 10–11)

1 $4 \times 12 = 48$ inches. The desk does not fit.
2 $\frac{1}{4}$ **3** 66 feet **4** 10
5 (a) 70 (b) 240 (c) 3 (d) 2.5
(e) 1.5 (f) 80 (g) 5.250 (h) 0.435
(i) 70 (j) 34 200 (k) 20 (l) 0.56
6 11.2 **7** 16.2
8 (a) 75 (b) No, too short.
9 (a) 5564 km (b) 8848 m or 8.848 km (c) 6695 km
10 3.3
11 3
12

Depart Exeter	1237	1336	1530
Arrive Reading	1416	1534	1716
Length of journey	1 hr 39 min	1 hr 58 min	1 hr 46 min

13 (a) 210 (b) 246 (c) 72 (d) 3
14 (a) 40 mph (b) $6\frac{2}{3}$
15 (a) 14.4 (b) 2
16 (a) 40 minutes (b) 24 minutes

Activity 1 Jamie weighs 8 stone 9 pounds. He is 11 pounds overweight.

Activity 2 (a) 96 (b) $21\frac{1}{3}$

2 Algebraic expressions (pages 20–21)

1 (a) $5a$ (b) $14b$ (c) $5c$ (d) $10d - 3$
(e) $4e + 2$ (f) $7f + 7g$ (g) $h + 4j$ (h) $k - m$
(i) $5n^2$ (j) $10p^2$ (k) $7q^2$ (l) r^2
2 (a) $2a$ (b) $4b$ (c) c^2
(d) $12d$ (e) $10e$ (f) $6f^2$
(g) $40g^2$ (h) $9hjk$ (i) $30m^2n$
3 (a) $3j + 2s$ (b) $3j + 40$ (c) £15
4 (a) 170 (b) 371 (c) 234 (d) 1577 (e) 4860 (f) 72 534
5 (a) (i) 3^3 (ii) 3^2 (iii) 3^4 (iv) 3^7
(b) (i) 27 (ii) 9 (iii) 81 (iv) 2187
6 (a) (i) 3^3 (ii) 27 (b) (i) 3^4 (ii) 81 (c) (i) 3^7 (ii) 2187
(d) (i) 3^4 (ii) 81 (e) (i) 3^4 (ii) 81 (f) (i) 3^3 (ii) 27
7 (a)

p^{10}
p^6 p^4
p^3 p^3 p

(b)

q^{14}
q^6 q^8
q q^5 q^3

8 (a) $5a + 10$ (b) $7b + 21$ (c) $6c + 8$ (d) $8d - 24$
(e) $10 - 2e$ (f) $12f - 20$ (g) $x^2 + 7x$ (h) $x^2 + xy$
(i) $6x^2 + 4x$ (j) $6x^2 + 15xy$ (k) $ax + bx$ (l) $x^3 + x^2 + x$
9 (a) $5(a + 2)$ (b) $4(2b + 1)$ (c) $11(3c - 2)$
(d) $4(4d - 1)$ (e) $6(3 + 4e)$ (f) $12(3f - 5)$
(g) $3(p + 2q)$ (h) $x(x + 9)$ (i) $2(3x + 4y)$
(j) $2x(2x + y)$ (k) $2x(x + 2y)$ (l) $x(x^2 + 2x + 1)$
10 (a) $5x + 7y$ (b) $7x + y$ (c) $7x - y$
(d) $7(a + 2b)$ (e) $4a + 5b$ (f) $5(a + 8b)$

3 Shape (pages 30–31)

1–6 Ask your teacher to check your diagrams.
4 Square-based pyramid
6 (a) 2 (b) 3 (c) 2
7 $V = 492\ m^3$, $S = 392\ m^2$
8 It depends on exactly what type of triangular end the prism has: if it is scalene, then Paul is correct; if it is isosceles, then Jenny is correct; if it is equilateral, then Ankur is correct.

4 Doing a survey (pages 42–43)

1 (a) 0.4°
(b) Pie chart with angles as follows:

Education services	232°
Social services	80°
Environmental services	44°
Other services	4°

2 (a) 15–29
(b) 60–74 and 75+. This is because women generally live longer than men.
3 (a) Stephen

Time	Tally	Frequency
$0 \leqslant t < 15$		0
$15 \leqslant t < 30$	||	2
$30 \leqslant t < 45$	卌	5
$45 \leqslant t < 60$	卌 卌 |	11
$60 \leqslant t < 75$	卌 卌	10
$75 \leqslant t < 90$	||	2
$90 \leqslant t < 105$	|	1
	Total	31

Anna

Time	Tally	Frequency
$0 \leqslant t < 15$	卌 |||	8
$15 \leqslant t < 30$	||||	4
$30 \leqslant t < 45$	||	2
$45 \leqslant t < 60$	|||	3
$60 \leqslant t < 75$	||	2
$75 \leqslant t < 90$	卌 ||	7
$90 \leqslant t < 105$	||||	4
$105 \leqslant t < 120$	|	1
	Total	31

(b) Two frequency charts to show above data.
(c) Stephen has used the computer every day, Anna hasn't. However, when Anna does use the computer she often spends a long time on it.
4 (a) Ask your teacher to check your scatter graph.
(b) In general, the shorter the time taken to run 100 m, the longer the distance jumped.
(c) Ask your teacher to check your line of best fit.
(d) About 1.8 metres

5 Ratio (pages 52–53)

1 (a) 3 : 2 (b) 2 : 1 (c) 1 : 9 (d) 1 : 4
(e) 2 : 1 : 5 (f) 9 : 19 (g) 10 : 1 (h) 1 : 4
2 (a) 18 (b) 4 (c) 1 (d) 3
3 (b) (i) AB = 5 cm, AC = 13 cm
(ii) PR = 6 cm, PQ = 2.5 cm
(c) (i) 5 : 2.5 (ii) 13 : 6.5 (iii) 12 : 6
(d) 2 : 1 for all the ratios
(e) The ratios are the same
4 (a) (i) 5 : 12.5 = 2 : 5 (ii) 2.4 : 6 = 2 : 5
(b) The ratios are equal
(c) area A = 12 cm^2, area B = 75 cm^2
(d) area A : area B = 4 : 25
(e) $4 : 25 = 2^2 : 5^2$
5 (a) area face A : area face B = 4 : 9
(b) volume A : volume B = 8 : 27
(c) area A : area B = $2^2 : 3^2$ = (ratio of edges)^2
volume A : volume B = $2^3 : 3^3$ = (ratio of edges)^3
6 (a) 50 000 cm = 500 m
(b) 1500 m = 1.5 km
7 6400 m = 6.4 km
8 Mrs Shah = £4500 Miss Kachinska = £10 500
9 left 32, right 20, headed 8, own goal 4
10 (a) rice flour 40 g, fat 60 g
(b) flour 300 g, rice flour 100 g
(c) flour 360 g, rice flour 120 g, fat 180 g, sugar 240 g, currants 120 g, candied peel 30 g

6 Equations (pages 62–63)

1 (a) $x = 3$ (b) $x = 4$ (c) $t = 3$ (d) $x = 3$
(e) $x = 3$ (f) $s = 7$ (g) $x = 5$ (h) $x = 4$
(i) $p = 8$ (j) $x = 0$ (k) $x = 4$ (l) $x = 2$

2 (a) (i) $21 = 4x + 9$ (ii) $x = 3$
(b) (i) $2x + 7 = 3x + 5$ (ii) $x = 2$

3 (a) $x = 3$ (b) $x = 14$ (c) $x = 2$ (d) $x = 4$
(e) $x = 3$ (f) $x = 5$ (g) $x = 1$ (h) $x = 2$
(i) $x = 8$ (j) $x = 3$ (k) $x = 2$ (l) $x = 1$
(m) $x = -11$ (n) $x = 4$ (o) $x = 7$

4 (a) $x = 9$ (b) $x = 6$ (c) $x = 2\frac{1}{2}$
(d) $x = 12$ (e) $x = 3$ (f) $x = 3\frac{1}{2}$

5 (a) (i) $10 - 2x = 4$ (ii) $x = 3$
(b) (i) $13 - 3x = 1$ (ii) $x = 4$

6 (a) $12x = 108$
(b) (i) $x = 8$ cm, riser = 8 cm (ii) tread = 16 cm

7 $x \approx 4.50$

8 (a) $x = 5$ (b) $x \approx 3.59$

7 Angles and polygons (pages 72–73)

1 A = 41°, B = 44°, C = 96°, D = 96°, E = 84°,
F = 60°, G = 44°, H = 31°, I = 40°

2 107°

3 A = 23°, B = 67°

5 (a) (i) 6° (ii) 174° (b) 10 440°

6 13

7 (a) 111° (b) Corresponding angles (c) 900° (d) 99.5°

8 (a) 150° (b) 30° (c) 30°
(d) ∠DAJ = ∠BAL − ∠BAD − ∠LAJ
= 150° − 30° − 30°
= 90°
By symmetry, ADGJ is a square.

8 Directed numbers (pages 80–81)

1 (a) 0 (b) 2 (c) (−2) (d) 5 (e) 5
(f) (−21) (g) 14 (h) (−26) (i) (−14)

2 (a) (−6) (b) 0 (c) 6 (d) (−3) (e) (−3)
(f) 3 (g) (−8) (h) 8 (i) (−2) (j) 25
(k) (−5) (l) 5 (m) 20 (n) 20 (o) 8

3

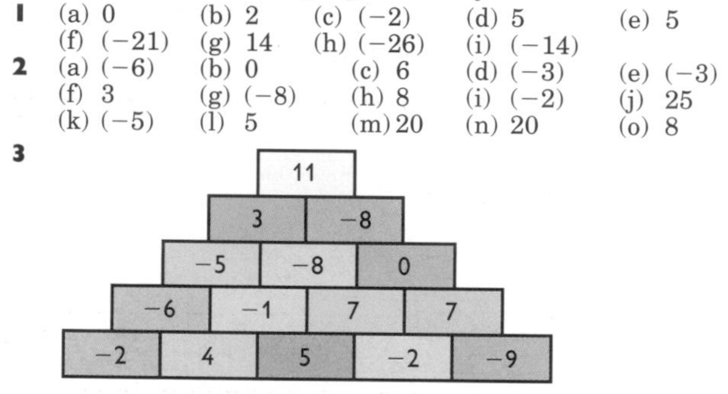

4 (a) 9 (b) 16 (c) 13 (d) (−3) (e) 1 (f) (−18)
(g) Examples of correct solutions:
0 − 5, 1 − 6, 2 − 7, (−5) − 0, (−6) − (−1).

5 (a) (−3) (b) 0 (c) (−1)

6 (a) (−4) + (−3) = (−7) (b) (−4) + 3 = (−1)

7 (a) 9 (b) 19 (c) (−13) (d) 2 (e) 0 (f) (−1)

8 (a) Left (b) No movement (c) Moves left

9 (a) (−100) (b) 100 (c) (−100)
(d) (−3.6) (e) 1.2 (f) (−1.2)
(g) 5 (h) (−8) (i) (−4)

10 (a) (−3) (b) (−21) (c) 12 (d) 3 (e) (−3) (f) 3

Activity

Examples of correct solutions:
(−5) and (−3), (−6) and (−2), (−7) and (−1), (−8) and 0,
(−9) and 1, (−10) and 2

Activity

6	−1	4
1	3	5
2	7	0

−1	−6	1
0	−2	−4
−5	2	−3

9 Sequences and functions (pages 90–91)

1 (a) (i) 3, 5, 7, 9 (ii) 2, 5, 8, 11 (iii) 1, 4, 9, 16
(iv) 2, 5, 10, 17
(b) (i) and (ii)

2 (a) $a = 4$ and $d = 4$ (b) $a = 3$ and $d = 3$
(c) $a = 100$ and $d = -5$ (d) $a = -10$ and $d = 2$
(e) $a = 5$ and $d = 5$ (f) $a = 2$ and $d = 2$

3 (a) 7, 9 (b) 10, 18 (c) 27, 21, 15
(d) 18, 20, 24 (e) 45, 40, 30, 20 (f) −18, −12, −10

4 (a) $3n$ (b) $4n$ (c) $2n + 1$
(d) $3n + 1$ (e) $2n - 1$ (f) $5n - 2$

5 (a) 57, 54 (b) −240

6 (a)–(b) 1, 3, 6, 10, 15
(c) Differences increase by 1 each time
(d) (i) 55 (ii) $\frac{1}{2}n(n + 1)$

7 (a) (i) rectangle 1: length 2, width 1;
rectangle 2: length 3, width 2;
rectangle 3: length 4, width 3
(b) (i) length 11, width 10
(ii) length 21, width 20
(iii) length 101, width 100
(c) (i) 110 (ii) 420 (iii) 10 100
(d) (i) length $n + 1$ (ii) width n (iii) $n(n + 1)$
(e) No. of dots in a triangle = $\frac{1}{2}$ no. of dots in rectangle

Investigation

1 Tends to 6 as $\frac{6}{3} + 4 = 6$

5 Results is solution to $n = \frac{n}{a} + b$

10 Percentages (pages 96–97)

1 (a) £12.50 (b) £37.50

2 (a) £0.20 (b) £10

3 (a) £8.50 (b) £170

4 (a) £60 (b) £180 (c) £240 (d) £40

5 £330

6 (a) £212.50 (b) £25.50 (c) £340

7 £12 750

8 £93 312

9 (a) 75% (b) £975

10 (a) 80%
(b) (i) jeans £0.60, trainers £0.45, T-shirt £0.40
(ii) jeans £60, trainers £45, T-shirt £40

11 (a) £1.50 (b) £150

12 (a) £90 (b) £52 (c) £5

13 (a) 8% (b) 7.5% (c) 12.5%

14 (a) £6000 (b) £500 (c) £1250

11 Circles (pages 102–103)

1 (a) 44 cm (b) $6\frac{2}{7}$ inches (c) 35.2 m (d) 8 m

2 (a) (i) 2.2 m (ii) 11 m
(b) (i) 4.55 (ii) 45.45 (iii) 4545.45

3 (a) (i) 100.48 inches (ii) 803.84 square inches
(b) (i) 81.64 inches (ii) 530.66 square inches
(c) (i) 62.8 inches (ii) 314 square inches

4 36.28 square feet

5 19.464 m

6 (a) 22.47 m^2 (b) 9.18 m^2

Investigation

1 (a) 0.2355 m s^{-1} (b) 0.26 m s^{-1}

2 (b) 741.62 m (c) 88.99 m^2

12 Fractions (pages 108–109)

1 (a) $\frac{7}{10}$ (b) $\frac{7}{12}$ (c) $\frac{1}{2}$ (d) $\frac{5}{8}$ (e) $\frac{5}{6}$ (f) $\frac{13}{16}$
(g) $\frac{13}{16}$ (h) $\frac{8}{9}$ (i) $\frac{7}{8}$ (j) $\frac{11}{30}$ (k) $\frac{11}{18}$ (l) $1\frac{1}{3}$
(m) $1\frac{1}{16}$ (n) $1\frac{3}{20}$

2 (a) $\frac{1}{6}$ (b) $\frac{1}{10}$ (c) $\frac{5}{21}$ (d) $\frac{7}{18}$
(e) $\frac{1}{12}$ (f) $\frac{11}{70}$ (g) $\frac{17}{120}$ (h) $\frac{7}{36}$

3 (a) $\frac{5}{3}$ (b) $\frac{9}{5}$ (c) $\frac{7}{3}$ (d) $\frac{12}{5}$ (e) $\frac{22}{7}$

4 (a) $1\frac{1}{3}$ (b) $1\frac{2}{7}$ (c) $2\frac{1}{3}$ (d) $3\frac{1}{4}$ (e) $3\frac{3}{7}$

5 (a) $\frac{1}{9}, \frac{1}{7}, \frac{1}{5}, \frac{1}{3}, \frac{1}{2}$ (b) $\frac{5}{12}, \frac{7}{12}, \frac{2}{3}, \frac{3}{4}, \frac{5}{6}$ (c) $\frac{17}{15}, 1\frac{3}{10}, \frac{7}{5}, 1\frac{3}{5}, 1\frac{2}{3}$

6 (a) $\frac{12}{35}$ (b) $\frac{1}{6}$ (c) $\frac{1}{7}$ (d) $\frac{1}{3}$
(e) 18 (f) 12 (g) 6 (h) 35

7 (a) $\frac{3}{2} = 1\frac{1}{2}$ (b) $\frac{5}{4} = 1\frac{1}{4}$ (c) $\frac{3}{7}$ (d) $\frac{3}{5}$ (e) $\frac{2}{7}$

8 (a) 45 (b) 36 (c) 45 (d) 35 (e) 10 (f) 8
(g) 4 (h) 6 (i) 5 (j) 13 (k) 5 (l) 14

9 $\frac{1}{5}$

10 $1\frac{23}{30}$ miles

11 Darren (John receives $\frac{3}{14}$ of the votes)

12 (a) $1\frac{7}{12}$ of an hour or 2 h 35 min (b) $\frac{5}{12}$ of an hour or 25 mins

Activity

1 $\frac{1}{4} + \frac{1}{8} = \frac{3}{8}$

2 $\frac{3}{8} - \frac{1}{16} + \frac{1}{2} = \frac{13}{16}$

3 $\frac{3}{16}$

4 64 people

5 20 people

13 Graphs (pages 120–121)

1

x	0	1	2	3	4	5	6
$2x$	0	2	4	6	8	10	12
$+4$	4	4	4	4	4	4	4
$y = 2x + 4$	4	6	8	10	12	14	16

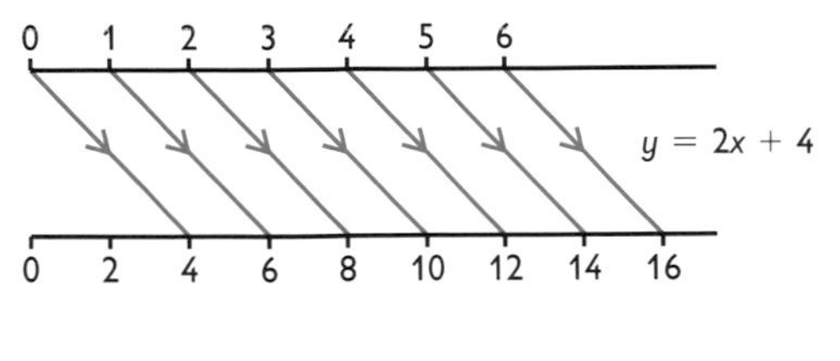

2 (a)

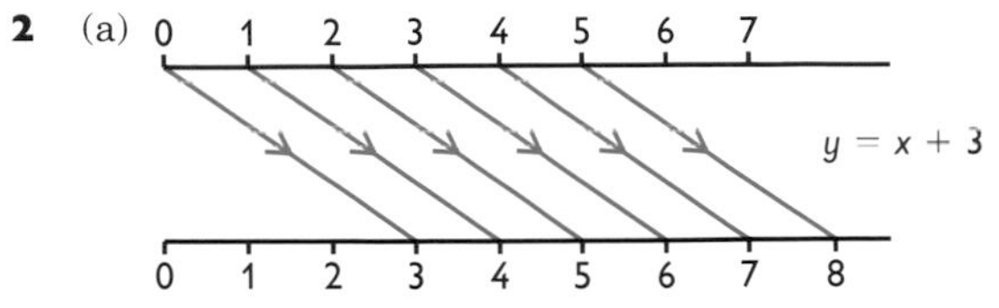

(b) Set of arrowed lines are parallel.

3 (i) $y = x - 5$ (ii) $y = \dfrac{x}{9}$

4 (a)

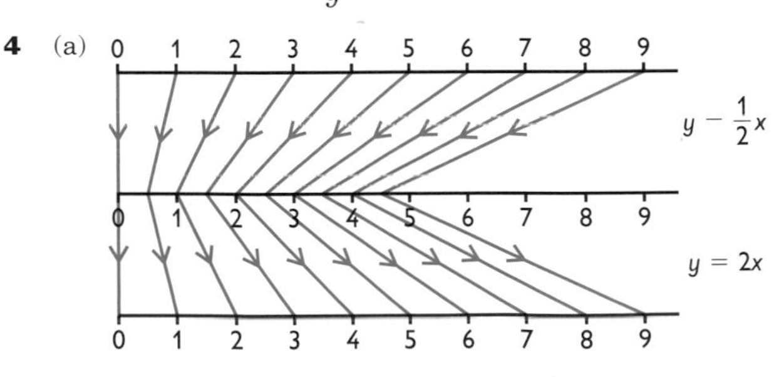

(b) $y = 2x$ is the inverse mapping of $y = \frac{1}{2}x$ and vice-versa.

5 (a)

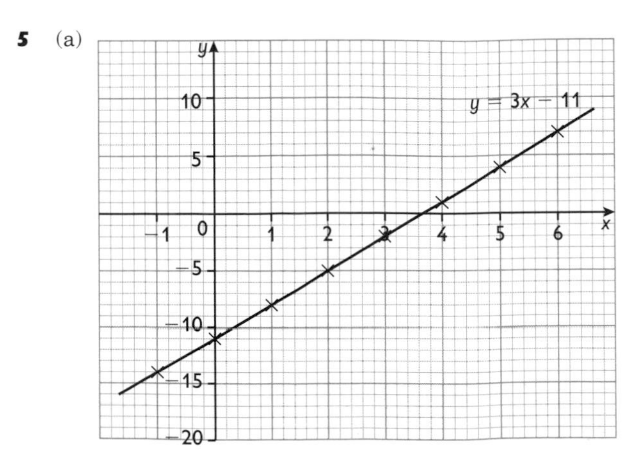

(b) $x = 3\frac{2}{3}$. This is the intercept of the line on the x–axis.

6

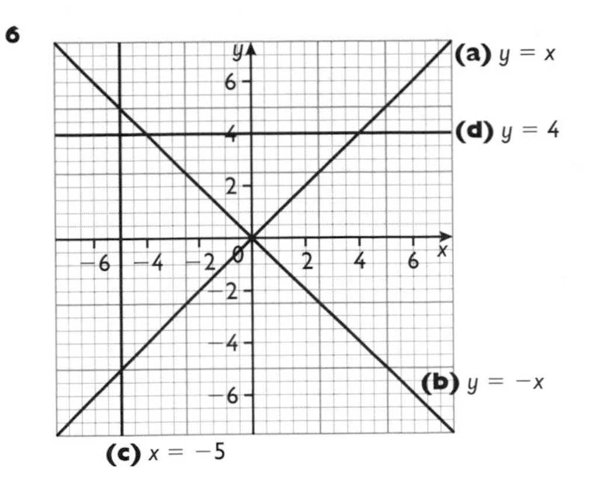

7 (a)

x	−3	−2	−1	0	1	2	3
$4x^2$	36	16	4	0	4	16	36
$+9$	9	9	9	9	9	9	9
$y = 4x^2 + 9$	45	25	13	9	13	25	45

(b) (c)

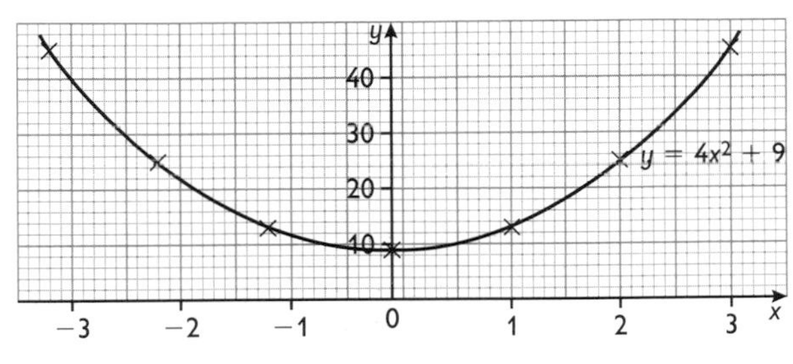

(d) Minimum value of $y = 9$.

8 (a)

x	0	1	2	3	4	5	6	7	8
$x - 4$	−4	−3	−2	−1	0	1	2	3	4
$y = (x - 4)^2$	16	9	4	1	0	1	4	9	16

(b) (c)

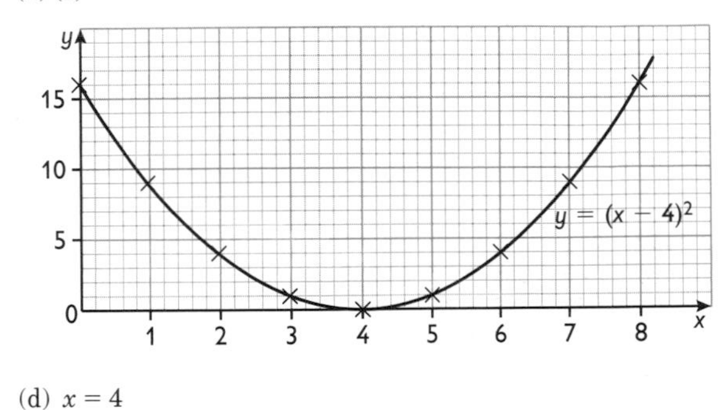

(d) $x = 4$

9 (a)

x	−1	0	1	2	3	4	5	6	7	8
$7x$	−7	0	7	14	21	28	35	42	49	56
$-x^2$	−1	0	−1	−4	−9	−16	−25	−36	−49	−64
$y = 7x - x^2$	−8	0	6	10	12	12	10	6	0	−8

(b) (c)

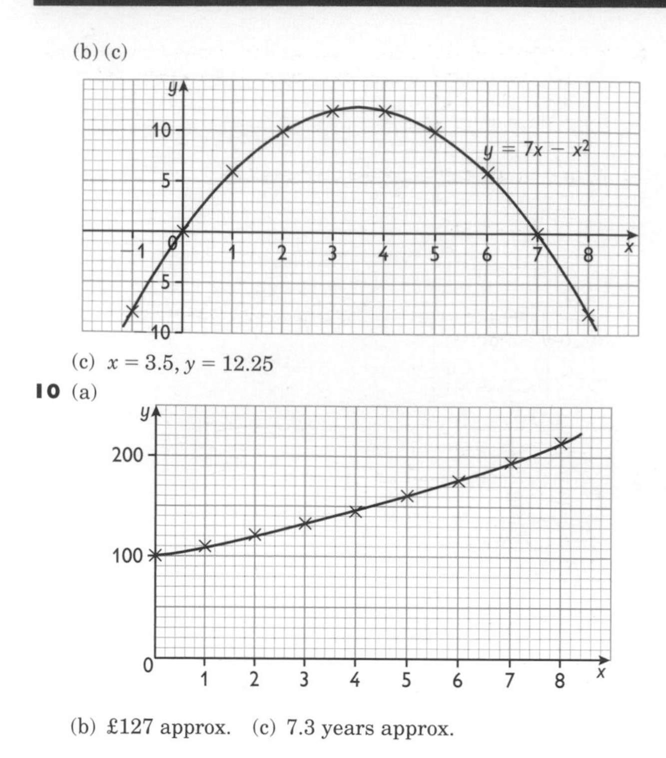

(c) $x = 3.5$, $y = 12.25$

10 (a)

(b) £127 approx. (c) 7.3 years approx.

Investigation

Michael's claim is true.

14 Drawing and construction (pages 128–129)

1 (a) 4.8 m (b) 73°
2 (a) (b) Ask your teacher to check your drawing.
(c) Angle bisector
3 (a) Ask your teacher to check your drawing.
(b) (i) 87° (ii) 7.9 cm
(c) (i) 8.64 cm² (ii) 20.64 cm²
4 (a) (b) Ask your teacher to check your drawing.
(c) Perpendicular bisector
5 (a) (b) Ask your teacher to check your drawing.
6 (a) Circumference of a circle, centre C, radius 5 cm (in 2-D)
(b) Ask your teacher to check your drawing.
(c) (i) Common chord
(ii) AB bisects CD at right angles
(d) Rhombus. Each side is 5 cm.
7 (a) (b) Ask your teacher to check your answer.
(c) 16.4 m (d) 13.8 m

Activity

1,2 Ask your teacher to check your diagram.
3 Square
4,5 Ask your teacher to check your diagram.
? AECGBFDH is a regular octagon

15 Working with data (pages 138–139)

1 (a)

Number of pints	Tally	Frequency
1	𝍸 |	6
2	𝍸 𝍸	10
3	𝍸 ||	7
4	𝍸 ||	7
5	||	2

(b) Bar chart to illustrate above data
(c) 2 (d) 2.66 (e) 2.5
2 (a) Boys 53.8, girls 58.9 (b) Boys
(c) Boys 38, girls 52 (d) Girls
3 1.48
4 (a) (i) 14 (ii) 15.5
(b) (i) 11 (ii) 15
(c) Girls (d) Boys

5 (a) 16 | 8 represents 168 cm

15*	
15	6 9
16*	4 3 2 2 1 3
16	8 5 9 7 8 6 5 9
17*	2 3 0 2 4 4 3 0 3 4
17	9 5 8 8 5 9 7 6 7 8 6
18*	2 1 4 0 2 4 0
18	5 8 6 5
19*	1 4
19	

(b) 16 | 8 represents 168 cm

15*	
15	6 9
16*	1 2 2 3 3 4
16	5 5 6 7 8 8 9 9
17*	0 0 2 2 3 3 3 4 4 4
17	5 5 6 6 7 7 8 8 8 9 9
18*	0 0 1 2 2 4 4
18	5 8 6 8
19*	1 4
19	

(c) Median = 174 cm
(d)

Height (cm)	Frequency
155–159	2
160–164	6
165–169	8
170–174	10
175–179	11
180–184	7
185–189	4
190–194	2

(e) Modal class is 175–179 cm

16 Index notation (pages 146–147)

1 (a) 3.67×10^2 (b) 2.5×10^4 (c) 3.4×10^7 (d) 1.23×10^6
(e) 6×10^{-2} (f) 3.76×10^{-3} (g) 1.4×10^{-5} (h) 3.214×10^{-4}
2 (a) 400 (b) 621 000 (c) 1 000 000
(d) 65 000 000 000 (e) 0.3 (f) 0.0042
(g) 0.000 1 (h) 0.000 000 005 32
3 (a) 3.6×10^3 (b) 2.7×10^2 (c) 1.23×10^6 (d) 8×10^3
(e) 4.5×10^7 (f) 1.2×10^5 (g) 2.4×10^9 (h) 6×10^5
(i) 2×10^{-3} (j) 2.5×10^{-4} (k) 4×10^{-3} (l) 3.5×10^{-6}
4 (a) 10^6 (b) 10^5 (c) 10^9 (d) 10^9 (e) 10^{-3} (f) 10^{-6}
5 (a) 1.9×10^8 (b) 5.7×10^6 (c) 3×10^{-4}
(d) 1.23×10^{-6} (e) 2 570 000 (f) 1.9×10^{10}
(g) 0.004 89 (h) 0.000 006 7
6 Ask your teacher to check your answer.
7 (a) 0.000 005 (b) 0.000 000 5 (c) 553
(d) 0.000 000 000 000 000 000 000 000 000 000 911
(e) 6 370 000
8 (a) 3.7×10^7 (b) 2.39×10^5
(c) 2.2×10^{-4} (d) 1×10^{-6} (e) 1.36×10^{10}
9 (a) 2000 metres (b) 5 000 000 tonnes
(c) 0.000 000 008 secs (d) 0.000 000 000 004 metres
(e) 0.000 008 grams (f) 3 000 000 000 000 joules
10 510 seconds or 8 mins 30 seconds
11 15 seconds

17 Formulae (pages 152–153)

1 (a) 108° (b) 135°
2 (a) $x = y - 5$ (b) $x = 5 - y$
(c) $f = e + 2 - v$ (d) $e = r + f - 2$
(e) $b = \dfrac{a}{3}$ (f) $d = 4c$
(g) $x = \dfrac{y - 6}{3}$ (h) $t = \dfrac{v - u}{a}$
(i) $q = \dfrac{s + 4}{t}$ (j) $u = \dfrac{s}{t}$
(k) $y = x - z$ (l) $b = \dfrac{a - c}{3}$
3 (a) $c = 15h + 25$ (b) (i) £55 (ii) £77.50
(c) $h = \dfrac{c - 25}{15}$
(d) (i) 6 hours (ii) 8 hours (iii) $3\frac{1}{2}$ hours

4 (a) (i) $t = n - 2$ (ii) $n = t + 2$ (b) 8 (c) 12

5 (a) $c = 8p + 75$

(b) $p = \dfrac{c - 75}{8}$

(c) 43 people

6 (a) (i) 225 mins or $3\frac{3}{4}$ h (ii) 165 mins or $2\frac{3}{4}$ h (iii) 3 h

(b) 5

Investigation

(a) Ask your teacher to check your table of values for N, A and R.

(b)(c) $R = A - N + 2$, $N = A - R + 2$, $A = N + R - 2$

18 Accuracy (pages 160–161)

1 (a) 12.3 (b) 4.84 (c) 0.136 (d) 0.03
(e) 0.048 (f) 5.0 (g) 4.00 (h) 1.000

2 (a) 10 (b) 1.4 (c) 104 (d) 0.51
(e) 0.35 (f) 0.605 (g) 0.002 (h) 1000

3 The approximate answers are the result of very simple rounding. If your answers are different, ask your teacher to check them.
(a) 100, 114 (b) 8, 5.67 (c) 25, 32.3 (d) 1, 1.21
(e) 140 000, 173 000 (f) 3, 3.09 (g) 0.04, 0.0406
(h) 20, 15.5 (i) 10, 11.9 (j) 10 000, 16 300 (k) 7, 7.26
(l) 5, 5.30 (m) 0.37, 0.355 (n) 30, 32.8 (o) 20, 19.5
(p) 1, 0.934

4 18 106 500 ≤ Australia < 18 107 500
5 228 500 ≤ Denmark < 5 229 500
58 285 500 ≤ France < 58 286 500
125 155 500 ≤ Japan < 125 156 500
58 305 500 ≤ United Kingdom < 58 306 500
263 562 500 ≤ United States < 263 563 500

5 Your estimates should be about 21 cm^2 and 800 cm^2.

6 1 roll will provide approx. 5 full lengths. $12 \div 2.3 \approx 5$
Each long side requires approx. 8 lengths. $4 \div 0.53 \approx 8$
Side containing door requires approx. 5 lengths.
Window side requires approx. 2 full lengths and 4 half lengths.
Total lengths required = 25.
Number of rolls required = 25 ÷ 5. 5 rolls are required.

7 Volume of balloon $= \frac{4}{3}\pi \times 15^3 \approx 14\,000\ cm^3$.
Volume of inside of car $\approx 300 \times 150 \times 120 \approx 5\,400\,000\ cm^3$
Number of balloons = 5 400 000 ÷ 14 000 = 400
However allowing for space between the balloons, an answer between 300 and 350 is more reasonable.

Investigation

1 3.65 cm to 3.75 cm

2 Minimum circumference = 11.46681... cm

3 Maximum circumference = 11.78097... cm

4 Freda's = 11.6 cm, minimum = 11.5 cm, maximum = 11.8 cm

5 11.6 cm

19 Real life graphs (pages 166–167)

1 (a) (i) Noon (ii) 30 minutes
(b) (i) 1 hour (ii) 20 miles (iii) 20 miles per hour
(c) 2.30 pm
(d) (i) 1 hour (ii) 63.33 miles per hour

2 (a)

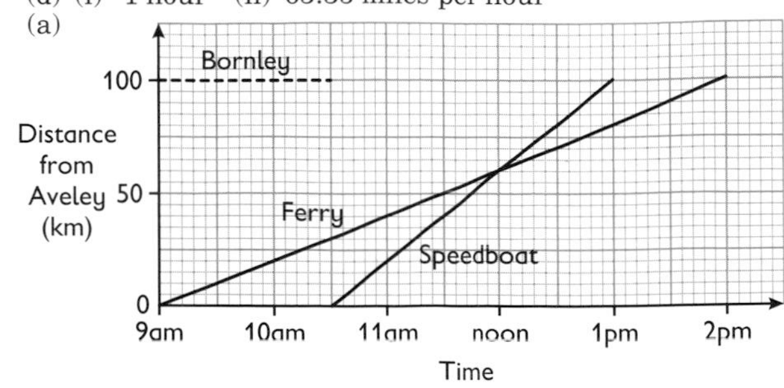

(b) 20 km per hour (c) 1 pm (d) Noon (e) 20 km

3 (a) Approx. 85 pence
(b) From 0–2 days and 4–7 days
(c) Day 4
(d) Day 7
(e) True

4 (ii) Because initially the cross-section is small but as time elapses the cross-sectional area increases thus slowing down the height.

5 (ii) is the correct shape.

20 Transformations (pages 174–175)

1 (a) Reflection in y axis
(b) Reflection in $y = x$
(c) Translation $\begin{pmatrix} 4 \\ -2 \end{pmatrix}$
(d) Rotation through 90° anticlockwise about the origin
(e) Reflection in $y = -x$
(f) Translation $\begin{pmatrix} 0 \\ -2 \end{pmatrix}$
(g) Rotation through 180° about the origin
(h) Reflection in $x = 1$

2 (a) A to C (or B to F, or F to I)
(b) E to F
(c) C to G (or G to C)
(d) D to F (or F to D)

3 (a)(b) Ask your teacher to check your diagram.
(c) Translation $\begin{pmatrix} 3 \\ 5 \end{pmatrix}$

4 (a)(b) Ask your teacher to check your diagram.
(c) Ask your teacher to check your answer.
(d) (3, 13)
(e) 2
(f) (i) Times by 2 (ii) Times by 4 ($= 2^2$)

5 (a) Ask your teacher to check your diagram.
(b) 52°, 56°, 72°
(c) Ask your teacher to check your diagram.
(d) Corresponding angles are equal.

Activity

For A4 297 ÷ 210 = 1.414 ... and so on.

21 Probability (pages 182–183)

1 (a) 0.1 (b) 40 (c) 6

2 (a)

HH	DH	CH	SH
HD	DD	CD	SD
HC	DC	CC	SC
HS	DS	CS	SS

(b) (i) $\frac{1}{4}$ (ii) $\frac{1}{4}$ (iii) $\frac{1}{2}$ (iv) $\frac{1}{8}$

3 (a) 0.15
(b) (i) 60 (ii) 20 (iii) 30

4 (a) (i) 0.36 (ii) 0.1 (iii) 0.3
(b) Red 6, yellow 5, green 2, blue 7

5 (a) 4 (b) 1

(c)

Red die (columns) / Blue die (rows)

	1	2	3	4	5	6
1	1	3	5	7	9	11
2	0	2	4	6	8	10
3	−1	1	3	5	7	9
4	−2	0	2	4	6	8
5	−3	−1	1	3	5	7
6	−4	−2	0	2	4	6

(d) (i) $\frac{1}{12}$ (ii) $\frac{1}{6}$ (iii) $\frac{1}{12}$ (iv) $\frac{1}{18}$ (v) 0

22 Inequalities (pages 190–191)

1 (a) $63 \leq S \leq 175$
(b) $13 \leq F \leq 37$
(c) $11 \leq G \leq 39$

2 $12 \leq A < 21$

3 (a) $<$ (b) $>$ (c) $<$ (d) $>$
(e) $<$ (f) $>$ (g) $<$ (h) $>$

4 Any values which fit the inequalities below. Ask your teacher to check your answers.
(a) $7 \leqslant x \leqslant 15$ (b) $0 < x \leqslant 4$ (c) $-2 \leqslant x \leqslant 2$
(d) $-7 \leqslant x < -1$ (e) $-5 \leqslant x \leqslant 0$ (f) $17 < x \leqslant 23$

5 (a) $-2 \leqslant x < 3$ (b) $-3 < x < 0$ (c) $-1 \leqslant x < 4$
(d) $0 < x < 5$ (e) $-4 < x < 1$ (f) $-3 \leqslant x < 3$

6 (a) −5 3 (b) 0 5
(c) −3 0 (d) −4 4
(e) $2\frac{1}{2}$ 5 (f) −2 $3\frac{1}{2}$
(g) −4 6 (h) 0 5

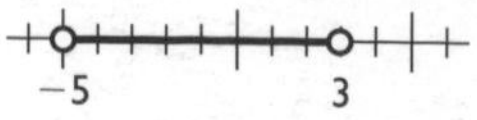

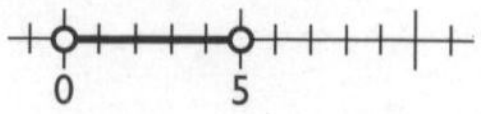

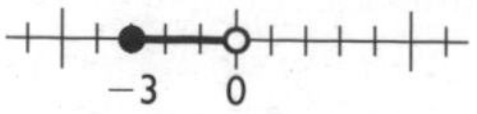

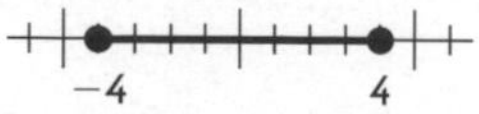

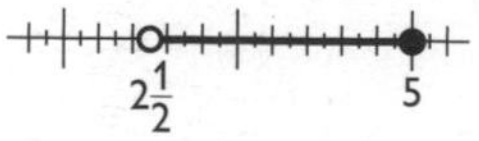

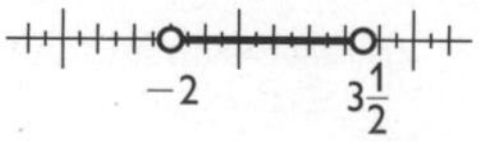

7 (a) $x > 8$ (b) $x \geqslant 4$
(c) $x < 4$ (d) $x > 7\frac{1}{2}$
(e) $x < 5$ (f) $x \geqslant 4$
(g) $x > 2$ (h) $x < 3$

8 (a) (i) $<$ (ii) $>$ (iii) $>$
(b) (i) $<$ (ii) $>$ (iii) $<$
(c) (i) $<$ (ii) $>$ (iii) $>$
(d) (i) $<$ (ii) $<$ (iii) $<$
(e) (i) $>$ (ii) $>$ (iii) $>$

23 Pythagoras (pages 196–197)

1 $a = 13.6$ cm $b = 26.9$ cm $c = 9.2$ feet
$d = 7.1$ m $e = 15.6$ m $f = 5.1$ inches

2 $a = 14.4$ yards $b = 3.5$ miles $c = 53.7$ cm
$d = 11.3$ cm $e = 3000$ m $f = 0.604$ m

3 (a) $x = 10, y = 8$
(b) $p = 7, q = 25$
(c) $r = 36, s = 45$

4 26.25 feet

5 1000.80 m

6 5610 m = 5.610 km

26 Investigations (page 220)

1 Useful formulae
$1 + 2 + 3 + \ldots + n = \frac{1}{2}n(n + 1)$
$1^2 + 2^2 + 3^2 + \ldots + n^2 = \frac{1}{6}n(n + 1)(2n + 1)$
$1^3 + 2^3 + 3^3 + \ldots + n^3 = \left[\frac{1}{2}n(n + 1)\right]^2$